Mathematics and Modelling

Mathematics and Modelling

Edited by
A. Bazykin and Yu. Zarkhin

Nova Science Publishers, Inc.

Nova Science Publishers, Inc.
6080 Jericho Turnpike, Suite 207
Commack, New York 11725

Book Production Manager: F. Columbus
Graphics: Elenor Kallberg, Christopher Concannon

Library of Congress Cataloging-in-Publication Data available upon request

ISBN 1-56072-104-9

Printed in the United States of America

Table of Contents

Mathematics
and Modelling

ISBN 1-56072-104-9

GLOBAL ANALYSIS OF THE PHASE PORTRAIT OF THE NONLINEAR PARABOLIC EVOLUTION EQUATION

Yu.S. Il'yashenko

1. MAIN RESULTS AND RELATIONSHIPS BETWEEN THEM

1.1. Formulation of Results. The study is devoted to the analytical investigation of the so-called Kuramoto-Sivashinsky equation

$$u_t = -((u_x)^2 + u_{x^2} + \nu u_{x^4}), \quad \nu > 0, \quad u_{x^n} = \partial^n u/\partial x_n$$

with the periodic boundary conditions $u(t,x + 2\pi) = u(t,x)$. This equation is encountered in a series of physical problems [1,2,3]. Its numerical study shows a chaotic behavior of the solutions for $\nu \ll 1$ [4]. The phase space of this equation is a set of functions on the circumference $S^1 = \mathbb{R}/2\pi Z$. The mean value of the right-hand side is negative if U $\neq$ const. The difference between the function f and its mean value $\bar{f}$ will be called the centered f function. The main results of the present study consist in the following (the detailed formulations will be given later on):

1. Boundedness of the centered solutions of the Kuramoto-Sivashinsky equation and all their derivatives on the interval $t \in [0,+\infty)$.

2. Energy transfer from low to high harmonics.

3. Existence, Liapunov stability, and finite dimensionality of the maximum attractor.

1.2. Equation for Centered Solutions and Euristic Considerations. Denote by P an operator transforming each function $f \in L^2(S^1)$ into the centered f funtion; P is projecting of $L^2(S^1)$ along the vector $f_o \equiv 1$ onto a hyperplane orthogonal to the vector f_o. The right-hand side of the Kuramoto-Sivashinsky equation depends on Pu alone. Therefore the centered solution of this equation, w = Pu, satisfies the equation

$$u_t = -(P(w_x)^2 + w_{x^2} + \nu w_{x^4})$$

or, since $P(w_x) = (Pw)_x$, we substitute w by u to obtain

$$u_t = -(P(u_x)^2 + u_{x^2} + \nu u_{x^4}), \quad \bar{u} = 0. \tag{1}$$

In what follows we shall consider the above equation. Its right-hand side consists of two terms. The first one corresponds to the linear equation

$$u_t = -(u_{x^2} + \nu u_{x^4})$$

and the second one to the equation

$$u_t = -P(u_x)^2 \qquad \text{(HJ)}$$

Somewhat arbitrarily, it will be called the Hamilton-Jacobi equation since it is satisfied by the centered solution of the real Hamilton-Jacobi equation $u_t = -u_x^2$. A linear equation has an immobile point O: $u \equiv 0$ which for $\nu < 1$ is an infinite dimensional saddle. The respective operator has eigenvectors e^{ikx} with eigenvalues $\lambda_k = k^2 - \nu k$. If $k \le \nu^{-1/2}$, then $\lambda_k \le 0$ (such k values correspond to "low harmonics"). If $k \ge \nu^{-1/2}$, then $\lambda_k \ge 0$ ("high harmonics"). V.I. Arnold, who attracted my attention to the Kuramoto-Sivashinsky equation, noted that the "linear terms swing the low harmonics and quench the high ones, whereas nonlinear terms mix them both". The main part of the present study (sections 3-5) is devoted to the substantiation of this heuristic picture. Let us proceed to exact formulations.

1.3. Boundedness of Solutions. The phase space of equation (1) is the space $X = C|^{\infty}(S^1)$ of infinitely differentiable functions with the zeroth mean on S^1. In this space various scalar products and the respective norms are considered; it is not complete in any of them. Let

$$f = \sum_{-\infty}^{\infty} a_k e^{ikx}, \quad a_k = \bar{a}_k, \quad a_0 = 0.$$

By definition, for any real s, the sth Sobolev norm, n_s, of the function f is

$$n_s(f) \overset{\text{def}}{=} \|f\|_s = \left[\sum k^{2s} |a_k|^2\right]^{1/2}.$$

For integer $s = m$, the m-norm, $n_m(f)$, is equal to L^2, the norm of the derivative $f^{(m)}$:

$$\|f\|_m = \|f^m\|_0.$$

Consider also the C^m norm in the space X:

$$\|f\|_{C^m} = \max_{S^1} |f^{(m)}|.$$

Since $\bar{f} = 0$, this norm is equivalent to the conventional one that takes into account not only the mth derivative but also all the preceeding ones.

Below the term smoothness always signifies the infinite smoothness.

Theorem 1. For any s there exists such R_s that all the solutions of equation (1) with the smooth initial condition enter the sphere $n_s \leq R_s$ within some positive time and never leave it.

1.4. Properties of Norms and Universal Constants. For any given $u \in C|^{\infty}(S^1)$ the function $s \dashrightarrow n_s(u)$ is monotonic and logarithmically convex: its logarithm is a downward convex function [11]. Whence it follows that the functional n_s possesses the properties:

$$\frac{n_s}{n_{s-ka}} \quad \left[\frac{n_s}{n_{s-a}}\right]^k, \ (a > 0); \ n_s^2 \leq n_{s-a}\, n_{s+a}\,.$$

According to the Sobolev enclosure theorem, for any natural m and s > m + 1/2 there exists a constant C such that $\|u\|_{C^m} \leq C|u|$. For brevity, let us take that the inequality A $<$ B in which the left-hand side is a function or a funtional and the right-hand side can be constant signifies that there exists such a constant C > 0 that everywhere A ≤ CB. For example,

$$\|u\|_{C^m} \lesssim \|u\|_s \text{ for } s > m + 1/2.$$

1.5. Energy Transfer. In distinction from the following text, in this Section we call an energy any squared Sobolev norm. If f = A sin x, then all the energy is concentrated in the right harmonics; if $f = \sum a_k e^{ikx}$, $g = \sum_{k>100} a_k e^{ikx}$, and $\|g\|_s^2 \geq f_s$, then more than half of the energy n_s^2 for function f is concentrated in harmonics of the order exceeding 100.

For the exact formulation of the theorem of energy transfer let us denote by E_N the space of trigonometric polynomials of degree not higher than N having the zeroth mean and by P_N the operator of orthogonal projecting $L^2(S^1) \dashrightarrow E_N$ (reject of the higher-order terms of the Fourier series), and by $P_N^{\perp}$ projecting onto the orthogonal complement of E_N (reject of lower-order terms up to the order N).

Theorem 2 (of energy transfer). For any N and any $\lambda \in (0,1)$ there exists such R and for any function $\varphi \in X$ lying outside the sphere $n_1 \leq R$ there exists such moment of time $T_\varphi > 0$ that the solutions of equation (1) with initial condition φ at the moment of time T_φ take a value ψ for which $\|P_N^{\perp}\psi\|_3^2 \geq \lambda\|\psi\|_3^2$.

Thus, for example, if $\lambda = 0.99$ and N = 100, the theorem states that more than 99% of the energy n_3^2 for function ψ is concentrated in harmonics of the order exceeding 100. In this case, φ could be equal to A sin x for A > R, and the Fourier expansion of the initial condition has no higher-order harmonics.

1.6. Generalized Dissipative Systems and Their Attractors. I did not manage to reveal for equation (1) an absorbing domain at the boundary of which the vector field of the equation is directed inward (systems with such a domain are called dissipative). It seems that such a domain does not exist at all. But it was possible to reveal a "generalized absorbing domain" which was "not worse" than the absorbing one, i.e., which dictated to the equation mainly the same properties as the absorbing domain. It is defined as follows.

Definition 1. Let a semiflow $\{g^t | t>0\}$ with trajectories continuously depending on the initial conditions be given in the space X; B and $\tilde{B} \supset B$ are the closures of the domains in X. Let, for example, the domain B be a generalized globally absorbing domain for a semiflow $\{g^t\}$ with periphery $\tilde{B}$ and delay T, if:

1°. The trajectory of any point $\varphi \in B$ returns to the domain B not later than for time T without leaving the periphery $\tilde{B}$.

2°. The trajectory of any initial condition φ comes into domain B for some positive time. A system determining a semiflow with the generalized globally absorbing domain which is compact with the periphery is called a generalized dissipative system.

Thus, for example, a trajectory γ of a semiflow is said to be defined on the whole time axis if for any point $y \in \gamma$ and any moment of time $t > 0$ the point x is defined: $g^t x = y$.

Definition 2. The maximum attractor of a generalized dissipative system is called the closure of the union of all its trajectories defined on the whole time axis and belonging to the periphery.

It turns out that the attractor of a generalized dissipative system can be defined by the same formula as the attractor of a dissipative system and that it is also Liapunov stable.

Proposition. Let $\{g^t | t > 0\}$ be a generalized dissipative semiflow with the globally absorbing domain **B**, periphery $\tilde{B}$, and delay T, and A be its attractor in the sense of definition 2. Then

$$A = \bigcap_{n>0} g^{Tn} \tilde{B}.$$

Also, the attractor A is non-void and Liapunov stable in the following sense: for any neighborhood U of the set A there exists $t(U) > 0$ such that $g^t \tilde{B} \subset U$ at $t > t(U)$.

▹ Let us prove the nonvoidness of the set A. From definition 1 it follows that each point $x \in \tilde{B}$ enters the domain $\tilde{B}$ for time T and then does not leave $\tilde{B}$ any more. Thus, $g^T \tilde{B} \subset \tilde{B}$. Let us take

$$A_1 = \bigcap_{n>0} g^{Tn} \tilde{B}.$$

It is the enumerable intersection of imbedded compacts; it is nonvoid. The set A_1 is just the attractor of the semiflow $\{g^t\}$ in the sense of definition 2. Indeed, the set A_1 consists of all the trajectories defined on the whole time axis and belonging to $\tilde{B}$, moreover it is closed. The attractor nonvoidness has been proven.

Now let us prove its Liapunov stability. Let us define on $\tilde{B}$ a function t, the "time of leaving $\tilde{B}$ in the time reversal":

$$t(x) = \max\{t \geq 0 \mid (g^{\tau})^{-1}x \in \tilde{B} \text{ for } \tau \in [0,t]\}.$$

Note that $t(x) = \infty$ if and only if the trajectory of point x is defined on the whole time axis, i.e., if $x \in A$. The function t is semicontinuous from above: if $x = \lim x_n$, then $t(x) \geq \limsup t(x_n)$. This follows from the continuous dependence of the semiflow trajectories on the initial conditions and the closure of the set $\tilde{B}$. Now let us assume that the attractor A is Liapunov unstable. Then for each n there exists a point $x_n \in \tilde{B}\backslash U$ such that $t(x_n) \geq n$. Since $\tilde{B}$ is compact, it is possible to take that $x_n \dashrightarrow x$. Since the function t is semicontinuous, $t(x) = \infty$. Therefore $x \in A$, i.e., we arrive at contradiction.▹

<u>1.7. Finite Dimensionality of the Attractor.</u> As usual, denote by $\mathring{H}_s$ the Sobolev space obtained by complementing the space $\mathring{C}$ by the norm n_s. Below it will be proven that in each of the spaces $\mathring{H}_s$ equation (1) sets a generalized dissipative system at $s \geq 4$; the standard part of the proof is given schematically in Sect.1.8; a more complicated part is considered in Sects.2-5. Thus, in each of the spaces $\mathring{H}_s$ equation (1) has the attractor A_s. Similar to the heat equation, equation (1) smoothes the initial conditions; therefore all the A_s attractors consist of smooth functions and coincide. Below we shall write A instead of A_s.

<u>Theorem 3</u>. Consider the metrics of the space L^2 in space $X = \mathring{C}^{\infty}$. The Hausdorff dimesionality, $\dim_H A$, of the attractor A defined with the aid of this metrics is finite. Moreover, $\dim_H A \lesssim \nu^{-36.5}$.

<u>1.8. The Main Theorem and Reductions</u>

Theorem 1a. The Kuramoto-Sivashinsky equation defines a semiflow having a globally absorbing domain B given by the inequality

$n_1 \leq R,\ R \lesssim \nu^{-72}$. Here R is ν-dependent: $R \lesssim \nu^{-72}$.

The above statement is the main theorem that will be proven in details later on. Theorems 1 and 3 are schematically derived from this theorem in the remaining part of Sect.1. Another conclusion without the statement about the Liapunov stability of the attractor can be found elsewhere [3]. Theorem 2 will be proven in details later on.

In the present Section we outline the derivation of Theorem 1 from Theorem 1a. It is proven that for each m there exists such R_m that the below defined domain B_m is globally absorbing for equation (1):

$$B_1 = \{n_1 \leq R_1\},\ B_3 = B_1 \cap \{n_3 \leq R_3\},\ B_m = B_{m-1} \cap \{n_m \leq R_m\},\ m > 3.$$

Consider the case m = 3. Let us put $R_1 = R$, R is taken from theorem 1a, B_1 is the generalized globally absorbing domain, in accordance with theorem 1a. Now let us prove that in the domain $B_3 \setminus B_1$ the functional n_3 decreases along the solutions. It will be proven in Sect.4.2 that

$$\frac{d(n_3^2)}{dt} \lesssim n_3^3 + n_4^2 - \nu n_5^2,$$

where, in virtue of equation (1), d/dt is a derivative. In the sphere $n_1 < R_1$ the following inequalities are valid in virtue of the logarithmic convexity of the norm n_s

$$n_3^3 \leq n_1^{3/2}\, n_5^{3/2} \leq R_1^{3/2}\, n_5^{3/2};\ n_4^2 \leq n_1^{1/2}\, n_5^{3/2} \leq R_1^{1/2}\, n_5^{3/2}.$$

Hence

$$\frac{d(n_3^2)}{dt} \lesssim 2R_1^{3/2}\, n_5^{3/2} - \nu n_5^2 < 0 \text{ for } n_5 > 4R_1^3,\ R_1 > 1.$$

If $R_3 > 4R_1^3$, then, since $n_5 \geq n_3$, we have $n_5 > 4R_1^3$ in the domain

$B_1 \backslash B_3'$, i.e., in this domain $dn_3/dt < 0$.

Now let us prove that the trajectories of equation (1) can emerge from domain $B_1 \backslash B_3$ only by entering the sphere B_3, if $R_3 > 2v^{-1}R_1$. Indeed, Proposition 4 of Sect.3.6. states that at the intersection of the sphere $n_1 = R_1$ with the domain $R_3 > 2v^{-1}R_1$ the trajectories of equation (1) are directed inside the sphere $n_1 \leq| R_1$.

Thus, according to theorem 1a, all the trajectories enter the sphere B_1 and then, in accordance with the above proven, the intersection $B_3 =$

$B_1 \cap \{n_3 \leq R_3\}$.

The case m > 3 is considered in a similar way.

1.9. Upper Estimate of the Hausdorff Attractor Dimensionality. We outline here the derivation of Theorem 3 from Theorem 1a. The below considerations allow the estimate of the Hausdorff dimensionality of the Galerkin approximation attractors to equation (1) of sufficiently high dimensionality N of the phase space; this estimate is N-independent. In order to obtain the same estimate for the attractor of the complete equation (1) additional substantiations are needed. After publications of studies [5–8] they seem to be standard.

In order to obtain the needed upper estimate, suffice it to consider the quadratic form F_u in the periphery $\tilde{B}$ of the generalized globally absorbing domain of equation (1)*/

$$F_u(\xi) = (dw_u(\xi), \xi),$$

where w is the right-hand side of equation (1). Then it is necessary to find the quadratic form G such that $G(\xi) \geq F_u(\xi)$ at $u \in \tilde{B}$. Let $\lambda_1 > \dots$ be the eigenvalues of the form G. If the number k possesses the property

$\sum_1^k \lambda_j < 0$, then $\dim_H A \leq k$.

The form F_u is:

$$F_u(\xi) = (2u_x\xi_x, \xi)_{L^2} + \|\xi\|_1^2 - \nu\|\xi\|_2^2.$$

---

*/The results [9,10] relate to dissipative systems. But after proposition in Sect.1.6 they can be fully used for generalized dissipative systems.

Let us estimate it for $\|u\|_1 \leq 2R$.

$$|(u_x\xi_x, \xi)| \lesssim \|u\|_1 \; \|\xi\|_1^2 \lesssim R_1\|\xi\|_1^2 \leq R_1\|\xi\|_0\|\xi\|_2 \leq$$

$$3\nu^{-1} R_1^2 \; \|\xi\|_0^2 + \frac{\nu}{3} \|\xi\|_2^2 ;$$

$$\|\xi\|_1^2 \leq 3\nu^{-1}\|\xi\|_0^2 + \frac{\nu}{3} \|\xi\|_2^2.$$

Now G can be taken as:

$$G(\xi) = 3\nu^{-1}(R_1^2 + 1)\|\xi\|_0^2 - \frac{\nu}{3} \|\xi\|_2^2.$$

The eigenvalues of this form are $\lambda_j = \mu - \nu j^4/3$, $\mu = 3\nu^{-1}(R_1^2+1)$, their sum is $\sum_k = \sum_{j=1}^{k}\lambda_j \approx \mu k - \nu k^5/5$. For $k \gtrsim (\mu\nu^{-1})^{1/4}$ we have $\sum^{k} <$ 0. In virtue of theorem 1a, $\mu \lesssim \nu^{-145}$. Therefore, $\dim_H A \lesssim \nu^{-36.5}$.

1.10. Earlier Results. The solution boundedness and the dimensionality of the Kuramoto-Sivashinsky equation was studied earlier in [3] whose authors gave the name to the equation and obtained the conditional theorems on the finite dimensionality of the attractor under the assumption that all the solutions enter the sphere $n_1 \le R$. It was also proven there that the even solutions really enter such a sphere and remain there. Our method based on the application of the theory of ordinary differential equations to infinite dimensional systems provides a similar estimate for all the solutions and not only even ones. The estimate of the attractor dimensionality in [3] for an equation in the phase space of even functions, $u(x) = u(-x)$, is better than ours; the exponent of ν is equal to -13/8 instead of -36.5. It seems to be probable that the method developed in our study would allow one to prove the boundedness of the solutions and the finite dimensionality of the attractors for more general equations than that considered here. The main results of this study have already been announced in [12].

I use this opportunity to thank V.I. Arnol'd, N.D. Vvedenskaya, L.R. Volevich, Yu.B. Radvogin, E.V. Radkevich, and A.V. Fursikova for fruitful discussions.

2. GLOBAL ANALYSIS OF THE PHASE PORTRAIT

2.1. Energy Transfer and Dissipation. In what follows we call the energy the functional n_1^2. The Hamilton-Jacobi equation, $u_t = -P(u)$, is characterized by the energy integral $E = n_1^2(u)$. Indeed,

$$\frac{d}{dt_{HJ}} E = 2(u_x, (-Pu_x^2)_x) = 2(Pu_{xx}, u_x^2) = 2 \oint u_x^2 \, u_{xx} = \frac{2}{3} \oint (u_x^3)$$

Hereinafter d/dt_{HJ} (d/dt_{ks} is differentiation in virtue of the (HJ) equation and, therefore, also equation (1)). Here $\oint f$ is the integral of the function f by the Lebesgue measure on S^1. In virtue of equation (1) the energy derivative is

$$\frac{d}{dt_{KS}}E = n_2^2 - \nu n_3^2.$$

For $\nu > 1$, the energy n_1^2 is the Liapunov function since $n_2 < n_3$, and the point O is globally attracted. Hereinafter it is taken that $\nu \leq 1$.

Definition. A pumping cone K_p, a dissipation cone K_d, and cones of slow and rapid dissipation, K_{sd} and K_{rd}, respectively, are defined by the following inequalities:

$$K_p:\ n_2^2 \geq \nu n_3^2;\ K_d:\ n_2^2 \leq \nu n_3^2;$$

$$K_{sd}:\ n_2^2 \geq \frac{1}{2}\nu n_3^2;\ K_{rd}:\ n_2^2 \leq \frac{1}{2}\nu n_3^2.$$

In virtue of equation (1), the energy derivative at the point u is nonnegative if and only if this point belongs to the pumping cone.

2.2. Global Behavior of Solutions.

Theorem 4. There exist such positive T, c, and α that any solution of equation (1) with the initial condition φ in the slow dissipation cone outside the sphere B, $n_1 \leq R(\nu) = C\nu^{\alpha}$, will enter the cone of rapid dissipation for time $T_\varphi > 0$ not exceeding T. The respective trajectory arc,

$$\Gamma = \{g^t\varphi \mid t \in [0, T_\varphi]\}, \qquad (*)$$

belongs to the sphere $n_1 \leq 2|\varphi|_1$. Moreover, $|\psi|_1 \leq |\varphi|_1^{-1}$,where ψ is the end point of the arc Γ. It can be taken that $T = \pi^2/R(\nu)$ and a = 7.

Let us derive Theorem 1a from Theorem 4 and prove that the sphere B is a globally absorbing domain for system (1) with the delay T and periphery $\tilde{B}$: $n_1 \leq 2R(\nu)$. The transition from the initial condition φ to the end point ψ of the arc Γ described in Theorem 4 will be called a step. Consider an arbitrary initial condition φ. If it lies within the rapid dissipation cone or inside the sphere B, then the consideration starts with the next paragraph. If $\varphi \in K_{sd}\backslash B$, then, in virtue of Theorem

4, after one step the trajectory of the point φ will enter the rapid dissipation cone and the energy decrease at least by 1. Then three cases are possible (ure 1):

A. The solution will remain in the rapid dissipation cone forever. In this case the solution energy decreases and the solution itself tends to a saddle singular point. It is the case of probability O: the initial condition belongs to the stable manifold of the saddle O of a finite dimensionality.

B and C. The solution goes out of the rapid dissipation cone at point φ_1; in case B: $\varphi_1 \in B$ and in case C: $\varphi_1 \notin B$.

Case B. In virtue of Theorem 4 the solution with the initial condition φ_1 either remains in the sphere B forever or leaves it but then returns back within the time not longer than T without leaving the sphere $\tilde{B}$.

Case C. In virtue of Theorem 4, the transition from point φ_1 to the rapid dissipation cone occurs for one step with the energy loss of at least 1. Thus for a finite number of steps the solution will enter the sphere B and will never go outside the limits of the sphere $\tilde{B}$.▹

2.3. Lemmas of Emergence and Energy Loss. Theorem 4 follows from two lemmas: that of emergence and that of energy loss; these lemmas will be proven in Sects. 3 and 4, respectively.

Lemma 1 (of emergence). There exists such a positive T and for each $\rho > \varphi^{-1}$ there exists a constant $R = R(\nu,\rho)$ having the following property. For each initial condition φ belonging to the slow dissipation cone outside the sphere $n_1 \leq R$ there exists such a moment of time $T_\varphi <$ T, that the arc (*) belongs to the sphere $n_1 \leq 2R$ and ends at point ψ for which

$$\|\psi\|_3 \geq \rho\|\varphi\|_1.$$

Corollary. $\psi \in K_{rd}$ for $\rho \geq 4\nu^{-1}$.

▹ Let us check whether ψ satisfies the inequality defining the rapid dissipation cone.

$$\|\psi\|_2^2 \le \|\psi\|_1 \|\psi\|_3 \le 2\|\psi\|_3^2/\rho \le \frac{1}{2}\nu\|\psi\|_3^2.$$

The first inequality follows from the logarithmic convexity of the norm $n_s(u)$ over s; the second one follows from the conclusion of the Lemma: $|\psi|_1 \le| 2|\varphi|_1 \le 2|\psi|_3/\rho$; the third inequality follows from the expression for ρ. ▷

Lemma 2 (of energy loss). Let φ, ψ, and ρ be the same as in Lemma 1. If $\rho > C\nu^{-2}$ and C is sufficiently large, then

$$\|\psi\|_1 \le \|\varphi\|_1 - 1$$

2.4. Energy Transfer. Theorem 2 follows from Lemma 1.

▷ Let φ and ψ be the same as in Lemma 1. Let us assume that $|P_N^\perp \psi|_3^2 \le \lambda|\psi|_3^2$ and bring this inequality to the contradiction for a sufficiently large ρ. The backward inequality provides the statement of Theorem 2. By assumption,

$$\|P_N\psi\|_3^2 \ge (1 - \lambda)\|\psi\|_3^2.$$

But, by definition of Sobolev norms in terms of the Fourier coefficients

$$\|P_N\psi\|_3 \le N^2\|\psi\|_1.$$

Therefore we have

$$\|\psi\|_3 \le (1 - \lambda)^{-1/2}N^2\|\psi\|_1 \le 2(1 - \lambda)^{-1/2}N^2\|\varphi\|_1.$$

For $\rho > 2(1 - \lambda)^{-1/2}N^2$ we arrive at the contradiction. ▷

It is sufficient to prove the analogues of Lemmas 1 and 2 for Galerkin approximations to equation (1) with estimates independent of the dimensionality Λ of the approximation phase space. This will be done in the following three sections forming the main part of this study. Then lemmas 1 and 2 are obtained by the transition to the limit for $N \to \infty$.

3. GALERKIN APPROXIMATIONS TO THE KURAMOTO-SIVASHINSKY EQUATION AS SMALL PERTURBATIONS OF THE HAMILTON-JACOBI EQUATION

3.1. Scale Change. Outside a sphere of a large radius the quadratic terms of equation (1) start playing a more important role than the linear ones. Let us change the scales of the variables and time so that it will transform a sphere of radius ε^{-1} into a unit sphere but preserve the quadratic terms: $w = \varepsilon u$ and $\tau = \varepsilon^{-1}t$. Then equation (1) is transformed into the equation

$$w_\tau = -(P(w_x^2) + \varepsilon(w_{x^2} + \nu w_{x^4})).$$

Below we shall write u instead of w and t instead of τ:

$$u_t = -(P(u_x^2) + \varepsilon(u_{x^2} + \nu u_{x^4})). \qquad (1\varepsilon)$$

The transformation of the phase flow of equation (1ε) for time t is denoted by g_ε^t.

Lemma 1a. For any $\rho > \nu^{-1}$ there exists $\varepsilon_o = \varepsilon_o(\nu,\rho)$ such that at $\varepsilon \in (0,\varepsilon_o)$ for each initial condition φ belonging to the slow dissipation cone and sphere $n_1 = 1$ there exists such $t_\varphi \in (0,\pi^2)$ that the arc

$$\Gamma_{\varepsilon,\varphi} = \{g_\varepsilon^t\varphi \mid t \in [0,t_\varphi]\}$$

belongs to the sphere $n_1 \leq 2$ and outcomes to the sphere $n_3 = \rho$. Lemma 1 follows from Lemma 1a.

▹ Let us consider $R(\nu,\rho) = \varepsilon_o^{-1}$. Let the following inequality be valid in Lemma 1: $|\varphi|\, R > R(\nu,\rho)$. Similar to the case of Sect.3.1, let us change the scale in equation (1) with the coefficient $\varepsilon = R^{-1}$. Then the arc $\Gamma_{\varepsilon,\varphi}$ from Lemma 1a is transformed into the arc (*) from Lemma 1, and the inequality of Lemma 1a is transformed into the inequalities of Lemma 1 in which it is possible to put $T = \pi^2$ $T = \pi^2/R(\nu,\rho)$. ▹

3.2. Galerkin Approximations. Let the space of real trigonometric polynomials E_N and projecting operators P_N and $P_N^\perp$ be the same as above in Sect.1.5. The Nth Galerkin approximation to equation (1ε) is obtained if E_N is taken to be the phase space and the vector field on it is the projection of the constraint on E_N of the right-hand part of equation (1ε) on E_N. Since differentiation is commutative with projecting, the Nth Galerkin approximation to equation (1ε) has the form:

$$u_t = -(P_N(u_x^2) + \varepsilon(u_{x^2} + \nu u_{x^4})), \quad u \in E_N \quad (\Gamma_{N,\varepsilon})$$

(here $\Gamma_{N,\varepsilon}$ is the Galerkin approximation to equation (1ε) with the phase space E_N).

Lemma 1b. For any $\rho > \nu^{-1}$ there exists such $\varepsilon_o > 0$ depending on ν and ρ but independent of N that at $\varepsilon \in (0,\varepsilon_o)$ for any initial condition

$\varphi \in E_N \cap K_{sd}$, $|\varphi|_1 = 1$ there exists $t_\varphi \in (0,\pi^2)$ such that the arc

$$\Gamma = \{g_{N\varepsilon}^t \varphi \mid t \in [0,t_\varphi]\}$$

belongs to the sphere $n_1 \leq 2$ and comes to the sphere $n_3 = \rho$. Here $g_{N\varepsilon}^t$ is the transformation of the phase flow of the equation ($\Gamma_{N\varepsilon}$) for time t.

Lemma 1a is obtained from Lemma 1b by passing to the limit $N \mid \infty$. Galerkin approximations will be considered as small perturbations of the Hamilton-Jacobi equation in the sphere $n_3 \leq\mid \rho$. Let us prove first the general lemma of the perturbation for dynamical systems in the finite or infinite dimensional space.

3.3. Lemma of Perturbation. Let us consider two equations in the pre-Hilbertian space X:

$$\dot{x} = v(x) \text{ and } \dot{x} = v(x) + w(x).$$

If space X is finite dimensional, then it is Euclidean; if it is infinite dimensional, then it can be incomplete but has the scalar product and the respective norm. The most widespread estimate of the divergence of the tragectories of fields v and v + w is based on the Gronwall inequality that includes the Lipshitz constant of the field v. But in the infinite dimensional case a typical situation is when the field v is not a Lipshitz field. Nevertheless, the divergence of the trajectories can be estimated if the unperturbed equation $\dot{x} = v(x)$ has been integrated and the upper estimate is the norm of the derived mapping of the respective phase flow.

Let us remind the definition of a tube of trajectories (with a flat bottom).

Let D be a closed subset of hyperplane Λ of the space X at all the points of which the field v is transverse to the hyperplane. The tube of the trajectories of the field v with the bottom D and the time interval T is called the set

$$\Omega = \{g_v^t x \mid x \in D,\ t \in [0,T]\}.$$

The sets

$$D^T = g_v^T D,\ \{g_v^T D \mid t \in [0,T]\},\ D^t = g_v^t D$$

are called a cover and a wall and t is called the tube partition, the latter is defined at any $t \in (0,T)$. Denote by π_v the projecting of the tube of trajectories onto its bottom along the trajectories of the field v: if $x \in D$ and $t \in [0,T]$, the following equality is valid $\pi_v(g_v{}^t x) = x$. Hereinafter $g_v{}^t$ is the transformation of the phase flow (semiflow) of the field v for the time t.

Lemma 3. Let v and w be two vector fields in the pre-Hilbertian space X and φ be an arbitrary point of X. If X is finite dimensional, then the fields v and w are smooth; if X is infinite dimensional, the following smoothness constraint should be met: the mapping

$$\Phi:\ (t,\tau) \to g_v^{\,t} g_{v+w}^{\,\tau}\varphi$$

is smooth everywhere where it is defined. Let Γ be the arc of the trajectory of the field v + w:

$$\Gamma = \{\Gamma(\tau) \mid \tau \in [0,\tau_0]\},\ \Gamma(\tau) = g_{v+w}^{\tau}\varphi,$$

Ω is the tube of trajectories with the bottom D lying on the hyperplane Λ: $\langle\psi - \varphi, v(\varphi)\rangle = 0$ and the time interval T. Let $\Gamma \subset \Omega$ and the following conditions be fulfilled:

1. $|w| < \alpha$ everywhere on Γ.
2. If $y \in \Gamma$ and $g_v^t y \in \Omega$, then $|dg_v^t(y)| \leq F$.
3. $|v| \geq C$ everywhere on D.
4. The angle formed by the vector $v(\psi)$ and the normal $v(\varphi)$ to the plane Λ is not larger than a certain given angle (e.g., $\pi/3$) everywhere on D.

Then:

1. The length of the projection $\pi_v\Gamma$ does not exceed $2\alpha FT$.
2. $|(t\circ\Gamma)' - 1| \leq 3\alpha F/C$.

Consider first the finite dimensional case. Then the smoothness of all the functions considered in the proof follows from the smoothness of the fields v and w.

<u>Proposition 1.</u> Under the conditions of the Lemma:

$$\|d\pi_v(y)\| \leq 2F \text{ for } y \in \Gamma.$$

This proposition will be proven below.

Let us out $\gamma(\tau) = \pi_v\Gamma(\tau)$. Then

$$\dot{\gamma}(\tau) = d\pi_v\dot{\Gamma}(\tau) = d\pi_v(v\circ\Gamma(\tau) + w\circ\Gamma(\tau)) = d\pi_v w\circ\Gamma($$

In virtue of condition 1 of the Lemma and Proposition 1 we have

$$\|\dot{\gamma}(\tau)\| \leq 2\alpha F,$$

whence follows the first statement of the Lemma.

2. Let us define the funtion t: $\Omega \to \mathbf{R}$ by the formula

$$t(g_v^t x) = \dot{t} \text{ for } x \in D.$$

The dt form is invariant with respect to the flow of the field v. Therefore

$$\frac{d}{d\tau}(t\circ\Gamma(\tau)) = (dt,\ (v+w)/y)) = 1 + (dt,\ w(y)),$$

where $y = \Gamma(\tau)$ and (ω,ξ) is the value of the form ω on the vector ξ. Let $x = \pi_v y$. Let us translate the vector $w(y)$ to point x by the flow of the field v; denote the thus obtained vector by w_1; $w_1 \in T_xX$. In virtue of condition 2 of the Lemma we have $|w_1| \leq F\alpha$. Let us expand the vector w_1 into a sum $w_1 = w' + w''$, $w' = \beta v(x)$, $w'' \in T_x\Lambda$, $\beta \in \mathbf{R}$. Since $dt = 0$ on $T_x\Lambda$ and $(dt, v(x)) = 1$, we obtain

$$(dt, w_1) = (dt, w\ \ ) = \beta.$$

But $w' = w_1 - w''$, $w'' = d\pi_v(x)w_1$. From elementary geometric considerations (stated below in the proof of Proposition 1) it follows that $|d\pi_v(x)| \leq 2$; hence, $|w''| \leq 2|w_1|$. Therefore

$$|\beta| = \|w'\| \ /\ \|v(x)\| \leq 3\|w_1\| \ /\ \|v(x)\| \leq 3\alpha F/c.$$

This proves statement 2.

3. For the infinite dimensional case, the proof of statement 1 is just the same as for finite dimensional. In order to transfer the proof of statement 2 to the infinite dimensional case it is necessary to check the smoothness of the boundedness of function t on the surface

$\sum = \{\Phi(t,\tau)\}$. Such smoothness follows from the smootness of mapping Φ and the transversality of the field v to the tube bottom. Then the previous computations are repeated without any changes.

Now in order to prove Lemma 2 we have to prove Proposition 1 alone.

▷ Proposition 1 follows from conditions 2 and 4 of the Lemma and inequality

$$\|d\pi_v(y)\| \leq \|g_v^{-t}(y)\|/\cos\angle(v(x), v(\varphi)) \text{ for } y = g_v^t x.$$

Let us prove this inequality. We have

$$d\pi_v(y) = d\pi_v(x)\circ dg_v^t(y);$$

in order to project this vector onto the tube bottom along the field trajectories it can be first translated by the flow to a point of the bottom and then the obtained vector should be projected onto the bottom along the respective vector of the field v. The norm of the operator of the projecting onto hyperplane Λ along the vector $v(x) \in T_x X$ for $x \in \Lambda$ does not exceed unity divided by the cosine of the angle formed by the vector $v(x)$ and the normal $v(\varphi)$ to the plane Λ. If the space X is two-dimensional and the plane Λ is one dimensional, this follows from elementary geometric considerations; the general case can be reduced to the two dimensional one by constructing the plane through the projecting and progected vectors. ▹

3.4. The Galerkin Approximations to the Kuramoto-Sivashinsky Equation. They are considered as small perturbations of the Hamilton-Jacobi equation. The proof of Lemma 1b is based on the following two considerations.

1. Each solution of the Hamilton-Jacobi equation lives for a finite time, then it collapses: the higher-order derivatives (in particular, of order 3) go to infinity. 2. So long as the solution of the Hamilton-Jacobi equation lies within the sphere $n_3 \le \rho$, the solution of the equation $(\Gamma_{N\varepsilon})$ with the same initial condition remains to be close to the first solution for sufficiently large N and small ε. The first solution attains the boundary of the sphere $n_3 \le \rho$ and then goes outside the sphere; therefore the second solution also goes to the boundary of the sphere $n_3 \le \rho$. Then the "solution paths" become different: the solution of the Hamilton-Jacobi equation collapses and the solution of the equation $(\Gamma_{N\varepsilon})$ enters the rapid dissipation cone and the high-order harmonics along this solution attenuate.

Now let us pass to the formal statement of the problem. Lemma 3 of perturbation will be applied to the unperturbed (HJ) and perturbed $(\Gamma_{N\varepsilon})$ equations. As a phase space X we take the space of infinitely smooth functions with the zero mean on S^1, having the scalar product and the norm of the space H_{-1}. Thus space is not complete (it is pre-Hilbertian and not Hilbertian).

Proposition 2. For each pair (α,ρ) there exist such N and ε, that the right-hand sides of equations (HJ) and $(\Gamma_{N\varepsilon})$ in the intersection of spheres $\tilde{B}$: $n_1 \le 2$ and Q: $n_3 \le \rho$ differ by norm H_{-1} less than by α.

Addendum. Suffice it to choose N and ε in such a way that

$$\alpha \geq \rho(2C/N + \nu\varepsilon + 2\varepsilon,$$

where C is the universal constant.

▷ It is necessary to prove the inequality

$$\|P_N^{\perp}(u_x^2) + \varepsilon(u_{x^2} + \nu u_{x^4})\|_{-1} \leq \alpha,$$

if $|u|_3 \leq \rho$ and $|u|_1 \leq 2$, N is sufficiently large, and ε is sufficiently small. Then we have

$$\|P_N^{\perp}(u_x^2)\|_{-1} \overset{1}{\leq} \frac{1}{N}\|u_x^2\|_0 \overset{2}{\leq} \frac{2\pi}{N}\|u_x\|^2_{W_4^1} \overset{3}{\leq} \frac{C}{N}\|u\|^2_{3/2} \overset{4}{\leq} \frac{C}{N}\|u\|_1\|u\|_3,$$

where C is the universal constant. The first inequality follows from the definition of the norm $\|\cdot\|_{-1}$ in terms of the Fourier coefficients, the second one from the Cauchy-Bouniakowsky inequality, the third one from the Sobolev classification theorem, and the fourth one from the logarithmic convexity and monotony of the norm $\|\cdot\|_s$ with respect to s. On the other hand

$$\|\varepsilon(u_{x^2} + \nu u_{x^4})\|_{-1} \leq \varepsilon(\|u\|_1 + \nu\|u\|_3).$$

The required inequality follows from the two last estimates. ▷

3.5. The Properties of the Hamilton-Jacobi Equation.

Let us enumerate the properties of nondegenerate (HJ) equation that will be proven in Sect.5. We use them here to derive Lemma 1b.

1°. The Moment of Collapse. The solution $g_{HJ}^t\varphi$ of the (HJ) equation ($\{g_{HJ}^t\}$ is the phase flow of the Hamilton-Jacobi equation) with a varying periodic initial condition φ experiences collapse during both an increase and a decrease of t. The moments of collapse are called such

T_φ^+ and T_φ^- values, $T_\varphi^- < 0 < T_\varphi^+$, for which the solution $g_{HJ}{}^+\varphi_+$ is smooth for $t \in (T_\varphi^-, T_\varphi^+)$ and looses smoothness at T_φ^- and T_φ^+. Property 1° is:

$$T_\varphi^+ = -\frac{1}{2\min\varphi''}; \quad T_\varphi^- = -\frac{1}{2\max\varphi''}; \quad |T_\varphi^\pm| < \frac{\pi^2}{\|\varphi\|_1}; \quad |T_\varphi^\pm| \geq \frac{1}{2\|\varphi\|_0}.$$

2°. The Law of Energy Conservation (Sect.2.1). The functional $n_1{}^2$ is the first integral of the (HJ) equation.

3°. The Upper and Lower Estimates of the Growth Rate of the Norm n_3 with the Approach to the Moment of Collapse. There exists a constant $C > 0$ such that for $t \in (T_\varphi^+/2, T_\varphi^+)$:

$$\frac{c|T_\varphi^\pm|}{|T_\varphi^\pm - t|^3} \leq \|g_{HJ}^t\varphi\|_3^2 \leq \frac{|T_\varphi^\pm|^5\|\varphi\|_3^2}{|T_\varphi^\pm - t|^5}.$$

4°. Equation in Variations. There exists a constant $C > 0$ such that for $t \in (T_\varphi^-, T_\varphi^+)$:

$$\|dg_{HJ}^t(\varphi)\|_{-1} \leq C\|\varphi\|_3^{3/2}\|\varphi\|_1^{-1}$$

Note. With the approach to the moment of collapse along the solution of the (HJ) equation the norms n_s for $s > 1$ increase, the the norm n_1 being constant. Therefore it is not astonishing that the norm n_{-1} does not "notice" the approach to the moment of collapse.

3.6. The Choice of the Tube of Trajectories of the Hamilton-Jacobi Equation. Let us pass to the proof of Lemma 1b. Consider two vector fields, v and v + w, in the space X defined in Sect.3.4 that set the equations (HJ) and $(T_{N\varepsilon})$, respectively. Let φ be the initial condition from Lemma 1b: $|\varphi|_1 = 1$, $\varphi \in K_{sd}$. The hyperplane Λ is given by the equation

$$\langle\psi - \varphi, v(\varphi)\rangle_{-1} = 0, \quad v(\varphi) = -P(\varphi_x^2).$$

The bottom of the tube of trajectories of the field v is chosen in such a way that conditions 3 and 4 of Lemma 3 be satisfied: the lower estimate of length v in the norm n_{-1} on the botoom D and the upper estimate by the angle formed by the vectors of the field v at different points of the bottom. In addition, we also require that the moment of collapse at different points of the bottom occurred almost simultaneously. The time interval T is chosen in such a way that the inequality $n_3 > \rho$ be valid at the tube cover. In other words, the sphere $Q : n_3 \leq \rho$ does not intersect the tube cover. Let us prove that such a choice of the bottom and the interval is really possible.

Proposition 3. There exist such constants $C_1 > 0$ and $C_2 > 0$ that for any φ and ψ satisfying the conditions

$$\|\varphi\|_1 = 1, \quad \varphi \in K_{sd}, \quad \|\varphi - \psi\|_{C^2} = \sigma, \quad \sigma \leq C_1 \nu^{1,2},$$

the following inequalities are valid

$$\|P(\psi_x^2)\|_{-1} \geq C_2 \nu^{1,2}, \quad \angle(P(\varphi_x^2), P(\psi_x^2)) \leq \pi/3.$$

The proposition is proven in Sect.5.

Proposition 4. If $|\varphi|_1 = 1$ and $\varphi \in K_{sd}$, then

$$\|\varphi\|_3 \leq 2\nu^{-1}, \quad \|\varphi\|_{C^2} \lesssim \nu^{-4/5}.$$

⊳ From the logarithmic convexity of the norm n_s as a function of s it follows

$$\|\varphi\|_3 / \|\varphi\|_1 \leq \|\varphi\|_3^2 / \|\varphi\|_2^2.$$

Now the first statement of the lemma follows from the definition of the slow dissipation cone. The second statement follows from the first one and the Sobolev classification theorem. ⊳

The bottom D of the tube of trajectories of the (HJ) equation is now defined by the equation

$$D = \{\psi \in \Lambda \mid \|\psi - \varphi\|_{C^2} \le \sigma\}.$$

The time interval T is chosen to be close to T_φ: $T = T_\varphi - \delta$.

<u>Proposition 5</u>. If $\rho \ge 2\nu^{-1}$, $\delta = \delta(\nu,\rho)$, and $\sigma = \sigma(\nu,\rho)$ are sufficiently small, then for $t \in [T - \delta, T]$ everywhere on the t partition of the tube Ω the inequality $n_3 > \rho$ is valid. For sufficiently small β_1 and β_2 one can take that $\delta = \beta_1\rho^{-2/3}\nu^{4/15}$ and $\sigma = \beta_2\delta\nu^{8/5} = \beta_1\beta_2\rho^{-2/3}\nu^{28/15}$.

▷ First estimate the difference $|T_\varphi^+ - T_\psi^+|$ for $\|\varphi - \psi\|_{C^2} \le$

From the formula for T_φ^+ in Sect.3.5 it follows that if $\sigma < T_\varphi^+/2$, then

$$|T_\varphi^+ - T_\psi^+| \lesssim \sigma (T_\varphi^+)^{-2}$$

In virtue of Proposition 4, $(T_\varphi^+)^{-2} \lesssim \nu^{-8/5}$. Therefore for $\sigma = \beta_2\delta\nu$ and sufficiently small β_2 the following inequality is valid

$$|T_\varphi^+ - T_\psi^+| < \delta/2.$$

Let $y = g_{HJ}^{+}\psi$ be an arbitrary point of the t-partition of the tube Ω for $t \in [T - \delta, T]$. Then

$$0 < T_\psi^+ - t < 2\delta$$

In virtue of property 3° of the (HJ) equation and Proposition 4, there exist $C_1 > 0$ and $C_2 > 0$ such that for sufficiently small $\delta = \delta(\nu,\rho)$ we have:

$$n_3^2(y) \ge c_1 T_\psi^+ \delta^{-3} \ge c_2 \nu^{4/5} \delta^{-3} \ge \rho^2;$$

The first part of the Proposition has been proven. The second one (the exact equalities for δ and σ) is proven by the direct checking of the inequalities $\delta < T_\varphi^+/2$ ($T_\varphi^+/2 \geq| \beta\nu^{4/5} \geq| \beta_1\beta_2\nu^{2/3+28/15} \geq| \sigma$; the latter inequality follows from the conditions that $\rho \geq \nu^{-1}$ and $C_2\nu^{4/5}\delta^{-3} \geq \rho^2$ the validity of which is provided by the choice of β_1.▹

Note. For $\rho > \nu^{-1}$ and sufficiently small β_1 and β_2, the value of σ from Proposition 5 satisfies the inequality of Proposition 3.

This concludes the construction of the tube Ω of trajectories.

3.7. A priori Estimation. Let φ be the same as in Sect.3.6. From property 1° of the (HJ) equation it follows that $T_\varphi^+ < \pi^2$. Consider an arc $\Gamma|$ of the trajectory of the equation (Γ) with the initial condition φ and the time length $\tau_o = T - \delta$. We denote by Γ the portion of the arc $\Gamma|$ that is located at the intersection $\Omega \cap Q$ $Q = \{n_3 \leq \rho\}$, from the origin to the point of the first emerge from $\Omega \cap Q$. Below we shall prove that such a point really exists. But at the time being, the possibility that $\Gamma = \Gamma|$ cannot be rejected. Note that $\varphi \in \Omega \cap Q$, since $|\varphi|_3 \leq 2\nu^{-1} \leq \rho$. The first inequality follows from Proposition 4, the second one from the constraint on ρ.

Proposition 6. For sufficiently small ε and large N and at $\rho > 2\nu^{-1}$ the following inequalities are valid $1/2 < n_1/T < 2$.

▹ Let us remind that $\tau_o < T < \pi^2$. Therefore suffice it to prove that the function $|dn_1^2/dt_{N\varepsilon}|$ is small on Γ. We have

$$dn_1^2/dt_{N\varepsilon} = \varepsilon(n_2^2 - \nu n_3^2) \leq \varepsilon\rho^2 \ll 1.$$

The penultimate inequality follows from the fact that in the sphere Q: $n_2 \leq n_3 \leq \rho$.▹

3.8. Enter to the Sphere $n_3 = \rho$. Here the proof of Lemma 1b is ended. Let Γ be the same arc as in Sect.3.7. Let us assume that it does go out to the sphere $n_3 = \rho$. Then there are three possibilities (ure 2) each of which can be brought to contradiction: the arc Γ is fully contained in the sphere Q: $n_3 \leq| \rho$ and

A. Γ goes out to the cover of the tube Ω.
B. Γ goes out to the wall of the tube Ω.
C. Γ belongs completely to the tube Ω.

Case A cannot exist that has already been proven: the sphere Q does not intersect the cover of the tube Ω in virtue of Preposition 5.

Cases B and C cannot exist either. This will be derived from Lemma 3. Let us see whether the conditions of the Lemma for fields v and w, the tube Ω, and the arc Γ are satisfied.
Condition 1 follows from Propositions 2 and 6:

$$\|w|_\Gamma\|_{-1} \leq \alpha \text{ for } \alpha = \rho(2C/N + \nu\varepsilon) + 2\varepsilon \lesssim \varepsilon\nu\rho;$$

the latter inequality is satisfied for sufficiently large N and $\rho > 2v$.

Condition 2 follows from property 4° of the Hamilton-Jacobi equation and Proposition 6:

$$\|\max_{u\ \Gamma} dg^t_{HJ}(u)\|_{-1} \overset{\text{def}}{=} F; \ F \lesssim \rho^{3/2} \text{ for } \|u\|_3 \leq \rho, \ \|u\|_1 \geq 1/2.$$

Conditions 3 and 4 follow from Proposition 3. The constant from condition 4 is taken to be equal to $Cv^{1,2}$.

The requirement of smoothness formulated at the beginning of Lemma 2 is obviously fulfilled: the initial condition $u = g_{v+w}{}^\tau\varphi$ on the trajectory Γ of the equation $(\Gamma_{N\varepsilon})$ smoothly depends on τ; the solution of the equation (HJ) with the initial condition dependent smoothly on the parameter τ also depends smoothly on the same parameter.

Now in order to prove that cases B and C are not possible let us apply the Lemma of perturbation. The case C is not possible since for sufficiently small ε and large N the trajectory Γ falls on the t-partition of the tube Ω for times not longer than τ_0, t being close to $T_\varphi{}^+$ in virtue of statement 2 of Lemma 3. The difference $|T_\varphi{}^+ - t|$ is so small and the moment $T_\varphi{}^+$ is so close to the moment of collapse of all the solutions forming the tube that on the t-partition, as on the tube cover, the inequality $n_3 > \rho$ is valid (Proposition 5).

The case B is also impossible because of the lagorithmic convexity of the norm n_s as a function of s. For more details, let us assume that the trajectory Γ goes out to the wall of the tube Ω but remains within the sphere Q: $n_3 \leq \rho$. Let φ_1 be the point of its emergence. Let us take that

$\psi_1 = \pi_v \varphi_1$ and $\xi = \psi_1 - \varphi$ (ure 2). The following three properties of the vector ξ will be brought to contradiction:

$$1.\ \|\xi\|_{-1} \le 2\alpha FT,\quad 2.\ \|\xi\|_{C^2} = \sigma,\quad 3.\ \|\xi\|_3 \lesssim \rho\nu^{-2}.$$

First, let us prove these properties. Property 1° follows from the statement of Lemma 3. Property 2° follows from the definition of the tube of trajectories. Property 3° of the vector ξ follows from property 3° of the solution of the (HJ) equation and will be substantiated somewhat later. The considerations of Sect.3.6 provide simultaneously the lower estimate for the moment of collapse in both the past and the future of the solutions of the equation with the initial condition ψ on the bottom of the tube Ω: there exists such $\beta > 0$ that $\beta\nu^{-4/5} \le |T_*^-|$. From property 1° of the (HJ) equation it follows that $|T_*^-| \le \pi^2$. Whence, in virtue of property 3° of the (HJ) equation (the right-hand inequality) we have

$$\|\Psi_1\|_3 \lesssim \rho\nu^{-2}.$$

Since $|\varphi|_3 \le 2\nu^{-1}$, the vector $\xi = \psi_1 - \varphi$ possesses property 3°. Properties 1°-3° of the vector ξ are proven. Now let us prove that they contradict one another if ε is sufficiently small.

In virtue of the Sobolev classification theorem, $\|\xi\|_{C^2} \lesssim \|\xi\|_{2,6}$. In virtue of the logarithmic convexity of the norm n_s we have

$$\left(\frac{\|\xi\|_3}{\|\xi\|_{2,6}}\right)^{10} \ge \frac{\|\xi\|_3}{\|\xi\|_{-1}} \quad \text{or} \quad \|\xi\|_{-1} \ge \|\xi\|_3^{9}\ \|\xi\|_{2,6}^{10} \ge c\rho^{-9}\nu^{18}\sigma^{1}$$

for sufficiently small C. Since α (and also ε) is arbitrarily small, the latter inequality contradicts property 1° for sufficiently small ε. This concludes the proof of Lemma 1b.

Note. From the previous inequality it follows that the case B is not possible for $\varepsilon = c\rho^{-11.5}\nu^{17}\sigma^{10}$ if c is small. Here the fact is taken into account that $|\xi|_{-1} \leq 2\alpha FT$, where T is of the order of unity, $\alpha \lesssim \varepsilon\nu\rho$, and $F \lesssim \rho^{3/2}$. From the formula for σ of Proposition 5 we obtain that if $\rho = C\nu^{-2}$, then the statement of Lemma 1b is valid for $\varepsilon = c\nu^{72}$ if C is sufficiently large and c is sufficiently small.

4. ENERGY LOSSES

In this paragraph Lemma 2 of Sect.2.3 will be proven.

4.1. The Upper Energy Estimate. Let Γ be the same arc as in Sect.3.8. Let us estimate the maximum energy n_1^2 on Γ in terms of a quantity depending on ε and ν. We have

$$\frac{dn_1^2}{dt_{\Gamma_{N\varepsilon}}} = \varepsilon(n_2^2 - \nu n_3^2).$$

Now consider the portion Γ' of the arc Γ which is located in the pumping cone. In virtue of the statement of Sect.3.7 we have that $n_1|\Gamma \leq 2$. Then, taking into account also from Proposition 4, we have $n_3|\Gamma' \leq 2\nu^{-1}$. In virtue of the logarithmic convexity of the norm n_s, we obtain $n_2^2|\Gamma' \leq 4\nu.|^1$. The time τ for which the arc Γ' is run over is not longer than the time for which the set Γ' is run over and it does not exceed the universal constant. Everywhere on $\Gamma\backslash\Gamma'$ the derivative $dn_1^2/dt_{\Gamma_{N\varepsilon}}$ is negative; $n_1^2(\varphi) = 1$. Therefore

$$\max_{\Gamma} n_1^2 \le 1 + 4\varepsilon\nu^{-1}\tau.$$

Hence

$$\max_{\Gamma} n_1 \le 1 + 2\varepsilon\nu^{-1}\tau.$$

4.2. Energy Dissipation. In the domain $K_{sd}\backslash K_{rd}$ the energy decreases; the rate of its decrease is not important for the further consideration. Let us estimate from below the rate of the energy decrease in K_{rd}. By definition of the rapid dissipation cone, the following inequality is valid everywhere in K_{rd}:

$$n_2^2 - \nu n_3^2 \le -\nu n_3^2/2.$$

This proves that the energy n_1^2 in K_{rd} decreases rapidly along the solutions of the ($\Gamma_{N\varepsilon}$) equation. The considerations of the final part of this section and Sect.4.3 show that the arc Γ at $\rho \ge Cv^{-2}$ and sufficiently large C lies within the cone K_{rd} for such a long time that the energy along this arc can decrease drastically.

Now, let us choose on Γ a subset Γ''' consisting of a finite number of arcs ordered in correspondence with the time orientation of Γ and possessing the following property: the values of n_3 at the initial points of each arc are equal to the value of n_3 at the end point of the previous one; at the beginning of the first arc $n_3 = 4v^{-1}$, at the end of the last one $n_3 = Cv^{-2} \le \rho$ (ure). Such a set of arcs can be taken in virtue of the Lemma of emergence; it is finite because of the analiticity of the right-hand side of the $\Gamma_{N\varepsilon}$ equation. Let us estimate from the above the rate of increasing of n_3 along the solutions of the $\Gamma_{N\varepsilon}$ equation.

$$\frac{dn_3^2}{dt}\bigg|_{\Gamma_{N\varepsilon}} = \frac{dn_3^2}{dt}\bigg|_{HJ} + \varepsilon(n_4^2 - \nu n_5^2).$$

(the simple justification of this statement is omitted here). In virtue of the logarithmic convexity of the norm n_s over s we have: $n_5/n_4 \geq n_3/n_2$. Therefore in K_{rd} (and even in K_{sd}) we have $n_4^2 - \nu n_5^2 \leq 0$. Now let us estimate $dn_3^3/dt\,|_{\dot{H}J}$. We have

$$\frac{dn_3^2}{dt}\bigg|_{HJ} = -((Pu_x^2)_{x^3}, u_{x^3}) = (P(u_x^2), u_{x^6}) =$$

$$= (u_x^2, Pu_x^6) = (u_x^2, u_{x^6}) = -((u_x^2)_{x^3}, u_{x^3}).$$

$$|dn_3^2/dt_{HJ}| \lesssim |\oint u_{x^2} u_{x^3}^2| + |\oint u_x u_{x^3} u_{x^4}|.$$

But $\oint u_x u_{x^3} u_{x^4} = -\oint u_{x^2} u_{x^3}^2/2.$

Therefore

$$|dn_3^2/dt_{HJ}| \lesssim |\oint u_{x^2} u_{x^3}^2| \lesssim n_3^3, \quad |dn_3/dt_{HJ}| \lesssim n_3^2.$$

<u>4.3. Change of Variables</u>. Below the following general proposition is used.

<u>Proposition 1.</u> Let x be a smooth function of t and $0 \leq dx/dt \leq f(x)$ and g be a nonnegative function on the segment $t \in [a,b]$, $x(a) = A$. $x(b) = B$. Then

$$\int_a^b g(x(t))dt \ge \int_A^B \frac{g(x)}{f(x)}dx.$$

▹ The proposition follows directly from the formula for the change of variables. ▹

4.4. Estimation of the Energy Dissipation. Let . Now we assume that the set Γ''' consists of one arc (in the general case Proposition 1 is applied to each arc of the set Γ''') separately. Let $g = n_1|\Gamma'''$, $x = n_3|\Gamma'''$ and $t = t_{T_{N\varepsilon}}$, a and b are the moments of time t that correspond to the beginning and the end of the arc Γ'''; $A = 4v^{-1}$, $B = Cv^{-2}$. In Sect.4.2 it was shown that $dx/dt \lesssim$. In virtue of Proposition 1, we have

$$n_1(b) - n_1(a) = \int_a^b \frac{dn_1}{dt}dt \le - \frac{\varepsilon \nu}{2}\int_a^b n_3^2 dt \le -\int_A^B \frac{\varepsilon\nu}{2}dn_3 \le - \frac{\varepsilon}{2}$$

(for large C). For sufficiently large C the energy increment on Γ'' is smaller than the energy loss on Γ'''. The energy loss on Γ is of the order εv^{-1}. Lemma 2 is proven. ▹

5. PROPERTIES OF THE HAMILTON-JACOBI EQUATION

In this Section the statements of Sect.3.5 and Proposition 3 from Sect.3.6 will be proven.

5.1. The Solution of the Hamilton-Jacobi Equation. Consider two initial problems

$$u_t = -Pu_x^2, \quad u|_{t=0} = \varphi, \quad \bar{u} = 0;$$
$$w_t = -w_x^3, \quad w|_{t=0} = \varphi.$$

Let us denote the constraint imposed on $\{t\} \times S^1$ in the solution of the first problem by $u^t = g^t\varphi$, and that of the second problem by w^t. As was indicated in Sect.1.1, $u^t = Pw^t$. The solution w^t is found by the well-known formulae ([13], Sect.8):

$$u^t = v_0(\chi^t)^{[-1]}, \text{ where } v^t = \varphi + t\varphi'^2, \quad \chi^t = \mathrm{id} + 2\varphi' t.$$

In distinction from the conventional exponent the superposition exponent is in square brackets: $X^{[-1]}$ is mapping reciprocal to X. The solutions $g^t v$ are defined and are smooth for all t for which X^t is diffeomorphic mapping. Smoothness of $(X^t)^{[-1]}$ is lost at such values of T_φ^+ and $T_\varphi^- < 0$ of the variable t for which

$$\chi^{t\prime} \equiv 1 + 2\varphi'' t = 0.$$

Whence

$$T^+_\varphi = -1/2\min\varphi'', \quad T^-_\varphi = -1/2\max\varphi'', \quad |T^\pm_\varphi| \ge 1/2\,\|\varphi\|_{C^2}$$

If $\max|\varphi''|$ or $\min|\varphi''|$ are small, then the norm $|\varphi|_1$ is also small. This consideration readily leads to inequality $|T_\varphi^+| \le \pi^2/|\varphi|_1$. Property 1° of Sect.3.5 is proven.

2°. Estimation of Norm n_3 Along the Solutions of the Hamilton-Jacobi Equations. Let us prove that at $\varphi \neq 0$, $t \in (T_\varphi^+/2, T_\varphi^+)$ there exists $C > 0$ such that

$$\|g^t\varphi\|_3^2 = \oint (\varphi^{(3)})^2/(1 + 2\varphi'' t)^5 \, d\xi \qquad (*)$$

$$cT^+_\varphi / (T^+_\varphi - t)^3 \le \|g^t\varphi\|_3^2 \le T^{+5}_\varphi / (T^+_\varphi - t)^5 \|\varphi\|_3^2.$$

Analogous inequalities are also valid at $t \in (T_\varphi^-/2, T_\varphi^-)$. The right-hand inequality is valid for $t \in (0, T_\varphi^+)$.

▹ The calculation with the mysterious abridgements yields

$$(g^t\varphi)_{x^3} \circ \chi^t = \varphi^{(3)}(\chi^{t'})^{-3}.$$

Then formula (*) is obtained after the change of the variable $\xi = X^t(x)$

in the integral $\oint (g^t\varphi)^2_{x^3} dx$.

2°. Let us prove the right-hand inequality (**) for $t \in (0, T_\varphi^+)$. Let us take $\psi = x^{t'} = 1 + 2\varphi''t$. Formula (*) takes the form: $\|g^t\varphi\|_3^2 = \oint(\varphi^{(3)})^2/\psi^5$. For $t \in (0, T_\varphi^+)$ we have

$$\min \psi = 1 + 2 \min \varphi'' \quad t = 1 - t/T_\varphi^+ = (T_\varphi^+ - t)/T_\varphi^+ .$$

The required inequality follows from the above equation and formula (8).

3°. Let us prove the left-hand inequality (**) at $t \in (T_\varphi^+/2, T_\varphi^+)$. In the notation of Sect.2°, $\varphi''' = \psi|/2t$. By formula (*) we have

$$\|g^t\varphi\|_3^2 = (4t^2)^{-1}\oint(\psi'^2/\psi^5)d\xi = (9t^2)^{-1}\oint (\psi^{-3/2})^2 d\xi$$

$$\overset{1}{\geq} (36\pi^2 T_\varphi^{+2})^{-1}(\oint (\bar\psi^{-3/2}) \quad d\xi)^2 = c_0 \quad (T_\varphi^+ \quad)^{-2} \quad (\mathrm{var}\ \bar\psi^{3/2}$$

$$\overset{2}{\geq} c_0(T_\varphi^+)^{-2}[(T_\varphi^+ - t)^{-3/2}\, T_\varphi^{+3/2} - 1]^2, \quad c_0 = (36\pi^2)$$

Here var f is the variation of function f, inequality 1 follows from the Cauchy-Bouniakowsky inequality; inequality 2 follows from the fact that $\min \psi = (T_\varphi^+ - t)/T_\varphi^+$ (see 2°); the function ψ takes the value 1 where $\varphi'' = 0$. Since $t \in (T_\varphi^+/2, T_\varphi^+)$, we have

$$[(T_\varphi^+ - t)^{-3/2}\, T_\varphi^{+3/2} - 1]^2 \geq (T_\varphi^+)^3/8(T_\varphi^+ - t)^3.$$

Whence the necessary inequality for sufficiently small C:

$$\|g^t\varphi\|_3^2 \geq cT_\varphi^+ / (T_\varphi^+ - t)^3 \blacktriangleright$$

5.4. Equation in Variations. Here property 4° from Sect.3. will be proven.
One more calculation with mysterious abridgements yields:

$$(dg^t)(\varphi)\psi \overset{\text{def}}{=} \frac{d}{d\varepsilon} g^t(\varphi + \varepsilon\psi)|_{\varepsilon=0} = \psi \circ \chi_\varphi^+$$

Now let us pass from the space $\mathring{H}_{-1}$ to the space $\mathring{H}_0$. On the space of functions with the zero mean the operator I is defined: f|F, F' = f, $\bar{f}$ = 0. The norm n_1 can be defined not only with the aid of the Fourier-expansion but also by the equality $\|f\|_{-1} = \|If\|_0$. To each bounded operator $A: \mathring{H}_{-1} \to \mathring{H}_{-1}$ defined everywhere on the dense subset $\mathring{H}_0 \subset \mathring{H}_{-1}$ there correspond an operator **B**: $\mathring{H}_0 \to \mathring{H}_0$ defined everywhere on the dense subset $\mathring{H}_1$ of the space $\mathring{H}_0$ and having the same norm: $\|A\|_{-1} = \|B\|_0$. Namely, Bf = IAf'. Let us prove the norm coincidence.

$$\|A\|_{-1} = \sup_{\psi\neq0} \frac{\|A\psi\|_{-1}}{\|\psi\|_{-1}} = \sup_{\psi\neq0} \frac{\|IA\psi\|_0}{\|I\psi\|_0} = \sup_{f\neq0} \frac{\|IAf'\|_0}{\|f\|_0},$$

where $f \in I\psi \in \mathring{H}_1$. In our case $A\psi = \psi_0 X^{t[-1]}$, $Bf = I(f'_0, \chi^{t[-1]})$. Let us find the upper estimate of the norm $\|B\|_0$. Let $f \in \mathring{H}_1 \neq 0$, $f_0\chi^{t[-1]}(x_0) = 0$

$g(x) = \int_{x_0}^{x} f\circ|\chi^{t[-1]} d\xi$, $x \in [x_0, x_0 + 2\pi]$, $g = Bf + a(x - x_0)$,

where a is the mean value of the funtion $f'\circ\chi^{t[-1]}$. Then $Bf = g - a(x - x_0)$.
Let us prove the following inequalities

$$\|g\|_0 \lesssim \|f\|_0 \, (\|\varphi\|_3 + \|\varphi\|^{3/2}_{C^2}) \, \|\varphi\|_1^{-1}; \quad \|a\| \lesssim \|f\|_0 \|\varphi\|_3 \|\varphi\|_1^{-1} \qquad (*)$$

Property 4° follows from the above inequalities and the Sobolev classification theorem for the space C^2. For brevity, we put $\chi^t = \chi$. Then we have

$$g(x) = \int_{x_0}^{x} f' \circ \chi^{[-1]} = \int_{x_0}^{x} (f \circ \chi^{[-1]})' \; \chi' \circ \chi^{[-1]} = f \circ \chi^{[-1]}(x) \; \chi' \circ \chi^{[-1]}(x) -$$

$$- \int_{x_0}^{x} (f \circ \chi^{[-1]} \; \chi'' \circ \chi^{[-1]} \; \chi^{[-1]\prime} = (f\chi') \circ \chi^{[-1]}(x) - \int_{\chi^{[-1]}(x_0)}^{\chi^{[-1]}(x)} f\chi''.$$

Now let us prove the following general inequality: $\|h \circ \chi^{[-1]}\|_0 \leq |$

$\|h\|_0 \sqrt{\max |\chi'|}$. |. Indeed,

$$\|h \circ \chi^{[-1]}\|_0^2 = \oint h^2 \circ \chi^{[-1]} = \oint \frac{h^2 \circ \chi^{[-1]} (\chi^{[-1]})'}{(\chi^{[-1]})'} \leq \|h\|_0^2 \max|\chi'|.$$

Whence

$$\|f \circ \chi^{[-1]} \; \chi' \circ \chi^{[-1]}\|_0 \leq \|f \circ \chi^{[-1]}\|_0 \|\chi\|_{C^1} \leq$$

$$\|f\|_0 \, \|\chi\|^{3/2}_{C^1} \lesssim \|f\|_0 \, \|\varphi\|^{3/2}_{C^2} \, \|\varphi\|_1^{-3/2} .$$

The latter inequality uses the expression for χ and the upper estimate on $|I_\varphi^{\pm}|$, Sect.5.1. For any α, $\beta \in S^1$ we have

$$|\int_\beta^\alpha f\chi'' | \le \|f\|_0 \|\chi''\|_0 \le 2t\|f\|_0 \|\varphi\|_3 \lesssim \|f\|_0 \|\varphi\|_3 \|\varphi\|_1^{-1} .$$

The first of inequalities (*) has been proven; now let us prove the second one. Taking into account the 2π-periodicity of f and the definition of a, we arrive at

$$a = g(x_0 + 2\pi)/2\pi = \oint f\chi''/2\pi;$$

$$a \lesssim \|f\|_0 \|\varphi\|_3 \|\varphi\|_1^{-1} \quad ▸$$

5.5. Proof of Proposition 3 of Sect.3.6.

Proposition 1. Let function $f \in H_1(S^1)$ go to zero, $\max |f| \ge C$, $|f'| \le C_1$. Then: (a) if $f = 0$, we have:

$$\max|If| \ge C^2/2C_1 \quad b) \oint|f| \ge C^2/C_1 \quad c) \|f\|_0 \ge C^{3/2}/\sqrt{2C_1}.$$

▹ The proof of the statement (a) is illustrated by Figure 4. The function f can be considered as a 2π-periodic function on a line. By condition, the f function goes to zero. Therefore the dashed triangle of area C^2/C_1 in Figure 4 decreases under the graph of the function f, its base decreases between two neighboring zeroes of f, i.e. within one the period. Therefore, the oscillation of any primitive for f on S^1 is not smaller than C^2/C_1; in particular, osc If $\ge C^2/C_1$. Whence follows the statement a. The statement b is proven in a similar way. Now let us prove statement (c). We have: $\max |f^2| \ge C^2$, $|(f^2)'| \le 2CC_1$. Therefore,

in virtue of statement (b) we have $\oint f^2 \ge C^4/2CC_1 = C^3/2C_1$,. ▸

Proposition 2. Let $\bar{f} = 0$, $\|f\|_0 = 1$, $|f'| < C_1$. Then

$$\|Pf^2\|_{-1} \gtrsim C_1^{3/2}.$$

▹ Let us estimate from above max $|IPf^2|$, using Proposition 1a, then let us estimate it from below $\|IPf^2\|_0 - \|Pf^2\|_{-1}$, using Proposition 1c. Let max $|f| = C$. Then

$$\max|Pf^2| \geq C^2/2, \quad |(Pf^2)'| \leq 2CC_1.$$

In virtue of Proposition 1a applied to the function Pf^2 we obtain:

$$\max|IPf^2| \geq C^3/16C_1.$$

From Proposition 1c applied to the function IPf^2 satisfying the previous inequality and the estimate $|(IPf^2)'| = |Pf^2| \leq 2C^2$, we obtain

$$\|IPf^2\|_0 = \|Pf^2\|_{-1} \gtrsim C^{7/2}C_1^{-3/2}$$

Finally, from inequality $\|f\|_0 = 1$ it follows that $C \geq 1/\sqrt{2\pi}$. Therefore

$$\|Pf^2\|_{-1} \gtrsim C_1^{-3/2}$$ ▸

Note. From the logarithmic convexity of the norm n_s over s and condition $\|\varphi\|_1 = 1$, $\|\varphi\|_3 \leq 2\nu^{-1}$ the lower estimate on $\|\varphi_x\|_{-1}$ is obtained. But no estimate on $|P\varphi_x^2|_{-1}$ is obtained. Indeed, if the function φ_x is equal to 1 on one semicircumference and to -1 on the other, then $P\varphi_x^2 = 0$.

Let us prove the first statement of Proposition 3 from Sect.3.6: $\|P\varphi_x^2\|_{-1} \gtrsim \nu^{1,2}$. Let us put $\varphi_x = f$. By condition, we have $\|f\|_0 = 1$.

According to Proposition 4 of Sect.3.6, $|f'| \lesssim v^{-4/5}$. Now it follows from Proposition 2:

$$\|Pf^2\|_{-1} \gtrsim \nu^{1,2} \blacktriangleright$$

Prior to the proof of the second statement let us make the following note

$$P(\varphi_x^2 - \psi_x^2) = P\,[(\varphi_x - \psi_x)(\varphi_x + \psi_x)].$$

The function $\varphi_x - \varphi_x$ is small in H_{-1}, and the funtion $\varphi_x + \varphi_x$ is bounded, but multiplication by the bounded function with the subsequent porojecting P is not the bounded operator in $\mathring{H}_{-1}$. Therefore the following proposition is natural.

<u>Proposition</u> 3. Let $f \in H_{-1}$ and $g \in H_1$. Then

$$\|P(fg)\|_{-1} \le \|f\|_{-1}\|g\|_1.$$

▷ The formula for integration by parts in the space of functions with the zero mean on S^1 is:

$$I(P(fg)) = I\,[\,((If)g)' - P((If)g')\,] =$$
$$= P((If)g) - IP((If)g').$$

Let us estimate the first term

$$\|P((If)g)\|_0 \le \|(If)g\|_0 \le \|If\|_0 \max|g| \lesssim \|f\|_{-1}\|g\|_1$$

Let us estimate the second term

$$|IPh| \le 2\oint|h|$$

$$\|IP((If)g')\|_0 \lesssim \oint|(If)g'| \le \|If\|_0\|g'\|_0 = \|f\|_{-1}\|g\|_1 \blacktriangleright$$

Let us prove the second statement of Proposition 3 of Sect.3.6. Everywhere on D we have $n_1 \le 2$, if σ is sufficiently small. Moreover $n_0 \le 2$; $|\varphi_x + \psi_x|_{-1} = |\varphi + \psi|_0 \le 4$. Then

$$\|\varphi_x - \psi_x\|_1 = \|\varphi - \psi\|_2 \lesssim \|\varphi - \psi\|_{c^2} \le \sigma \lesssim \nu_{1,2}.$$

From Proposition 3 it follows now that for $\varphi,\psi \in D$ we have

$$\|P(\varphi_x^2 - \psi_x^2)\|_{-1} \lesssim \sigma \lesssim \nu_{1,2}.$$

Finally, if the difference between the vectors is small in comparison with their lengths, then the angle formed by these vectors is smaller than $\pi/3$. $\triangleright$

REFERENCES

1. D.J. Beney, "Long waves in liquid film." J. Math. and Phys., 1966, 5,150-155.
2. G. Sivashinsky and D.M. Michelson. "On irregular wavy flow of a liquid down a vertical plane." Prog. Theor. Phys., 63, 1980, 2112-2114.
3. B. Nikolaenko, B. Scheurer, and R. Temam. "Some global dynamical properties of the Kuramoto-Sivashinsky equations: nonlinear stability and attractors." Phisica, 16D, 1985, 155-183.
4. Y. Kuramoto. "Diffusion-induced chaos in reactions systems." Supp.Progr.Theor. Phys., 64, 1978, 346-367.

5. A.V. Gabin and M.I. Vishik. "Attractors of evolution equations with partial derivatives and the estimates of their dimensionalities." UMN, 1983, 38: 4, 133-187 (in Russian).
6. O.A. Ladyzhenskaya. "On finite dimensionality of bounded invariant sets for the Navier-Stokes and other dissipative systems." Zapiski Nauchnykh Seminarov LOMI, 1982, 115, 137-155 (in Russian).
7. Yu.S. Il'yashenko. "On dimensionality of the attractors of k-systems in an infinite dimensional space." Vestnik MGU, ser. mat. mekh., 1983, No.3, 52-59 (in Russian).
8. P. Constantin, C. Foias, and R. Temam. Attractors, representing turbulent flows. Memoir of AMS.
9. Yu.S. Il'yashenko. "Weakly compressing systems and attractors of the Galerkin approximations of the Navier-Stokes equation on a two-dimensional tore." Uspekhi Mekhaniki, 1982, 5, 1/2, 31-63 (in Russian).
10. A. Douady and J. Oesterle. "Dimension de Hausdorff des attracteurs." C.R.Acad.Sci., 1980, 290, 1135-1138.
11. R. Feller. Probability theory. vol.1. Moscow: Mir, 1967 (in Russian).
12. Yu.S. Il'yashenko. "Weakly compressive dynamical systems and attractors." UMN, 1982, 37, 1, 166 (in Russian).
13. V.I. Arnol'd. Additional chapters of the theory of differential equations. Moscow: Nauka, 1978 (in Russian).

---

*/The results [9,10] relate to dissipative systems. But after proposition in Sect.1.6 they can be fully used for generalized dissipative systems.

Mathematics
and Modelling

ISBN 1-56072-104-9

SYMPLECTIC ORBITAL EQUIVALENCE OF VECTOR FIELDS ON PLANE (Elementary Singular Points)

R.I. Bogdanov

A singular point of a vector field on a plane is called elementary if the linearization matrix of the vector field at the singular point is not nilpotent.

The main result of the paper is the following theorem:

In the vicinity of an elementary singular point of any vector field on a plane except of a set of codimensionality infinity in the space of all the vector fields there exists the change of coordinate of the class C^{∞}, preserving the phase volume on the plane, such that the first integral of the vector field in new variables is given by an elementary function of coordinates.

A similar theorem without the requirement of phase-volume preservation upon change of coordinates was obtained in [4]. In addition to the first integral, other invariants of elementary singular points are given [4]. The table of invariants given in [4] is valid also for the change of coordinates preserving the phase volume on a plane except in all but one case. This case, hereinafter called symplectically separated, is characterized by that the linearization of the vector field at the singular point is not zero-divergent and has the nontrivial homogeneous first integral explicitly dependent on both coordinates on the plane (in the notation of Ref. [4] $A_j^{\pm}$, $\lambda = -m/n < 0, \lambda \neq | -1$, a resonance saddle with linearization that is not a Hamiltonian vector field).

The above formulated theorem is more convenient for applications than the similar theorem from Ref. [4]. Indeed, under the action of the group of the change of phase variables preserving the phase volume, the transformation of the vector-field integrating factor on the vector-field space is simplified. The integrating factor is transformed under such an action as function. Thus it is possible to define in the vicinity of an elementary singular point a following pair of functions: the first integral and the integrating factor (more exactly, the algebra of the first integrals and the module with respect to multiplication over this algebra in the ring of functions of phase variables).

1. INTRODUCTION

1°. The action of the group of diffeomorphisms of the phase space on the vector-field space was systematically used by Poincaré [8]. Let us denote by Diff a group of diffeomorphisms that leaves a separated point of the phase space at its place. Studying the action of group Diff on the vector-field space with a separated point (a stationar of the vector field), Poincaré formulated a theorem of reduction of a vector field to a linear one. Since a linear vector field is explicitly integrable, all the invariants of singular points satisfying the conditions of the Poincaré theorem can be calculated.

2°. Extension of the group Diff with the aid of the multiplicative group in a ring of functions of two variables.

The multiplication of a vector field by a function that does not go to zero on the phase space does not change the first integral of the vector field. Therefore, studying the first integral of the vector field, it is possible to use both the action of the group Diff on the vector-field space and the multiplication of the vector field by functions not vanishing on the phase space. Let us denote the multiplicative group of positiove functions on the phase space by ε_2^+.

Now let Diff^+ denote the semidirect product of the group Diff by the group ε_2^+ with such multiplication law which obeys the above indicated action of the pairs (g,f), $g \in \mathrm{Diff}$, $f \in \varepsilon_2^+$ on the vector-field space (for details see [5]).

3°. The orbital equivalence of two vector field signifies, by definition, that these two vector fields lie in one orbit of the action of the group Diff^{+} on the vector-field space.

The classification of elementary singular points of vector fields with respect to the action of the group Diff^{+} on the vector-field space was suggested in [3].

4°. Symplectic Orbital Equivalence. Let us fix the area element on the plane, $dx_1 \wedge dx_2$. Then it is possible to single out in the group Diff a subgroup of diffeomorphisms called symplectomorphisms that preserve the area element $dx_1 \wedge dx_2$ (see [2]). We denote the group of symplectomorphisms by Sp. Similar to the procedure in Sect.2°, an extension of the group is constructed $Sp \subset S_p^{+} = S_p \times \varepsilon_2^{+}$. The group S_p^{+} acts on the vector-fields space and, similar to the procedure considered in Sect. 3°, defines the symplectic orbital equivalence of vector fields.

Thus two vector fields are symplectically orbitally equivalent if one of them is transformed into the other with the aid of an appropriate symplectomorphism and multiplication by an appropriate positive function on the phase space.

5°. Invariants of Symplectic Orbital Equivalence. Let two vector fields be symplectically orbitally equivalent. Then there exists a pair (g, f), $g \in S_p$, $f \in \varepsilon_2^{+}$ that provides the symplectic orbital equivalence. Further on, let H be the first integral of one vector field, and $1/\alpha$ be an integrating factor. Then $g^{*}(H)$ is the first integral and $1/fg^{*}\alpha$ is an integrating factor of the other vector field.

Another possibility of interpreting the function f is based on the fact that this function defines the normalization of the velocity of motion along the phase curves of one vector field along symplectically orbitalally equivalent one. The normalization obeys the condition that $1/g^{*}(\alpha)$ is an integrating factor of the normalized vector field.

6°. For applications using numerical methods it is useful to have asymptotic functions at singular points of the vector fields of both first integrals and invariant measure. The use of the computers in calculations is very efficient if there is a priori information about the asymptotic behavior of these first integrals. In particular, it is necessary to know the character of singularities treated in the computations. From this standpoint, the knowledge of the first integral with the allowance for the nonlinearity of differential equations provides the computations in the vicinity of elementary singular points. The invariant measure is reconstructed with due regard of Sect.5° by the pair of funtions - f and $g^{*}(\alpha)$. Note also that f is a smooth function and simgularity of $g^{*}(\alpha)$ is defined by singularity of the function α.

1.1. Formulation of Results

1.1.1. Definition. A singular point of a vector field on plane is called symplectically finite modal if its class of symplectic orbital equivalence is of a finite codimensionality*/ in the space of all vector fields.

1.1.2. Lemma. Let a singular point of a vector field on a plane be symplectically finite modal. Then this simgular point is elementary (i.e., the linearization matrix of the vector field at the singular point is not nilpotent).

1.1.3. Lemma. Let a singular point of the vector field on a plane be elementary. Then such a singular point is symplectically finite modal in the space of all vector fields except of a set of vector fields of codimension infinity.

1.1.4. Theorem. Symplectically finite modal singular points of vector fields on a plane are classified in Table 1 (i.e. variouos classes listed in Table 1 are symplectically orbitally nonequivalent).

1.1.5. Definition. An elementary point of the vector field on a plane is called symplectycally separated, if the linearization of the vector field at the singular point has a homogeneous first integral of the form $H = (ax_1 + bx_2)^m(cx_1 + dx_2)^n$ with $m,n, \in Z$, $m > 0$, $n > 0$, $m \neq n$, $a,b,c,d \in \mathbf{R}$, $ad - bc \neq 0$.

1.1.6. Corollary. The normal form (see Table 1) of an elementary singular point which is not symplectically separated with respect to the symplectically orbital equivalence coincides with the normal form with respect to the orbital equivalence of vector fields.

Symbol	V	Φ	$1/\alpha$	ξ	Notation
	$sx_1\partial_1 + \lambda x_2\partial_2$	$x_1^{-\lambda}x_2^{s}$	$1/x_1x_2$	e	nonresonance (s,λ)
	$\pm x_2\partial_1 + se$	$e^{u}x_2^{\mp s}$	$1/x_2^2$	e	$u = x_1/x_2$
	$\pm IdH/2 + ce$	$e^{\mp 2cU}H$	$1/H$	e	$c \neq 0$, $u = \tan^{-1} x1/x2$, $H = x_1^2 + x$
$SA_j^{\pm}$	$\alpha_0 Id\ln U + Q(u)Id\alpha_0$	$\frac{\alpha_0^{(s/U^j+c)^{f/j}}}{e^{s/jU^j}}$	$\frac{1}{\alpha_0 Q(u)}$	$\frac{Q(u)Id\alpha_0}{\{\alpha_0, \ln U\}}$	$\alpha_0 = x_1 x_2$, $H_0 = x_1^m x_2^n$, $m \neq n$, $U = H_0\exp(\sum_{l=1}^{j} c_l H_0^l$
$A_j^{\pm}$	$IdH + Q(H)e$	$\frac{U^j e^{sV}}{(c+sV)^c}$	$\frac{1}{x_1x_2Q(H)}$	$Q(H)e$	$H = x_1x_2$, $U = x_1/x_2$, $V = H^{-j}$
$C_j^{\pm}$	$IdH/2 + Q(H)e$	$\frac{e^{2jU \pm sV}}{(c+sV)^{\pm c}}$	$\frac{1}{HQ(H)}$	$Q(H)e$	$H = x_1^2 + x$, $V = H$, $U = \tan^{-1} x_1/x_2$
$B_j^{\pm}$	$Q(H)x_1\partial_1 + px_2\partial_2$	$\frac{u^{pj}e^{sV}}{(c+sV)^c}$	$\frac{1}{x_1x_2Q(H)}$	$x_2\ 2$	$p = \pm 1$, $H = x_1$, $V = H^{-j}$, $U = x_2$
$D_j^{\pm}$	$s(IdH + W)$	$\frac{e^U}{H^{p/(j+1)}}$	$\frac{1}{x_1^{j+1}}$	W	$W = x_1\partial_1 + jx_2\partial_2$, $p=$ $H = -px_1^{j+1}/(j+1)$, $U=x$

$c = x_1\partial_1 + x_2\partial_2$; $Q(H) = sH^j + cH^{2j}$; j is the index in the series, s
Φ is the first integral, $\alpha(1/\alpha)$ is the integrating divisor (factor)
ξ is a generator of stationary algebra $\varepsilon_2(x_1, x_2)\ V \ni \varepsilon_1(\Phi)\xi$

$SA_j^{\pm}$ is symplectically separated case. Equivalent form

$$\alpha_0 = x_1x_2\prod_{k=1}^{j} 1_k,\quad U = H_0\prod_{k=1}^{j} 1_k^{-1/k},\quad 1_k = 1 - c_k k H_{k-1}^{k}/\alpha_{k-1},$$

$$H_k = H_0\prod_{l=1}^{k} 1_l^{-1/l},\quad \alpha_k = x_1x_2\prod_{l=1}^{k} 1_l,\quad \{\alpha_0, \ln U\} = \{x_1x_2, \ln H_0\}\prod_{k=1}^{j} 1_k^{\bullet}.$$

$(1-\varepsilon)^{\bullet} = 1 + \varepsilon$, $\{,\}$ - Poisson braces

Table 1. Symplectic orbital normal forms of elementary singular points of vector fields on a plane.

1.2. Plan of the Paper

Below we state in an abridged form the proof of Theorem 1.1.4. The necessary calculations are made till the end only for symplectically separated elementary singular points of vector fields on a plane. At the same time, we shall also prove Lemma 1.1.3. The proof of Lemma 1.1.2 follows from the proof of a similar lemma for the action of the group $Diff^+$ on the vector-field space considered elsewhere*/ [3]. Indeed, the group $S_p^+ \subset Diff^+$ and therefore symplectically finite modal singular points are finite modal with respect to the action of the group $Diff^+$ on the vector-field space.

In Sect.2 we shall remind the main method for studying the orbital equivalence of vector fields on a plane: the transition from the action of the Lie group to the respective Lie-algebra representation and then reduction to the calculations in bivector fields (for details see [3]). Then the modification with respect to the action of the S_p^* group on the vector-field space is given.

In Sect.3 we give the necessary calculations described earlier in Sect.2 in the application to the symplectically separated elementary singular points of vector fields on a plane.

The main abridgement made in the proof of Theorem 1.1.4 given below reduces to the omission of the algebraic formalism associated with homologies of short complexes and the technique of the spectral sequences for their calculations. Such an abridgement was possible because of existing voluminous texts to the effect (see, e.g., [1], [3], and [7]).

1.3. Main Notation

$\mathbf{R}$, Z	is the ring of real (integer) numbers;
$C^\infty(x_1,x_2)$	is a ring of germs at the point $x_1 = x_2 = 0$ of the functions of class C^∞;
$\varepsilon_j(x)$	is a ring of the germs of functions of j variables at point $x = 0$
$\varepsilon_2^+(x_1,x_2)$	is multiplicative subgroup $C^\infty(x_1,x_2)$:
$\{f \in C^\infty(x_1,x_2)\colon f(0,0) > 0\}$;	
Diff, Sp	is a group (or subgroup) of germs at point (0) $\mathbf{R}^2$ of (0) $\in \mathbf{R}^2$ diffeomorphisms (symplectomorphisms) of the phase plane $\mathbf{R}^2$;
$Diff^+ = Diff \times \varepsilon_2^+$	is a semidirect product of groups Diff and ε_2^+;
$Sp^+ = Sp \times \varepsilon_2^+$	is a semidirect product of groups S_p and ε_2^+;
$V = \alpha \operatorname{Id} \Phi$	is a germ at point $(0) \in \mathbf{R}^2$ of the vector field on the phase plane $\mathbf{R}^2$;
α	is an integrating factor;
Φ	is the first integral;

Id(•)	is the Hamiltonian vector field with Hamiltonian (•) $\in \varepsilon^2$; div V = $\{\alpha,\Phi\}$ is the divergence of the vector field V;
{•,•}	are the Poisson braces of two functions (•) $\in \varepsilon_2$;
•∧•	is the outer product of two vector fields (•);
L_V	is the Lie derivative along the vector field V;
$\{1\} \subset C^\infty(x_1,x_2)$	is the group of identity in the ring $C^\infty(x_1,x_2)$;
$(\bullet)^*$	is conjugacy in the group $\{1\}:(1 + h)^* = 1 - h$, $h(0,0) = 0$;

2. LENEARIZATION OF THE ACTION OF GROUP S_p^+ ON SPACE OF VECTOR FIELDS

2.1. Reduction to Bivector Fields

2.1.1. The Lie Algebra of the S_p^+ Group is formed by the pairs (ξ,f) where ξ is the zero-divergence vector field on a plane, and f is a smooth function (not necessarily definite) on the plane. The Hamiltonian of the zero-divergence field is denoted by S where $S(0) = 0$, $(0) \in \mathbb{R}^2$, $\xi = \mathrm{Id}S = \partial S_{/\partial x_2} \bullet \partial_{/\partial x_1} - \partial S_{/\partial x_1} \bullet \partial_{/\partial x_2}$.

2.1.2. In the group S_p^+ a single parametric subgroup corresponding to an algebraic element $(\xi,f) \in J\,Sp^+$ acts on the vector field V. The rate of the change of the vector field V in the unit element of the group S_p^+ under the action of the single-parametric subgroup is given by the expression

$$L_\xi(V) + f\bullet V. \tag{1}$$

The image of linear operator (1) when (ξ,f) runs over the algebra JS_p^+ yields the tangential space at the point V to the orbit of the vector field V under the action of the S_p^+ group on the vector-field space.

2.1.3. Multiplying external (1) by the vector field V, we arrive after several simple transformations (see, e.g., [3]) at

$$L_V(V \wedge \xi). \tag{2}$$

Thus, the vector field η is at the point V of the space tangential to the action orbit of the S_p^+ group on the vector-field space if

$$\eta \wedge V + L_V(\xi \wedge V) = 0. \tag{3}$$

2.1.4. Note. Thus, the problem of reducing the vector field to the "simplest form" with respect to the symplectic orbital equivalence is divided into three stages:

(a) the calculation of the "difference" in bivector fields $\eta \wedge V$ (where η is infinitesimal deformation and V is an arbitrary vector field) which we want to "kill" by an appropriate symplectomorphism and multiplication by an appropriate funcion;

(b) the calculation of the projection of the Lie algebra of the symplectomorphisms in the bivector-field space $\xi \wedge V$ and $\xi \in JS_p^+$;

(c) the calculation of the image of the Lie derivative along the vector field V that acts on the spaces of bivector-field space described in (b).

2.2. Obstructions to Trivialization of the Symplectic Normal Form (Smooth Integrating Divisor and First Integral)

2.2.1. Note. The projection of zero-divergence vector fields onto the space of bivector fields obtained with the aid of outer multiplication by the vector field V is generated by the action of the Lie derivative along the vector field V on a space of smooth functions of phase variables. Indeed,

$$\xi \wedge V = \mathrm{IdS} \wedge V = L_V(S)\, \partial_1 \wedge \partial_2.$$

2.2.2. Lemma. Let $V = \alpha \mathrm{IdH}$. Then

(1) $\{\alpha,H\} = \partial\alpha/\partial x_1 \cdot \partial H/\partial x_2 - \partial\alpha/\partial x_2 \cdot \partial H/\partial x_1 = \operatorname{div} V$;

(2) $L_V(G(\alpha,H)\partial_1\wedge\partial_2) = (\alpha\, \partial G/\partial\alpha - G)\, \{\alpha,H\}\, \partial_1\wedge\partial_2$;

(3) $\mathrm{IdS} \wedge V = \alpha\, \{S,H\}\, \partial_1\wedge\partial_2 = \alpha\, \partial S/\partial\alpha \cdot \{\alpha,H\}\, \partial_1\wedge\partial_2$.

2.2.3. Note. The proof of formulae (1)-(3) of Lemma 2.2.2 is formally obvious. We leave the formal checking of it to the reader. The meaning of these formal transformations is not obvious. The reader can

make certain that for the vector fields listed in Table 1 formulae (1)-(3) have concrete meaning. To have a useful guide, we refer the reader to [3] where these formulae are used systematically. All the notions of homological algebra in application to calculation of orbital normal forms of vector fields on a plane necessary to justify the used formalism are also stated in [3].
This note is true to even greater extent for Lemma 2.2.4.

2.2.4. Lemma. Let $V = \alpha \mathrm{Id} H$, where α and H are smooth functions. Then

$$\mathrm{Coim}L_V \Big|_{\cdot \Lambda V} \cong \left[\alpha \epsilon_1(H)\{\alpha,H\} + \epsilon_1(H)\left[\alpha\{\{\alpha,H\},H\} - \{\alpha,H\}^2\right]\right] \partial_1 \Lambda \partial_2, (2,1)$$

where $(\bullet) \in J\,Sp$.

2.2.5. Lemma. The obstructions to trivialization of the symplectic orbital normal form of the vector field V in the vector-field space have the following form in the notation of the lemma 2.2.4:
$\alpha\, \varepsilon_1(H)\, \mathrm{Id}(\mathrm{div}\, V_{/\alpha}) + \varepsilon_1(H)\, \mathrm{Id}\alpha$

2.2.6. Corollary. Deformation of the vector field V along the vector field $\eta \in \alpha\varepsilon_1(H)\mathrm{Id}(\mathrm{div}V)/\alpha)$ or $\eta \in \varepsilon_1(H)\mathrm{Id}\alpha$ generates the vector field integrable in quadratures.
Indeed, in the first case

$$V + \eta = \alpha \mathrm{Id} H + \alpha f(H) \mathrm{Id}(\mathrm{div}\, V_{/\alpha}) = \alpha f(H) \mathrm{Id} \left[\int \frac{dH}{f(H)} + \frac{\mathrm{div}\, V}{\alpha}\right] \quad (2.2)$$

in the second case

$$V + \eta = \alpha \mathrm{Id} H + f(H) \mathrm{Id}\alpha = \alpha f(H) \mathrm{Id} \left[\int \frac{dH}{f(H)} + \ln \alpha \right] \quad (2.3)$$

2.2.7. Under the conditions of Corollary 2.2.6, the integrating deformation factor is expressed algebraically (without quadratures) in terms of the integrating factor of the vector field V and integrating factor of deformation.

2.2.8. Elucidation. The terms in Lemma 2.2.4 are of different nature. The first one is associated with

$$\operatorname{Ker} L_V \Big|_{\varepsilon_2(x_1,x_2)\partial_1\Lambda\partial_2} = \alpha\, \varepsilon_1(H)\partial_1\Lambda\partial_2.$$

The second one is associated with JSp $\in$ JDiff and, in particular,

$$(\text{J Diff}\ \Lambda\ V)_{/(\text{J Sp}\ \Lambda\ V)} \cong \varepsilon_1(H)\{\alpha,H\}\partial_1\Lambda\partial_2.$$

2.3. Obstructions to Trivialization of the Symplectic Orbital Normal Form (Smooth Integrating Divisor and the First Singular Integral)

2.3.1. Lemma. Let $V = \alpha \operatorname{Id} \ln H + cH^r \operatorname{Id}\alpha$, where α and H are smooth functions, $c \in \mathbf{R}$ and $c \neq 0$. Then the obstructions to trivialization of the symplectic orbital normal form are
$\mathbf{R} < H^{2r}\operatorname{Id}\alpha >$.

2.3.2. Corollary. In the notation of Lemma 2.3.1 the symplectic orbital normal form is

$$\alpha \operatorname{Id} \ln H + (cH^r + aH^{2r}) \operatorname{Id}\alpha =$$

$$= \alpha(cH^r + aH^{2r}) \operatorname{Id} \left[\int \frac{dH}{H^{r+1}(c + aH^r)} + \ln\alpha \right] \qquad (2.4)$$

2.3.3. Note. In the notation of Lemma 2.3.1, the vector field V has a smooth integrating divisor αH^r and the plane first integral on the "curve" $H = 0$. The proof of this note follows from formula (2,4) at $a = 0$.

The note explains the title of Section 2.3.

2.3.4. The Proof of Lemma 2.3.1. The direct calculation yields

$$L_V(G(\alpha,H)\partial_1\Lambda\partial_2) =$$

$$= \left[\alpha\frac{\partial G}{\partial \alpha} - cH^{r+1} \frac{\partial G}{\partial H} - G(1 - crH^r) \right] \frac{\{\alpha,H\}}{H}\partial_1 \Lambda\partial_2.$$

Putting $G = \alpha^m H^n$, we arrive at

$$L_V(G(\alpha,H)\ \partial_1\Lambda\partial_2) = \alpha^m\ H^{n-1}\ [m - cnH^r - (1 - crH^r)]\ \{\alpha,H\}\partial_1\Lambda\partial_2.$$

Thus we obtain the kernel $I_V \mid_{\varepsilon_2(\alpha,H)\partial_1\Lambda\partial_2}$ $\alpha H^r\partial_1 \Lambda \partial_2$ and coimage $L_V \mid_{\varepsilon_2(\alpha,H)\partial_1\Lambda\partial_2} - \mathbf{R} < \alpha H^{2r-1}\{\alpha,H\}\partial_1 \Lambda \partial_2 >$. The vector field corresponding to such coimage is

$$\mathbf{R} < H^{2r} I d\alpha >.$$

3. CALCULATIONS OF NORMAL FORMS IN SYMPLECTICALLY SEPARATED CASE

3.1. Obstructions to Trivialization of Normal Form Determined by Linearization of Symplectically Separated Elementary Singular Point of a Vector Field

3.1.1. Lemma. A vector field in the vicinity of a symplectically separated elementary singular point is symplectically orbitally equivalent to the following field

$$V = (nx_1\ \partial_{/\partial x_1} - mx_2\ \partial_{/\partial x_2} + o(|x|), \tag{3.1}$$

where $m,n \in Z$, $m > 0$, $n > 0$, $m \neq n$, $(m,n) = 1$.

3.1.2. The proof of Lemma 3.1.1 follows from the consideration of the action of the group $GL(2,\mathbf{R})$ on a space of linear vector fields. The group $GL(2,\mathbf{R})$ is factorized over the subgroup $SL(2,\mathbf{R})$. The scalar factor that arises in this operation is "killed by the absolute value" due to the action of the second multiplier ε_p^+ in the group $S_p^+ = S_p \times \varepsilon_2^+$.

3.1.3. Corollary. To linearization of the vector field (3.1) there corresponds the following representation

$$V_0 = \alpha_0 \mathrm{Id} \ln H_0, \tag{3.2}$$

where $\alpha_0 = x_1 x_2$, $H_0 = x_1^m x_2^n$.

3.1.4. Corollary. Except of the cases of codimension | in the space of all vector fields the vector field in the vicinity of symplectically separated elementary singular point is symplectically orbitally equivalent to the vector field of the form

$$V_0 + \sum_{j=1}^{r} c_j \frac{H_0}{\alpha_0} e + (sH_0^r + cH_0^{2r})e, \tag{3.3}$$

where $e = x_1 \partial/\partial x_1 + x_2 \partial/\partial x_2$, $s = \pm 1$, $c_j \in \mathbf{R}$, $c \in \mathbf{R}$.

3.1.5. The proof of Corollary 3.1.4 follows from Lemma 2.2.4 with due regard for the following note. The terms of the form $H_0^j/\alpha_0 e$ in the outer multiplication by V_0 generate the first term from the coimage considered in Lemma 2.2.4. The second term from the coimage indicated in Lemma 2.2.4 has the corrresponding vector field of the form $H_0^r e$ in the sum (3.3). It should be noted that the obstructions to trivialization of the symplectic orbital normal form vanish if we take into account the term $cH_0^{2r}e$ in (3.3). This fact will be considered in more detail later on.

3.2. Explicitly Integrable Symplectic Normal Form with Multiplicatively Modified α and H

3.2.1. Lemma. Except of the cases of codimension ∞,| in the space of all vectors the vector field in the vicinity of symplectically separated elementary singular point is symplectically orbitally equivalent to

$$\alpha \mathrm{Id} \ln U + (sU^r + cU^{2r}) \mathrm{Id}\alpha,$$

where $s = \pm 1$, $c \in \mathbb{R}$;

$\alpha = \alpha_0 | 1$, $U = H_0 \tilde{} 1$, 1 and $\tilde{1}$ are the unit elements of the ring $C^\infty(x_1,x_2) \cap \varepsilon_2(H_0,\alpha_0)$, in this case

$$1 = \prod_{j=1}^{r} 1_j, \quad \tilde{1} = \prod_{j=1}^{r} 1_j^{-1/j}, \ \text{div}\ \alpha\ \text{Id}\ \ln U =$$

$$\prod_{j=1}^{r} 1_j^* \ |\text{div}\alpha_0 \text{Id}\ \ln H_0,$$

$$1_j = 1 - c_j j\ H_{j-1}{}^j \alpha_{j-1},\ H_j = H_0 \prod_{l=1}^{j-1} 1_l^{-1/l}, \qquad \alpha_j = \alpha_0 \prod_{l=1}^{j-1} 1_l.$$

3.2.2. Lemma 3.2.1 is proved by induction. The initial induction step is: $\alpha_0 = x_1 x_2$, $H_0 = x_1{}^m x_2{}^n$, $V_0 = \Lambda x = \alpha_0 \text{Id}\ \ln H_0$.
The first (auxilliary) induction step . Let us assume that $V_1 = V_0 + \eta_1$, where $\eta_1 = c_j \alpha_0 H_0{}^j \text{Id}(\text{div}\ V_0/\alpha_0)$.
Then in virtue of (2.2)

$$V_1 = \alpha_0 \text{Id}\ \ln H_0 + c_j \alpha_0 H_0^j \text{Id}(\text{div}\ V_0/\alpha_0) = \alpha_0 H_0^j \text{Id}\left[\int \frac{H_0}{H_0^{j+1}} + c_j \frac{\text{div}\ V_0}{\alpha_0}\right] =$$

$$= \alpha_0 H_0 \text{Id}\left[- \frac{1}{jH_0^j} + c_j \frac{(n-m)}{\alpha_0}\right] =$$

$$= \alpha_0\left[-\frac{1}{j} + c_j\frac{(n-m)H_0^j}{\alpha_0}\right]\frac{\left[-\frac{1}{jH_0^j} + c_j\frac{(n-m)}{\alpha_0}\right]}{\left[-\frac{1}{jH_0^j} + c_j\frac{(n-m)}{\alpha_0}\right]} =$$

$$= -\alpha_0\left[\frac{1}{j} - c_j\frac{(n-m)H_0^j}{\alpha_0}\right] \mathrm{Id}\ \ln\frac{1}{H_0^j}\left[-\frac{1}{j} + c_j\frac{(n-m)H_0^j}{\alpha_0}\right] =$$

$$= \alpha_0\left[\frac{1}{j} - c_j\frac{(n-m)H_0^j}{\alpha_0}\right] j\ \mathrm{Id}\ \ln \mathbf{H_0}\left[\frac{1}{j} - c_j\frac{(n-m)H_0^j}{\alpha_0}\right]^{-1/j}$$

$$= \alpha_0(1 - c_j(n-m)jH_0^j/\alpha_0)\ \mathrm{Id}\ \ln H_0[1/j - c_j(n-m)H_0^j/\alpha_0]^{-1/j} =$$

$$= \alpha_0(1 - c_j(n-m)jH_0^j/\alpha_0)\ \mathrm{Id}\ \ln H_0[1 - c_j(n-m)jH_0^j/\alpha_0]^{-1/j} =$$
$$= \alpha_0\ 1(H_0,\alpha_0)\ \mathrm{Id}\ \ln H_0\tilde{1}(H_0,\alpha_0).$$

Thus

$$\left.\begin{array}{l}\alpha_1 = \alpha_0|1(H_0,\alpha_0)\\ \\ H_1 = H_0|1|(H_0,\alpha_0)\end{array}\right\}\begin{array}{l}V_1 = \alpha_1 \mathrm{Id}\ \ln H_1\\ \\ 1| = 1^{-1/j}\end{array}$$

Induction hypothesis. Let us put

$$V_i = \alpha_i \mathrm{Id}\ \ln H_i,$$

where

$$\alpha_i = \alpha_0|1(H_0,\alpha_0) \tag{3.4}$$

$$H_i = H_0|1|(H_0,\alpha_0).$$

Let us perform the induction step under the assumption that

$$V_{i+1} = V_i + \eta_i,\ \eta_i \in \alpha_i \varepsilon_1(H_i) Id(\mathrm{div}\ V_i/\alpha_i).$$

Without reducing the generality, we put

$$\eta_i = c_i \alpha_i\ H_i^{k_i}\ Id(\mathrm{div}\ V_i/\alpha_i).$$

Then, in virtue of (2.2) we have

$$V_{i+1} = \alpha_i Id\ \ln H_i + c_i \alpha_i\ H_i^{k_i}\ Id(\mathrm{div}\ V_i/\alpha_i) =$$

$$= \alpha_i\ H_i^{k_i}\ Id\left[\int \frac{dH_i}{H_i^{k_i+1}} + c_i\ \ \mathrm{div}\ V_i/\alpha_i\right] =$$

i_1

$$= \alpha_i\ H_i^{k_i}\ Id\left[-\frac{1}{k_i H_i^{k_i}} + c_i\ \ \mathrm{div}\ V_i/\alpha_i\right] =$$

$$= -\alpha_i\left[\frac{1}{k_i} - c_i\ \ \mathrm{div}\ V_i\ H_i^{k_i}\ /\alpha_i\right] Id\ \ln\ \frac{1}{H_i^{k_i}}\left[\frac{1}{k_i} - c_i\ \ \mathrm{div}\right.$$

$$\left. V_i\ H_i^{k_i}\ /\alpha_i\right] =$$

$$= \alpha_i(1 - c_i k_i \mathrm{div}\ V_i\ H_i^{k_i}\ /\alpha_i)\ Id\ \ln H_i\left[\frac{1}{k_i} - c_i\ \ \mathrm{div}\ V_i\ H_i^{k_i}\ /\alpha_i\right] =$$

$$= \alpha_i(1 - c_i k_i \mathrm{div}\ V_i\ H_i^{k_i}\ /\alpha_i)\ Id\ \ln H_i(1 - c_i k_i \mathrm{div}\ V_i\ H_i^{k_i}\ /\alpha_i)^{-1/k_i}.$$

Note also that in virtue of (3.4):

$$\operatorname{div} V_i = \frac{1}{H_i}\{\alpha_i, H_i\} = \frac{1}{H_i}\left\{\alpha_0\, 1(H_0,\alpha_0),\ H_0\,\tilde{}1(H_0,\alpha_0)\right\} =$$

$$= \left[1\ \tilde{}1\{\alpha_0,H_0\} + H_0\, 1\{\alpha_0, 1\} + \alpha_0\tilde{}1\{1,H_0\}\right.\ \frac{1}{H_i} =$$

$$= \left[1\ \tilde{}1 + H_0\, 1\ \frac{\partial\tilde{}1}{\partial H_0} + \alpha_0 H_0\left(\frac{\partial 1}{\partial\alpha_0}\frac{\partial\tilde{1}}{\partial H_0} - \frac{\partial 1}{\partial H_0}\frac{\partial\tilde{1}}{\partial\alpha_0}\right) + \alpha_0\tilde{}\,1\frac{1}{\alpha_0}\right]\cdot\frac{\{\alpha_0, H_0\}}{H_0\ \tilde{}1} \in$$

$$\varepsilon_2(H_0,\alpha_0),\ \text{because}\ \{\alpha_0,H_0\}/H_0 \in \mathbb{R}$$

Therefore

$$V_{i+1} = \alpha_{i+1}\mathrm{Id}\ \ln H_{i+1},$$

where

$$\left.\begin{aligned} \alpha_{i+1} &= \alpha_i\, 1_i(H_0,\alpha_0) \\ H_{i+1} &= H_i\, 1_i(H_0,\alpha_0) \end{aligned}\right\} \qquad (3.5)$$

Whence it follows, in particular, that

$$\left.\begin{aligned} l_i(H_0,\alpha_0) &= 1 - c_i k_i(\operatorname{div} V_i\ )H_i^{k_i}\ /\alpha_i \\ \tilde{l}_i(H_0,\alpha_0) &= \ l_i^{-1/k_i}\ (H_0,\alpha_0) \end{aligned}\right\}, \qquad (3.6)$$

In distinction from (3.2.1) in (3.6) the quantities l_i and $l|_i$ denote quantities l_{k_i} and $\tilde{l}_{k_i}$ obtained at the ith step.
The induction step is completed.

The (r + 1)st induction step should be modified. Indeed, on the (r + 1)st step we have

$$V_{r+1} = V_r + \eta_r = \alpha_r \mathrm{Id} \ln H_r + f(H_r)\mathrm{Id}\alpha_r,$$

where

$$\left.\begin{aligned} \alpha_r &= \alpha_0 | l(H_0,\alpha_0) \\ H_r &= H_0 \tilde{l}(H_0,\alpha_0) \end{aligned}\right\}$$

$$f(H_r) = c_r H_r^r,\ c_r \in \mathbb{R},\ c_r \neq 0.$$

Whence, in virtue of 2.3, we obtain the symplectic normal form indicated in Lemma 3.2.1 with due regard for the following.

3.2.3. Note. In performing induction in 3.2.2, we can restricts outselves in the choce of η_i to the main terms alone, i.e., to take

$$\eta_i = c_i \alpha_i\ H_i^{k_i}\ \mathrm{Id}(1/\alpha_i).$$

Then

$$\operatorname{div} \mathbf{V}_{i+1} = \operatorname{div}[\mathbf{V}_i + \eta_i] = \operatorname{div} \mathbf{V}_i + \operatorname{div} \eta_i = \frac{\{\alpha_i, H_i\}}{H_i} - c_i k_i \frac{H_i}{\alpha_i}\{H_i, \alpha_i\} =$$
$$= (1 + c_i k_i H_i^{k_i}/\alpha_i) \operatorname{div} \mathbf{V}_i = 1_i^* \operatorname{div} \mathbf{V}_i .$$

Thus, by induction, we obtain the relationship between the multiplicative unities deforming α_o, H_o, and div V_o indicated in Lemma 3.2.1.

3.3. Modification (3.3) to the Integrable Form with Preservation of Integrating Factor

3.3.1. Lemma. With the aid of the symplectic orbital equivalence the normal form (3.3) can be reduced to the form

$$\alpha_0 \mathrm{Id} \ln U + (sU^r + cU^{2r})\mathrm{Id}\alpha_0, \tag{3.7}$$

where $\alpha_0 = x_1 x_2$, $U = H_0 \exp\left(\sum_{j=1}^{r} c_j H_0^j / \alpha_0\right)$ $s = \pm 1$, $c \in \mathbf{R}$.

3.3.2. Lemma 3.3.1 is proved by changing the form of the obstructions H_o^j/α_o e. It can readily be checked whether the indicated generator can be substituted by

$$H_0^j/\alpha_0\, \xi_j,$$

where

$$\xi_j = (jn - 1)x_1 \partial/\partial x_1 + (1 - jm)x_2 \partial/\partial x_2.$$

Indeed,

$$H_0^j/\alpha_0 | \xi_j \wedge \Lambda x = H_0^j(n(jm - 1) + m(1 - jn))\partial_1 \wedge \partial_2 =$$
$$= (m - n)H_0^j \partial_1 \wedge \partial_2 \neq 0.$$

Further on,

$$H_0^j/\alpha_0|\xi_j = \alpha_0 \mathrm{Id}(H_0^j/\alpha_0).$$

Therefore,

$$\alpha_0 \mathrm{Id} \ln H_0 + \sum_{j=1}^{r} c_j \frac{H_0^j}{\alpha_0} \xi_j$$

$$= \alpha_0 \mathrm{Id} \left[\ln H_0 + \sum_{j=1}^{r} c_j - \frac{H_0^j}{\alpha_0} \right] = \alpha_0 \ \mathrm{Id} \ln U,$$

where

$$U = \exp \left(\ln H_0 + \sum_{j=1}^{r} c_j - \frac{H_0^j}{\alpha_0} \right) = H_0 \ \exp \left(\sum_{=1}^{r} c_j - \frac{H_0^j}{\alpha_0} \right)$$

Finally, in virtue of 2.3, we arrive at symplectically normal form (3.7).

3.4. Integration in Elementary Functions of Symplectic Orbital Normal Forms

3.4.1. Lemma. Let

$$V = \alpha \ \mathrm{Id} \ln U + (sU^r + cU^{2r})\mathrm{Id}\alpha, \ s = \pm 1, \ c \in \mathbf{R}$$

Then: (a)

$$V = \alpha \ \mathrm{Id} \ln \Phi,$$
$$\alpha = \alpha(sU^r + cU^{2r}),$$
$$\Phi = \alpha(s/u^r + c)^{c/r} \ e^{-s/rU^r}.$$

(b) the generator of the stationary subalgebra of the phase portrait of the vector field V has the form

$$\xi = \frac{sU^r + cU^{2r}}{\mathrm{div}(\alpha \ \mathrm{Id} \ln U)} \mathrm{Id}\alpha.$$

3.4.2. Section (a) of the Lemma is proved by the following calculation.

$$V = \alpha(sU^r + cU^{2r})\mathrm{Id}\left[\int \frac{dU}{U^{r+1}(s+cU^r)} + \ln\alpha\right]$$

$$\int \frac{dU}{U^{r+1}(s+cU^r)} = -\frac{s}{rU^r} + \frac{c}{r}\ln\left(\frac{s}{U^r} + c\right).$$

Whence we obtain α and, exponetiating it, also Φ.
The proof of (b) follows from Sect.2.2.8 and the calculation

$$V \wedge \xi = \frac{(sU^r + cU^{2r})\alpha}{\mathrm{div}(\alpha\,\mathrm{Id}\ln U)}\,\mathrm{Id}\ln U \wedge \mathrm{Id}\alpha = -\alpha\partial_1\wedge\partial_2.$$

*/The definition of codimension is well known and can be found elsewhere (e.g., in [2]).

REFERENCES

1. V.I. Arnold. "Spectral sequences for reducing functions to the normal forms." In: Problems of mechanics and mathematical physics. Moscow: Nauka, 1976, 7-20 (in Russian).
2. V.I. Arnold and A.B. Givental. "Symplectic geometry." In: Modern problems of mathematics (Advances in science and technology. VINITI AN SSSR) v.4. Moscow, 1985, p.5-139 (in Russian).
3. R.I. Bogdanov. "Local orbital normal forms of vector fields on a plane." Tr. Seminara im.I.G. Petrovskogo, 1979, 5, pp.51-84 (in Russian).
4. R.I. Bogdanov. "Invariants of elementary singular points on a plane." UFN, 1985, 3(243), pp.199-200 (in Russian).
5. R.I. Bogdanov. "Singularities of vector fields on the plane with pointed direction." Inv. Math., 1979, 54, pp.247-259 (in Russian).
6. H. Broer. "Formal normal form theorems for vector fields and some consequences for bifurcations in the volume preserving case." In: Dynamical systems and turbulence, Warwick 1980. Lecture notes in mathematics. Springer-Verlag, 1981, No.898, pp.54-75.
7. S.MacLane. Homology. Moscow: Mir, 1966 (in Russian).

8. H. Poincaré. Sur les proptiétés des fonctions, définies par des équations aux differentielles partielles. Thése, 1879. Oevres, t.1, Paris, 1928.
9. S. Sternberg. Lectures on differential geometry. Moscow: Mir (in Russian).

Mathematics
and Modelling

ISBN 1-56072-104-9

A COMPLICATED SINGULAR POINT OF THE CENTER-FOCUS TYPE AND THE NEWTON DIAGRAM

F.C. Berezovskaya and N.B. Medvedeva

From the standpoint of Newton diagrams the approach to the study of singular points of vector fields on a plane is rather traditional. In 1970 A.M. Molchanov formulated the problem: from all the germs of vector fields with the given Newton diagrams to single out the germs of the general position whose asymptotic and topological properties were determined by the polynomial vector field corresponding to the Newton diagram and to indicate the algorithm for the analysis of such a singular point. This problem was solved for the case where at least one trajectory entered such a singular point with a known tangent [1,2,3].

The algorithm for distinguishing the center and focus of the vector field homogeneous in its main part is given in [1].

But up to the recent time it was impossible to distinguish between the center and the focus of the Newton diagram having more than one edge. In the present study the coefficient of the smooth (linear [4]) term of the asymptotic monodromy transformation was calculated for "typical" vector fields with the monodromic singular point and the given Newton diagram; for quite a large set of diagrams this coefficient was different from identical zero. The requirement of typicalness singles out a dense set in the space of all the vector fields with the given

Newton diagram which is open everywhere. If the logarithm of the above mentioned coefficient is nonzero, the singular point is a focus.

The results obtained in this study are also valid for analytical and infinitely smooth fields. For convenience, the consideration is for analytical vector fields.

The main results of the study were announced elsewhere [5].

The present study was performed under the guidance of Yu.S. Il'yashenko. The authors express him their most sincere gratitude.

1. MONODROMIC NEWTON DIAGRAMS

1.1. Newton Diagrams, edge and vertex indices. A singular point of a vector field on a plane is monodromic if the Poincaré implication function (monodromy transformation) for such a point has been defined. The rigorous definition can be found elsewhere ([4], p.89).

An analytical vector field V in the vicinity of the singular point (x,y) = (0,0) can be written in the form

$$V = \sum a_{\mu\nu} x^{\mu} y^{\nu-1} \frac{\partial}{\partial x} + \sum b_{\mu\nu} x^{\mu-1} y^{\nu} \frac{\partial}{\partial y}, \qquad (1)$$

where μ,ν are nonnegative integers, $a_{\mu o} = b_{o\nu} = 0$.

Definitions. 1. A carrier of vector field (1) is called the set $M = \{(\mu,\nu): |a_{\mu\nu}| + |b_{\mu\nu}| \neq 0\}$. The vector $(a_{\mu\nu}, b_{\mu\nu})$ is called the vector coefficient of the point $(\mu,\nu) \in M$. 2. The Newton diagram of vector field (1) is called a union of all the segments lying at the boundary of a convex shell of the set $\bigcup_{(\mu,\nu)\in M} \{(\mu,\nu + \mathbb{R}_+^2)$ where $\mathbb{R}_+^2$ is the positive quadrant*/. Such segments are called the edges of the Newton diagram and their ends are called vertices. 3. A vertex of the Newton diagram is called a boundary vertex, if it belongs to one of the coordinate axes, and an inner vertex otherwise. 4. Let (a,b) be a vector coefficient of a Newton diagram vertex. The index of this vertex is a number

$$\beta = b/a, \text{ if } a \neq 0$$

$$\beta = \infty, \text{ if } a = 0.$$

5. The index of an edge of a Newton diagram is called a rational number α equal to the tangent of an angle formed by the negative direction of the ν axis and the edge. 6. Let us define the functions X^α, Y^α, and F^α for an edge γ with the index α by the formulae:

$$X^\alpha(x,y) = \sum_{(\mu,\nu)\in\gamma\cap\mathbb{Z}^2} a_{\mu\nu} x^\mu y^{\nu-1}, \quad Y^\alpha(x,y) = \sum_{(\mu,\nu)\in\gamma\cap\mathbb{Z}^2} a_{\mu\nu} x^\mu y^{\nu-1},$$

$$F^\alpha(x,y) = -\alpha y X^\alpha(x,y) + x Y^\alpha(x,y).$$

The function F^α is called the α-characteristic of the field V. 7. For any quasiuniform polynomial R(x,y) with weights n and m of variables x and y, respectively, the following expansion is valid

$$R(x,y) = A x^{s_1} y^{s_2} \prod_i (y^n - b_i x^m)^{k_i},$$

where b_i are various complex numbers. The factor of the form $y^n - b_i x^m$, $b_i \neq 0$ is called a simple factor of the polynomial R(x,y), the number b_i is called its multiplicity. A vector field with the Newton diagram Γ is called Γ-nondegenerate, if: (1) all the polynomials F^α have no simple multiplicity factors exceeding unity; (2) the index of any inner vertex differs from the indices of the adjacent edges.

Definitions. A set of Γ-nondegenerate vector fields is denoted by N_Γ. Obviously that this set is open and dense everywhere in the space of all the vector fields with the Newton diagram Γ. The set of Γ-nonndegenerate vector fields with a zero monodromic singular point is denoted by M_Γ.

Lemma 1.1. A set M_Γ is open in N_Γ.

Definition. The Newton diagram Γ is called monodromic if the set M_Γ is nonvoid.

1.2. Quasiuniform polynomials corresponding to an edge of a Newton Diagram. Consider an edge of a Newton diagram with the index α. Let $\alpha = m/n$ Wbe an is irreducible fraction and N + 1 be the number of integer points on the edge, (p,q) be the coordinates of the vertex of a right angle in the triangle formed by the edge and the segments parallel to the coordinate axes. The polynomials X^α, Y^α, and F^α are represented in the form

$$X^\alpha(x,y) = x^p y^{q-1} P^\alpha(x,y),\ Y^\alpha(x,y) = x^{p-1} y^q Q^\alpha(x,y),$$
$$F^\alpha(x,y) = x^p y^q G^\alpha(x,y), \quad (2)$$

where P^α, Q^α, and G^α are quasiuniform polynomials of degree nmN with weights n and m of variables x and y, respectively.

Let R(x,y) be a quasioniform polynomial of an even degree with weights n amd m of variables x and y, respectively. Then:
1. If n is even and m is odd, then $R(x,-y) = R(x,y)$. (3)
2. If n is odd and m is even, then $R(-x,y) = R(x,y)$. (4)

1.3. Newton Diagrams Corresponding to Monodromic Singular Points. The vertices and edges of the Newton diagram are enumerated from the ordinate to the abscissa axes, $\alpha_1, \ldots, \alpha_k$ are the edge indices.

Lemma 1.2. Let $V \in N_\Gamma$. Then $V \in M_\Gamma$, if and only if :(a) the Newton diagram Γ has at least one edge; (b) all the polynomials $F^\alpha(x,y)$ have no real simple factors, (c) polynomials $F^{\alpha_1}(x,y)$ and $F^{\alpha_k}(x,y)$ are not divisible by x and y, respectively, and (d) for any inner vertex $\beta \notin [\alpha.\alpha|]$ where β is the vertex index, α and $\alpha|$ are the indices of the adjacent edges.
Lemma 1.1 follows from Lemma 1.2 proved in Sect.3 .

Theorem 1. The Newton diagram is monodromic then and only then when it has two boundary vertices and when the projection lengths of each p edge onto the μ and ν axes are integers.

Proof. Necessity. Let Γ be a monodromic diagram, $V \in M_\Gamma$. If the first condition is not fulfilled, then either x = 0 or y = o is a trajectory of field V. We arrive at the contradiction with monodromy. Let the second condition be unsatisfied. The lengths of projections of the edge with the index α on the μ and ν axes are equal to mN and nN, respectively, i.e., to the degrees of the polynomial $G^{\alpha}(x,y)$ in x and y, respectively. The oddness of one of these degrees results in the existence of a simple real factor for polynomial $G^{\alpha}(x,y)$ and, hence, $F^{\alpha}(x,y)$, since the simle factors of these polynomials are the same. We arrive at the contradiction with monodromy.

Sufficiency. Consider a vector field V_0 whose carrier is a set of the vertices of the diagram Γ. The vector coefficient of a vertex lying at the ν axis is equal to $(-1/\alpha_1, 0)$. The remaining vector coefficients are equal to (0, 1). It can readily be shown that $V_0 \in M_\Gamma$.

2. ASYMPTOTIC BEHAVIOR OF MONODROMY TRANSFORMATION

2.1. Formulation of Basic Theorems.

Notation. $0^x(1)$ is a continuous function of $x \in \mathbf{R}^n$ which is infinitesimal for $x \to 0$; $O_\varepsilon^x(1)$ is the function equal to $0^x(1)$ for each ε; the trend of $0_\varepsilon^x(1)$ to zero is not necessarily uniform with respect to ε.

Definition. Let $\alpha = m/n$ be an irreducible fraction. An edge with the index α is called even if one of the numbers m and n is even and odd otherwise.

Let $\alpha_1, \ldots, \alpha_k$ be the edge indices and $\beta_0, \ldots, \beta_k$ be the vertex indices for the Newton diagram Γ.

Theorem 2. Let Γ be a monodromic Newton Diagram, $V \in M_\Gamma$. Then, if a transversal is chosen in a proper way, and, probably upon time reversal, the monodromy transformation of the vector field V takes the form $\rho \to (\exp F_1(V)\rho(1 + O^{\rho}(1))$, where

$$F_1(V) = -4\pi \mathrm{Im} \sum_{i=1}^{k} \varepsilon_i \; k \sum_{\mathrm{Im} z>0}^{\bullet} \mathrm{res}_z \frac{X^{\alpha_i}(1,v)}{F^{\alpha_i}(1,v)} ;$$

$$k_1 = 1, \text{ and for } i > 1 \ k_i = \begin{cases} \dfrac{\beta_{i-1} - \alpha_i}{\beta_{i-1} - \alpha_{i-1}} \ k_{i-1} & \text{for } \beta_{i-1} < \infty \\ k_{i-1} & \text{for } \beta_{i-1} = \infty. \end{cases}$$

$\varepsilon_i = 0$ if the ith edge is even and $e_i = 1$ otherwise.

Theorem 3. Let Γ be a monodromic Newton diagram. Then $F_1(V) \equiv 0$ on M_Γ if and only if all the edges of the diagram Γ are even.

Thus Theorem 2 provides the sufficient condition for the focus for Γ-nondegenerate vector fields with Newton diagrams having at least one odd edge.

Sections 3-5 are devoted to the proof of Theorem 2, Section 3 also includes the proof of Lemma 1.2, and Section 5 has the proof of Theorem 3.

3. SECTORS AND RECTANGLES

3.1. Sectors. Let us consider a Γ-nondegenerate vector field in the first quadrant $I = \{(x,y): x \geq 0, y \geq 0\}$ intersected by the neighborhood of zero. Let us divide the first quadrant into sectors corresponding to the edges and vertices of the Newton diagram and make the exponential change of coordinates transforming our vector field into a vector field with elementary singular points.

Let α be an index of an edge of the Newton diagram and let α' and $\tilde{\alpha}$ be indices neighboring α, namely, $\alpha' < \alpha < \tilde{\alpha}$. Let $\alpha' = m'/n'$, $\alpha = m/n$, and $\tilde{\alpha} = \tilde{m}/\tilde{n}$ be irreducible fractions. Let us put $r = \tilde{m}n - m\tilde{n}$. For the first edge (with the index α_1) we put $\alpha \quad \alpha'_1 = 0/1 = 0$.

Let us establish the correspondence between the index α and a sector

$$S_\alpha = \{(x,y): \varepsilon^r X^\alpha \le y \le \varepsilon^{-n(\alpha-\alpha')} x^\alpha,\ 0 \le x \le \delta\}$$

at $\alpha \neq \alpha_k$ (α_k is the index of the last edge).

$$S_\alpha = \{(x,y): 0 \le y \le \varepsilon^{-n(\alpha-\alpha)},\ 0 \le x \le \delta\} \text{ for } \alpha = \alpha_k.$$

The quantity δ is bounded from above by the dimensions of the neighborhood in which the initial vector field is defined.

Now let us establish the correspondence between the vertex of the Newton diagram adjacent to the edges with indices α and $\alpha|$ and the sector

$$S_{\alpha\tilde{\alpha}} = \{\varepsilon^{-n(\tilde{\alpha}-\alpha)} \le y \le \varepsilon^r X^\alpha\}$$

The upper vertex (that with the maximum ordinate) is put in correspondence with the sector $S_1 = \{(x,y): 0 \le x \le \varepsilon y^{1/\alpha_1},\ y \le \delta\}$.

<u>Note</u>. The neighborhood of zero in the first quadrant is the union of sectors corresponding to the vertices and edges of the Newton diagram located at such distances from y and x axes that correspond to the location of the vertices and edges of the Newton diagram. This follows from the formulae for sectors.

3.2. <u>Rectangles</u>. Let us make the analytical change of coordinates in the interior of sectors S_a, $S_{\alpha\tilde{\alpha}}$ and S_1:

$$x = z^n,\ y = wz^m; \tag{5}$$

$$x = u^n v^{n\tilde{n}},\ y = u^m v^{n\tilde{m}}; \tag{6}$$

and

$$x = u_1 y_1^{n_1},\ y = y_1^{m_1}, \tag{7}$$

respectively. Here $\alpha_1 = m_1/n_1$ is the index of the first edge.

Upon changes (5), (6), and (7) the interior of sectors S_α, $S_{\alpha\tilde{\alpha}}$, and S_1 is transformed into the interior of rectangles

$$P_\alpha = \{(z,w): 0 \le z \le \delta^{1/n},\ \varepsilon^r \le w \le \varepsilon^{-n(\alpha-\alpha_j)}\},\ \alpha \ne \alpha_k,$$
$$P_{\alpha\tilde{\alpha}} = \{(u,v): 0 \le u \le \varepsilon,\ 0 \le v \le \varepsilon\} \text{ and}$$

$$P_1 = \{(y_1,u_1): 0 \le y_1 \le \delta^{-\frac{1}{m_1}},\ 0 \le u_1 \le \varepsilon\},$$

respectively.

Lemma 3.1. As a result of change (5), the Γ-nondegenerate field passes from sector S_α into the field of directions defined inside the rectange P_α set by the analytical vector field

$$\begin{cases} \dot{z} = \dfrac{1}{n} z(X^\alpha(1,w) + zX(z,w)) \\ \dot{w} = F^\alpha(1,w) + zW(z,w). \end{cases}$$

As a result of change (6) the Γ-nondegenerate vector field passes from sector $S_{\alpha\tilde{\alpha}}$ into the field of directions defined inside the rectangle $P_{\alpha\tilde{\alpha}}$ set by the vector field

$$\dot{u} = u(-\lambda + U(u,v)),\quad \dot{v} = v(1 + V(u,v)),$$

$$U(0,0) = V(0,0) = 0,\quad \lambda = \begin{cases} \tilde{n}\dfrac{\beta - \tilde{\alpha}}{\beta - \alpha}, & \beta < \infty \\ \tilde{n}, & \beta = \infty \end{cases};$$

where β is the index of the vertex ajacent to the edges with indices α and $\tilde{\alpha}$. Functions X, W, U, and V are analytically continued in the neighborhood of closures of the rectangles inside which they are defined.

Lemma 3.1 is proved by direct substitutions and division by the power functions.

3.3. The beginning of the Proof of Lemma 1.2. The parameter ε is taken so small that the rectangle P_α would contain all the positive roots of the polynomial $F^\alpha(1,w)$ at the boundary lying on the w axis and none of rectangles $P_{\alpha\tilde{\alpha}}$ and P_1 would contain at the boundary any singular points of the field of directions defined there, probably, except of point (0,0).

Lemma 3.2. Let $V \in N_\Gamma$. If the polynomial $F^\alpha(1,w)$ has a positive root, then vector field (8) in the rectange P_α has a singular point on the w-axis at sufficiently small ε, with the characteristic trajectory lying strictly inside P_α.

Proof. Any positive root of the polynomial $F^\alpha(1,w)$ is single in virtue of the condition of Γ-nondegeneracy. Therefore to each such root there corresponds an elementary singular point of vector field (8) lying on the w-axis. At least one characteristic trajectory of such a singular point lies strictly inside P_α (see [6]).

Lemma 3.3. Let $V \in N_\Gamma$. The vector field V has no characteristic trajectories in the first quadrant then and only then when: (a) the Newton diagram Γ has at least one edge; (b) the polynomials $F^\alpha(1,v)$ have no positive roots, (c) the polynomials $F^{\alpha_1}(u.1)$ and $F^{\alpha_k}(1,v)$ have no zero roots, and (d) for any inner vertex $\beta \notin [\alpha, \tilde{\alpha}]$, where β is the vertex index and α and $\tilde{\alpha}$ are the indices of the adjacent edges.

Proof. If the Newton diagram has only one vertex, then the singular point is either unisolated or else at least one of the coordinate axes is an invariant set.

Let the diagram Γ have at least one edge. According to Lemma 3.2, the condition (b) at $\alpha \neq \alpha_k$ is equivalent to the absence of characteristic trajectories in the sector S_α at $\alpha \neq \alpha_k$. Condition (c) plus condition (b) at $\alpha = \alpha_k$ are equivalent to the absence of singular points in rectangles

P_1 and $\mathbf{P}_{\alpha_1}$, and, hence. also to the absence of characteristic trajectories in sectors S_1 and $\mathbf{S}_{\alpha_k}$.

Note that in the rectangle $P_{\alpha\tilde{\alpha}}$ the vector field (9) at sufficiently small ε has the only singular point (u,v) = (0,0) which is nondegenerate in virtue of the Γ-nondegeneracy condition, and therefore it is either a saddle or a node. The condition (d) is equivalent to a saddle singular point, and, therefore, to the absence of characteristic trajectories in the sector $S_{\alpha\alpha|}$. Lemma 3.3 is proven.

3.4. Reflected Vector Fields. Let S^x and S^y be reflections of the plane (x,y) relative to the x and y axes, respectively, $S^{xy} = S^x \bullet S^y$. In addition to the vector field V = (X,Y) we consider also the vector fields $V_y = S_*^yV$, $V_x = S_*^xV$, and $V_{xy} = S_*^{xy}V$. The Newton diagram for these fields is the same as for the field V and, hence, the edge indices are also the same. The sets of the components of the reflected vector fields are denoted as (X,Y) with the respective subscripts. The α-characteristics of the vector fields V_y, V_x, and V_{xy} have the form
$F_y^\alpha = -F^\alpha \circ S^y$, $F_x^\alpha = -F^\alpha \circ S^x$, $F_{xy}^\alpha = F^\alpha \circ S^{xy}$. (10)

Lemma 3.4. (a) The vector fields V, V_y, V_x, and V_{xy} are characterized by the same vertex indices, (b) if the field V is Γ-nondegenerate, then the fields V_y, V_x, and V_{xy} are also Γ-nondegenerate.

Proof. Statement (a) follows from the fact that can readily be checked, i.e., that in reflection of a vector field relative to any axis the vector coefficient of any point of the carrier either does not change or changes to the opposite one. (b) Condition (2) of Γ-nondegeneracy for reflected vector fields follows from the coincidence of the edge and vertex indices, condition (1) follows from equations (10) and the expansion

$$F^\alpha(x,y) = Ax^py \sum_{j=0}^{N_\bullet} (y^n - q_jx^m),\ q_j \in \mathbf{C} - \{0\}. \tag{11}$$

Corollary. $V \in M_\Gamma \Rightarrow |V_x, |V_y, |V_{xy} \in MM_\Gamma$.

3.5. The End of the Proof of Lemma 1.2. Let $V \in N_{\Gamma}$. Obviously, the field V has a monodromic singular point at zero then and only then when for each of the four vector fields V, V_x, V_y, and V_{xy} the conditions (a) - (d) of Lemma 3.3 are satisfied. The conditions (a) and (d) of Lemma 3.3 for all the four vector fields are the same and they coincide with the conditions (a) and (d) of Lemma 1.2. According to formulae (10) and (11), the condition (c) of Lemma 3.3 is either satisfied or not satisfied for all four vector fields simultaneously whereas the condition (c) of Lemma 3.3. for field V is equivalent to the condition (c) of Lemma 1.2. It can readily be proven with the aid of formulae (10) and (11) that the fulfillment of the condition (b) of Lemma 3.3 for each of four vector fields, V, V_x, V_y, and V_{xy}, is equivalent to the condition that the polynomials $F^{\alpha}(1,v)$ and $F^{\alpha}(u,1)$ have no real roots having nonzero values. This latter condition is equivalent to the condition (b) of Lemma 1.2.

4. CORRESPONDENCE MAPPINGS

4.1. The composition of Correspondence Mappings. Let $V \in M_{\Gamma}$. The monodromy transformation Δ of the vector field V corresponding to the y axis can be divided into a composition of four mappings: $\Delta = \Delta_4 \circ \ldots \circ \Delta_1$ where Δ_1, Δ_2, Δ_3, and Δ_4 are the correspondence mappings for the vector fields V_y, V_{xy}, V_x, and V, respectively (Figure 1a). In accordance with the division of the first quadrant into sectors, each of the mappings Δ_1, Δ_2, Δ_3, and Δ_4 is divided into the correspondence mappings for sectors.

4.2. Correspondence Mappings in Sectors that Correspond to Edges. Mapping defined in disjoint union of rectangles P_{α}, $P_{\alpha\tilde{\alpha}}$, and P_1 by formulae (5), (6), and (7), respectively, is denoted by π. Let δ be a correspondence mapping along the phase curves of the field V for the sector S_{α} (Figure 1a). We denote by δ such a mapping of the sides of the rectangle P_{α} that $\pi \circ \delta = \delta' \circ \pi$; it is a mapping along the integral curves of the field of directions (see (8))

$$\frac{dz}{dw} = \frac{1}{n} z \frac{X^{\alpha}(1,w) + zX(z,w)}{F^{\alpha}(1,w) + zW(z,w)} . \qquad (12)$$

Choice of Parameters. As a coordinate in the image and also of preimage of mapping δ we take the restriction of the function Z on the respective side of the rectangle P_α.

Let us put $\kappa = \delta'(0)$. Mappings for vector fields v_x, v_y and v_{xy} analogous to δ are denoted by δ_x, δ_y, and δ_{xy}, and their derivatives at zero by κ_x, κ_y, and κ_{xy}, respectively.

Lemma 4.1. Let $v_\alpha = \lim_{\varepsilon\, 0} \ln \kappa\kappa_x\kappa_y\kappa_{xy}$. Then

$$\nu_\alpha = \begin{cases} -4\pi \mathrm{Im} \sum_{\mathrm{Im} z>0} \mathrm{res}_z \dfrac{X^\alpha(1,w)}{F^\alpha(1,w)} & \text{if the edge with index } \alpha \text{ is odd;} \\ 0 & \text{otherwise.} \end{cases}$$

0 otherwise.

Proof. Solving the equation in variations with the initial conditions along the solution z = 0 of equation (12), we obtain that

$$\ln \kappa = \int_{\varepsilon^r}^{\varepsilon^{-n}\,(\alpha-\alpha\ )} \frac{X^\alpha(1,w)}{F^\alpha(1,w)}\, dw.$$

Similarly,

$$\ln \kappa_x = \int_{\varepsilon^{-n}(\alpha-\alpha)^x}^{\varepsilon^r} \frac{X_x^\alpha(1,v)}{F_x^\alpha(1,v)}\, dv = \int_{\varepsilon^{-n}(\alpha-\alpha)}^{\varepsilon^r} \frac{X^\alpha(1,v)}{F^\alpha(1,v)}\, dv.$$

Whence

$$\lim_{\varepsilon\to 0} \ln \kappa\kappa_x = \text{v.p.}\int_{-\infty}^{\infty} \frac{X^\alpha(1,v)}{F^{\alpha}(1,v)}\, dv.$$

Similarly,

$$\lim_{\varepsilon\to 0} \ln \kappa_y\kappa_{xy} = -\text{v.p.}\int_{-\infty}^{\infty} \frac{X^\alpha(-1,v)}{F^{\alpha}(-1,v)}\, dv.$$

Whence

$$\nu_\alpha = \text{v.p.}\int_{-\infty}^{\infty} \left[\frac{X^\alpha(1,v)}{F^\alpha(1,v)} - \frac{X^\alpha(-1,v)}{F^\alpha(-1,v)} \right] dv. \quad (13)$$

Using quasiuniformity of the polynomials X^α, Y^α, and F^α and the conditions of Γ-nondegeneracy, it can readily be shown that the latter main value is

$$\text{v.p.}\int_{-\infty}^{\infty} \left[\frac{Y^\alpha(u,1)}{F^\alpha(u,1)} - \frac{Y^\alpha(u,-1)}{F^\alpha(u,-1)} \right] du. \quad (14)$$

From formulae (2) we obtain

$$\frac{X^\alpha(x,y)}{F^\alpha(x,y)} = \frac{P^\alpha(x,y)}{yG^\alpha(x,y)}, \quad \frac{Y^\alpha(x,y)}{F^\alpha(x,y)} = \frac{Q^\alpha(x,y)}{xG^\alpha(x,y)}.$$

Using these formulae and properties (3) and (4), we obtain that if m is even, the integrand is identical zero in formula (13), if n is even, the integrand is identical zero in formula (14).

For odd edges we derive from the residue theorem that the main value of (13) is equal to the sum of residues from the lemma formulation.

4.3. Gluings of Sectors. We shall call the gluing mapping φ of sectors $S_{\alpha\tilde{\alpha}}$ and S_α the mapping φ: $L \to T$ of the sides of rectangles $P_{\alpha\tilde{\alpha}}$ and P_α such that the image and pre-image are transformed by mapping π into one point (Figure 1b). In a similar way $\tilde{\varphi}$ is defined as a mapping of gluing of sectors $S_{\alpha\tilde{\alpha}}$ and S_α. In a similar way the mapping of gluing of sectors S_1 and $S_{\alpha 1}$ is defined.

4.4. Correspondence Mapping in Sectors That Correspond to Vertices. Let g be mapping of the sides of the rectangle $P_{\alpha\tilde{\alpha}}$ conjugated with the correspondence mapping along the phase curves of the field V for the sector $S_{\alpha\tilde{\alpha}}$ with the aid of mapping π. Let us put $f = \varphi \circ g \circ \tilde{\varphi}^{-1}$. The composition of the correspondence mapping for the rectangle P_1 with the glueing mapping for the sectors S_1 and $S_{\alpha 1}$ is denoted by h. In a similar way the f and h mappings for the reflected vector fields are denoted by f and h with the respective subscripts.

Lemma 4.2. For the parameters on the sides of rectangles P_α chosen in Sect.4.2 and the appropriately chosen parameters on the sides of the rectangle P_1 (1) the mappings f and f_{xy} have the form $\eta \to \zeta = \varepsilon^\mu(1 + o^\varepsilon(1))\eta^\gamma(1 + o_\varepsilon^\eta(1))$; and the mappings f_x and f_y have the form $\zeta \to \eta = \varepsilon(1 + o^\varepsilon(1))\zeta^{1/\gamma}(1 + o_\varepsilon^\zeta(1))$,

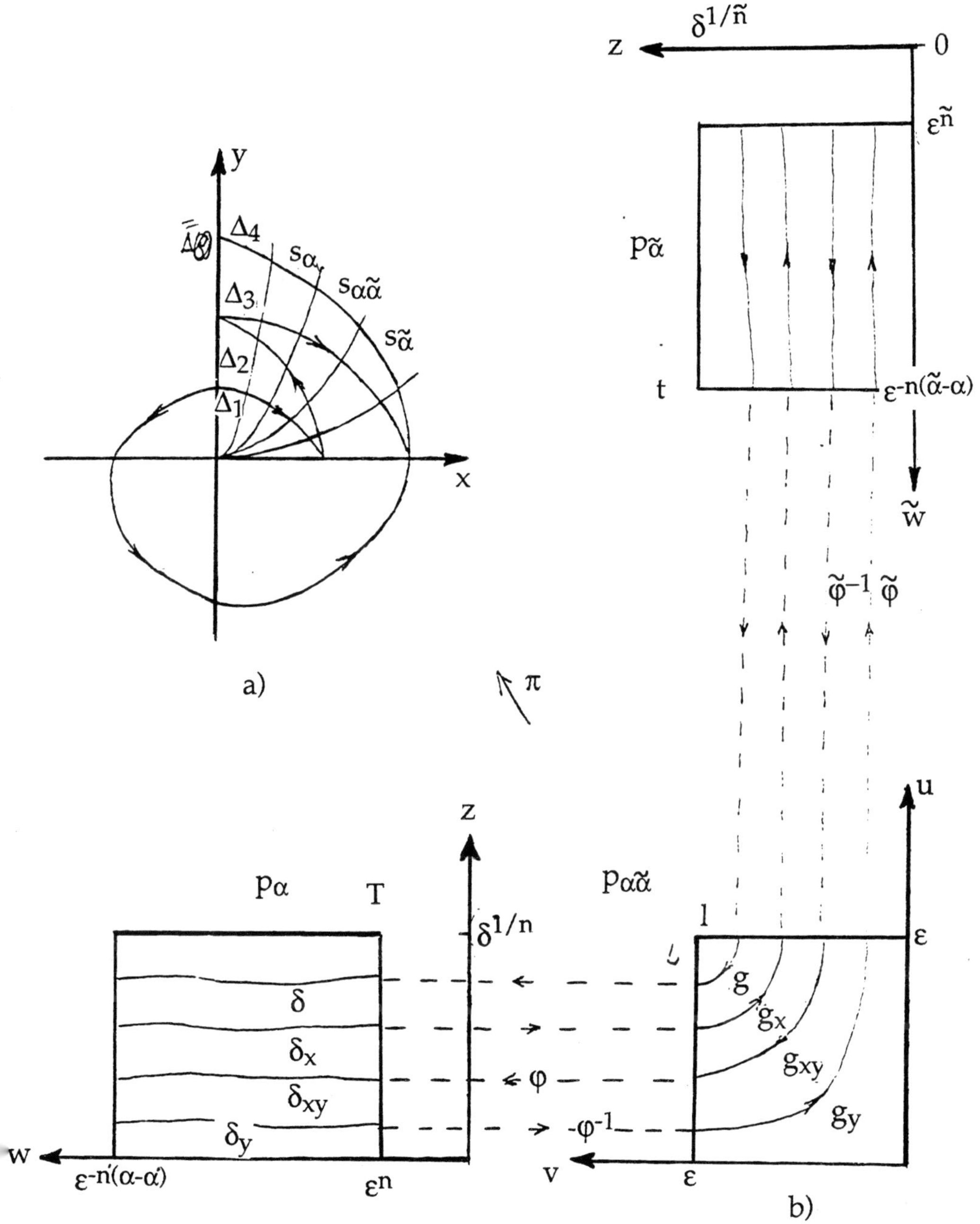

Figure 1. Point depiction of gene mapping. The direction of the mapping action is shown by arrows.

$$\text{where } \gamma = \begin{cases} (\beta - \tilde{\alpha}) / (\beta - \alpha), & \text{if } \beta < \infty \\ 1, & \text{if } \beta = \infty, \end{cases}$$

β is the index of the vertex adjacent to the edges with indices α and $\tilde{\alpha}$, γ and μ are real numbers such that $\nu\gamma + \mu = 0$; (2) the mappings h, h_x,h_y, and h_{xy} have a linear main term with the coefficient $1 + o^{\varepsilon}(1)$.

Corollary. Let $\varphi(\rho) = \kappa\rho(1 + o^{\rho}(1))$, then mappings $f|\varphi|f_x$and $f_{xy}|\varphi|f_y$ have the form $\rho \to \kappa^{\gamma}\rho(1 + o^{\varepsilon}(1)(1 + o_{\varepsilon}^{\varepsilon}(1))$.

Proof of Lemma 4.2. Consider the rectangles P_{α}, $P_{\alpha\tilde{\alpha}}$and $P_{\tilde{\alpha}}$. Transversals are denoted by t, l, L, and T, as is shown in Figure 1b. In sect.4.2 we choose a parameter $\zeta = Z^{n}|T$, on the transversal T and the parameter $\eta = Z^{\tilde{n}}|t$ on the transversal t. Let us consider parameters $u|_{L}$ and $v|_{l}$ on transversals L and l, respectively. From formulae (7) and (8) we obtain that the mappings φ and $\tilde{\varphi}^{-1}$ have the form v: $\zeta = \varepsilon^{n\tilde{n}}u^{\tilde{n}}$ and $\tilde{\varphi}^{-1} = \varepsilon^{-1/\tilde{n}}\eta^{1/n\tilde{n}}$.

Lemma 4.3 [7]. Mappings g and g_{xy} have the form

$$u = \varepsilon^{1-\lambda}(1 + o^{\varepsilon}(1))v^{\lambda}(1 + o_{\varepsilon}^{v}(1)).$$

Mappings g_x and g_y have the form

$$v = \varepsilon^{1-1/\lambda}(1 + o^{\varepsilon}(1))u^{1/\lambda}(1 + o_{\varepsilon}^{u}(1)).$$

Using Lemma 4.3 amd formulae for φ and φ^{-1} and making simple substitutions, we obtain statement (1) of Lemma 4.2.

(2) Consider as the parameter on the sides of the rectangle P_1 parallel to the y_1-axis the restriction of the function εy_1^{n1} on these sides. For such a choice of the parameters, the mapping of the gluing of sectors P_1 glues the points with the same values of the parameters,

whereas the correspondence mapping for the rectangle P_1 has the linear main term with the coefficient

$$\exp \int_0^{\varepsilon} \frac{cY^{\alpha_1}(u,1)}{F^{\alpha_1}(u,1)}\, du = (1 + o^{\varepsilon}(1).$$

5. The End of the Proof of Theorem 2

Theorem 2 follows from Theorem 4 and Lemma 4.1.

Theorem 4. Under the conditions of theorem 2 we have $E_1(V) = \sum_{i=1}^{k} k_i v_{\alpha i}$ where numbers k_i and $v_{\alpha i}$ are defined in terms of Theorem 2 and Lemma 4.1, respectively.

Proof. Let us enumerate the correspondence mappings and γ (from the formulation of Lemma 4.2) in accordance with the numeration of vertices and edges of the Newton diagram. We put $\kappa^j = (\delta^j)(0)$. We put also

$$\psi_k = \delta^k o \delta_x{}^k,\ \psi_k{}^y = \delta_{xy}{}^k o \delta_y{}^k, \text{ and for } 1 \le i \le k$$
$$\psi_i = \delta^i o f^i o \psi_{i+1} o f_x{}^i o \delta_x{}^i,\ \psi_i{}^y = \delta_{xy}{}^i o f_{xy}{}^i o \psi_{i+1}{}^y o f_y{}^i o \delta_y{}^i .$$

The monodromy transformation Δ is the following composition

$$\Delta = h o \psi_1 o h_x o h_{xy} o \psi_1{}^y o h_y. \tag{15}$$

Using the corollary of Lemma 4.2, it is easy to prove that the main term of the mapping ψ_1 is

$$\kappa^1\kappa_x{}^1(\kappa^2\kappa_x{}^2\ )^{\gamma_1} \ldots (\kappa^k\kappa_x{}^k\ )^{\gamma_1\gamma_2\cdots\gamma_{k-1}}\ (1 + o^{\varepsilon}(1))\rho. \tag{16}$$

In a similar way, we obtain that the main term of the mapping $\psi_1{}^y$ has the form (16) where κ^i and $\kappa_x{}^i$ are substituted by $\kappa_{xy}{}^i$ and $\kappa_y{}^i$, respectively. Multiplying the coefficients of the main terms of the mappings from composition (15) and passing in thus obtained product to the limit for $\varepsilon \mid 0$, we obtain that the mapping Δ has the following asymptotic function

$$\Delta(\rho) = (\exp \ v_{\alpha_1}) \prod_{i=2}^{k} (\exp \ v_{\alpha_i})^{\gamma_1 \cdots \gamma_{i-1}} \rho(1 + o_\varepsilon^{\rho}(1)).$$

Thus theorem 4 and also Theorem 2 have been proven.

<u>The Proof of Theorem 3.</u> The sufficiency is obvious. Now let us prove the necessity. Let $F_1(V) \equiv 0$ on M_Γ. We assume that the jth edge of the diagram Γ is odd. Consider the vector field W_o whose carrier consists of only one point - the second lower point on the jth edge, with the respective vector coefficient being $(1/\alpha_j, 1)$. Let V_o be the vector field from the proof of theorem I. It can readily be shown that $V_o + W_o \in M_\Gamma$, and also, with the help of Theorem 4 and formula (14), that in virtue of oddness of the jth edge and monodromy of the Newton diagram, $F_1(V_o + W_o) \neq 0$. Thus we arrived at contradiction.

REFERENCES

1. A.D. Bryuno. <u>Local method of nonlinear analysis of differential equations.</u> Moscow: Nauka, 1979 (in Russian).
2. F.S. Berezovskaya. <u>Complicated stationary point of the systems on a plane: the structure of the neighborhood and the index.</u> Preprint. Pushchino: ONTI NTsBI, 1978 (in Russian).
3. F.S. Berezovskaya. <u>Power asymptotic functions of the set of second-order differential equations.</u> Pushchino: ONTI NTsBI, 1978.
4. V.I. Arnol'd and Yu.S. Il'yashenko. "Ordinary differential equations. I. Nowaday problems of mathematics. Fundamental directions." vol.1. <u>Itogi nauki i tekhniki. VINITI AN SSSR.</u> Moscow, 1985 (in Russian).

5. F.C. Berezovskaya and N.B. Medvedeva. "On distinctions between the center and the focus of vector fields with the given Newton diagram." Uspekhi mat. nauk, 1986, 41, 198-299 (in Russian).
6. A.A. Andronov et al. Qualitative theory of second-order dynamical systems. Moscow: Nauka, 1966 (in Russian).
7. N.B. Medvedeva. "The first focus value of a complicated monodromic singular point." Tr. Seminara im.Petrovskogo, 1988, 13 (in Russian).

*/The Newton diagram of an infinitely smooth vector field can be constructed in a similar way using its infinite jet.[1]

Mathematics
and Modelling

ISBN 1-56072-104-9

ABELIAN VARIETIES AND LIE ALGEBRAS

Yu.G. Zarkhin

The paper consists of three parts. The first one contains the general information on Abelian varieties. The second one is devoted to the discussion of the Lie algebras corresponding to the Mumford-Tate and Galois groups acting in the Tate modules of the Abelian varieties. These sections are, in fact, a brief and incomplete review for nonexperts in the field that includes the definition of l-adic numbers and the formulation of the Mordell hypothesis proved by Faltings. For more detailed and clear statements the reader is sent to books written by Mumford [16] and Lang [14,32], a series of reviews written by Serre [37, 39, 40], and also articles written by Rosen and Miln for Arithmetic Geometry [44]. The third part is devoted to Abelian varieties over number fields. Here the finiteness theorem is proved for the dimensionality of the Abelian variety with the given dimensionality of the endomorphism algebra and the rank of a semisimple component of the l-adic Lie algebra of the Galois group acting in the Tate modules. We tried to state the material in the most elementary and concise form, sometimes at the expense of its generality and the strength of formulations.

The article is, in fact, the substantially revised and extended text of the paper delivered by the author at the Z.B. Vinberg and V.L. Popov Seminar on the theory of invariants held at the Mechanical--Mathematical Faculty of the Moscow State University in April 1986.

I am very grateful to J.-P. Serre and P. Deligne for fruitful discussions and their letters and also to G.A. Litvinov for reading the first version of the paper.

I dedicate this paper to Emmanuil El'evich Shnol who for many years have encouraged my mathematical studies and express my deep gratitude to him.

1. ABELIAN VARIETY

By definition, an Abelian variety is a connected algebraic group (over an algebraically closed field) which is a projective algebraic variety. The projectivity condition imposes rigid restrictions on the group structure, in particular, the group law on an Abelian variety is always commutative. The same situation is also typical for the theory of the Lie groups [16]: any connected compact complex Lie group is commutative. Such Lie groups are called comlex tores and can be represented in the form of the V/L factors of n-dimensional complex vector spaces $V \approx \underline{C}^n$ over the discrete lattices L of the maximum rank 2n [16, 26].

The place of Abelian varieties in the general theory of algebraic groups is determined by the Chevalley theorem: any algebraic group is an extension of an Abelian variety obtained with the aid of a linear algebraic group (i.e. a Zarisskii closed subgroup of the group of all nondegenerate matrices of a certain rank). Namely, any algebraic group contains the only closed normal divisor which is a linear algebraic group such that the group-factor over it is an Abelian variety. In this case there exist only zero algebraic homomorphisms between Abelian varieties and the linear algebraic groups. (At the same time, there exist nontrivial extensions of an Abelian variety obtained with the aid of a multiplicative group: their study is tightly related to the notion of a dual Abelian variety whose points parametrize such extensions).

Let X be an n-dimensional Abelian variety over the field of complex numbers $\mathbb{C}$. Then the group $X(\mathbb{C})$ of its complex points can be represented as a complex tore V/L. And vice versa, a complex tore V/L is a group of $\mathbb{C}$-points of a certain (n-dimensional) complex Abelian variety X then and only then when there exists a positively defined Hermitian bilinear form

$$H: V \times V \to \mathbb{C}$$

the imaginary part of which

$$\mathrm{Im}\, H: V \times V \to R,\ u,v \mid \longrightarrow \mathrm{Im}\, H(u,v)$$

takes the integer values on L x L; note that the form

$$\mathrm{Im}\, H: L \times L \to Z,\ u,v \mid \to \mathrm{Im}\, H(u,v)$$

is nondegenerate antisymmetric bilinear form. Hereinafter Z is a ring of integers. Such an H form is called polarization; the nonuniqueness of its choice is associated with the nonuniqueness of the embedding of the variety X into the projective space (with the aid of the theta-functions). (The lattice L admits the natural interpretation as the first group of integer homologies of the tore $X(\mathbb{C})$, and the space V as the first group of real homologies).

A ring End X is defined which is a ring of all (algebraic) endomophisms of the Abelian variety X. If X is a complex n-dimensional Abelian variety and $X(\mathbb{C}) = V/L$, then the ring End X is naturally identified with the ring of all the $\mathbb{C}$-linear mappings $V \to V$ transforming the lattice L into itself. As an additive group the ring End X is a free Abelian group of a finite rank not exceeding $2n^2$ for complex Abelian varieties. (In the p characteristic this estimate should be replaced by $4n^2$). Let us put

$$\mathrm{End}^{\circ}X = \mathrm{End}\, X \otimes Q.$$

The $\mathrm{End}^{\circ}X$ algebra is a finite dimensional semisimple algebra over the field Q of rational numbers. In the complex case it can be identified with the algebra of all $\mathbb{C}$-linear mappings $u: V \to V$ for which there can be found a natural number m (different for each u) such that $m\, u(L) \in L$, i.e., $mu \in \mathrm{End}\, X$. There is another description of the $\mathrm{End}^{\circ}X$ algebra. Let us denote by L_Q a subgroup in V consisting of all linear combinations of elements from L with rational coefficients or, which is the same, a vector Q-subspace in V spanned over L. We have

$$L \in L_Q \in V.$$

An element $x \in V$ lies in L_Q then and only then when $mx \in L$ for a certain natural m. In particular, torsion, i.e. a set of elements of the finite-order group $V/L = X(\mathbb{C})$ coincides with $L_Q/L \in V/L = X(\mathbb{C})$. The natural mappings

$$L \otimes Q \dashrightarrow L_Q,\ L \otimes \mathbb{R} \dashrightarrow V,\ L_Q \otimes {}_Q\mathbb{R} \dashrightarrow V$$

are isomorphisms. (Q-subspace of L_Q sets the Q structure on the $\mathbb{R}$-space V). Now it is possible to describe End°X as the Q-algebra of all C-linear mappings $V \rightarrow V$ transforming L_Q into itself or, which is the same, as Q-algebra of all the Q-linear mappings $L_Q \rightarrow L_Q$ which, being continued by R-linearity over all the V, turn out to be C-linear. Polarization H defines on the End°X algebra the following involution:

$$\mathrm{End}^\circ X \rightarrow \mathrm{End}^\circ X,\ u \mapsto u',$$

characterized (in the complex case) by the condition

$$\mathrm{Im}\, H(ux,y) = \mathrm{Im}\, H(x,u'y) \text{ for } u \in \mathrm{End}^\circ X;\ x,y \in L$$

and called the Rosati involution. Using this involution, it is possible to show that the center of the End°X algebra is the direct sum of number fields each of which is either quite real or is of the CM-type [16].

Let Γ be a lattice in space V lying in L_Q, i.e., $G \in L_Q \in V$ and V/Γ be a complex tore. Then there exists an n-dimensional complex Abelian variety Y such that $Y(\mathbb{C}) = V/r$. It is said that the Abelian varieties X and Y are isogeneous to one another. It is clear that for isogeneous Abelian variety End°X = End°Y. Here is an equivalent definition of isogeny that can readily be transferred to the p characteristic. The Abelian variety X and Y are isogeneous to one another if there exist a natural number m and (algebraic) homomorphisms $u: X \rightarrow Y$ and $v: Y \rightarrow X$ such that the composition $vu \in \mathrm{End}\, X$ is the multiplication by m (in X) and the composition $uv \in \mathrm{End}\, Y$ is the multiplication by m (in Y). The homomorphisms u,v are called isogenies. For complex Abelian varieties $X \subset X(\mathbb{C}) = V/L$ and $Y \subset Y(\mathbb{C}) = V/r$ with commensurate lattices

L and Γ it is possible to choose a natural number j such that $jL \in \Gamma$ and $j\Gamma \in L$ and to determine the isogenies u: $X \to Y$ and v: $Y \to X$ whose compositions are multiplications by $m = j^2$. These isogenies act at the complex points of the Abelian varieties X and Y according to the formulae:

$$u: V/L \to V/\Gamma, \quad u(z \bmod L) = jz \bmod \Gamma \quad \text{for } z \in V;e$$

$$v: V/\Gamma \to V/L, \quad v(z \bmod \Gamma) = jz \bmod L \quad \text{for } z \in V.$$

The Abelian variety X is called (absolutely) simple if $End^{\circ}X$ is a body. The condition of simplicity is equivalent to the absence of nontrivial (different from X and unity of the group law) Abelian subvarieties in X. Any Abelian variety is isogeneous to the product of primes [16, 26, 32] (see also [54]).

The "general" Abelian variety X of any dimensionality n has only trivial endomorphisms different from the multiplication by integers, i.e., End X = Z, $End^{\circ}X$ = Q. If End X = Z, it is said that X is an Abelian variety without multiplication. An opposite case is that of Abelian varieties with the complex multiplication (CM-type). An Abelian variety X of dimensionality n is of the SM-type if the $End^{\circ}X$ algebra includes a commutative semisimple Q-algebra of dimensionality 2n, i.e., the direct sum of number fields whose sum of degrees is equal to 2n. Since the dimensionalities and the Lie algebras End° for isogeneous Abelian varieties X and Y coincide, then X is a CM-type Abelian variety then and only then when Y is a CM-type Abelian variety. If the complex Abelian variety X is simple, it is a CM-type variety then and only then when $End^{\circ}X$ is the number field of degree 2n. Then EndX must be a CM-type field, i.e., an imaginary quadratic extension of quite a real number field (of degree n). Any CM-type complex Abelian variety X can be embedded into the projective space in such a way that the variety X is set by homogeneous equations with the coefficients which are algebraic numbers, and then the group law on X is given by the "formulae" whose coefficients are also algebraic numbers. In other words, any CM-type Abelian variety is defined over the field of algebraic numbers (the Shimura-Taniyama theorem) [26]. (The inverse is incorrect). In the p characteristic any Abelian variety set by equations whose coefficients lie within a certain finite field is a CM-type variety. In other words, any Abelian variety defined over a finite field is a

CM-type variety [23, 16, 20]. The inverse is "almost correct": in the p characteristic any CM-type Abelian variety is isogeneous to the Abelian variety defined over the finite field.

An n-dimensional Abelian variety X is that of the RM type (RM = real multiplication) if $\mathrm{End}^{\circ}X$ contains a commutative semisimple Q-subalgebra of degree n. (We do not require, as usual, that this subalgebra would be a direct sum of quite real fields). Any complex CN-type Abelian variety is also a RM-type variety. Any elliptic curve, i.e., a one-dimensional Abelian variety is an RM-type variety. (If X is the complex elliptic curve, then either End° = Abelian variety or X is a curve with the complex multiplication and $\mathrm{End}^{\circ}X$ is an imaginary quadratic field. (The p characteristic also includes the so-called supersingular elliptic curves for which $\mathrm{End}^{\circ}X$ is a quaternion algebra over Q branched only in p and at ∞). Any Abelian surface having a nontrivial endomorphism (different from multiplication by an integer) is an RM-type variety. The product of the RM (or CM)-type Abelian variety is also an RM (or CM) type variety; the same is true also for Abelian subvarieties.

Let X be a complex Abelian variety of dimensionality n and X(C) = V/L. For any natural m, X_m denotes the group of all the points in X(C) that are annihilated due to multiplication by m. We have

$$X_m = \frac{1}{m} L/L \in V/L = X(C), \quad \text{i.e.,} \quad X_m \approx (Z/mZ)^{2n}$$

is the direct sum of 2n cyclic groups of order m. Note here that multiplication by m establishes a natural isomorphism

$$X_m = \frac{1}{m} L/L \approx L/mL.$$

As to the whole torsion TORS(X(C)), we have

$$\mathrm{TORS}(X(C)) = L_Q/\ L \approx (Q/Z)^{2n} \quad = \oplus\ (Q_1/\ Z_1\ )^{2n}$$

where summation is carried out over all simle l: Q_l and Z_l denote the field and the ring of l-adic integer numbers, respectively. (The Appendix at the end of this Section gives the definition and lists the main properties of l-adic numbers). Hereinafter TORS(A) denotes torsion for any commutative group, i.e., the subgroup consisting of all the finite-order elements. (In the p characteristic for all m mutually simple with p, the group X_m of all the points of the Abelian variety X annihilated due to multiplication by m is also isomorphous to $(Z/mZ)^{2n}$, where n = 2 dim X. On the contrary, the group of points of period p is isomorphous to $(Z/pZ)^a$ where $0 \leq a \leq n$. For the Abelian variety of the "general position" (called ordinary Abelian varieties) a = n. Thus, an elliptic curve is an ordinary curve then and only then when it is not supersingular [16].

Let us assume that the Abelian variety X can be set by equations with the coefficients from the field $K \in C$ and that then the group law on X is given by the formulae with coefficients from field K, i.e., an Abelian subvariety in X x X x X consisting of the point of the form (s, t, s + t) "is carved up" from X x X x X by the additional equations with coefficients from the field K. Then the zero e of the group law on X is turned out to be a point with the coordinates from the field K, and one can consider a commutative group X(K) of K-rational points, i.e., a group of points in X(C) with coordinates lying in K.

Then all the points of the finite order from X(C) have coordinates lying in the algebraic closure K_s of the field K, i.e.,$X_m \in X(K_s)$ for all m. The Galois group G(K) = Gal (K_s/K) acts naturally on groups X_m transforming them into finite Galois modules. Of course, the Galois group G(K) acts naturally on the whole group $X(K_s)$ of K_s-rational points of the variety X. And a group X(C) = V/L is under the action of (discontinuous) automorphisms of the group Aut(C/K) of all the automorphisms of the complex number field not changing the place of all the elements of the K field.

If the field K is finitely generated over the field of rational numbers (e.g. is obtained by the adjunction to Q of all the coefficients of the equations setting the variety X in its certain projective embedding), then X(K) is a finitely generated commutative group (the Mordell-Weil-Neron equation [14, 15]), i.e., its torsion subgroup, TORS X(K), is finite and X(K) is isomorphous to the direct sum of the finite group and the free Abelian group Z^r of a certain finite rank r = r(X, K). The number r(X, K) is called the rank of the Abelian variety X over the field K or the K-rank of the variety X. If X and Y are the Abelian varieties isogeneous to one another and defined over the field K and if there exists an isogeny between them also defined over K, then for all the finite algebrain extensions K′ of field K the ranks r(X, K′) and r(Y, k′) coincide, in particular, r(X, K) = r(Y, K). And vice versa, let X and Y be Abelian varieties defined over the field K. Let us assume that r(X, K′) = r(Y, K′) for all finite algebraic extensions K′ of the field K. I do not know whether this condition results in isogeny of the Abelian variety X and Y over K, but I should like to note here that it results in the isomorphism of the representations of the Galois groups G((K) of the field K in the enumerable finite dimensional vector spaces

$$X(K_s)/(\text{TORS } X(K_s)) = X(K_s) \otimes Q, Y(K_s)/(\text{TORS } Y(K_s)) = Y(K_s) \otimes Q$$

over the field Q of rational numbers. Indeed, for any finite Galois extension K′ of the field K with the Galois group G′ it is possible to consider natural finite dimensional representations of the group G′ in the vector spaces

$$V = X(K') \otimes Q, \quad W = Y(K') \otimes Q$$

over the field of rational numbers. For any (cyclic) subgroup D of the group G′ the subspaces of D-invariants in the spaces V and W coincide with X(K″) x Q and Y(K″) x Q, respectively, and, hence, are of the same dimensionality equal to the number r(X,K″) = r(Y,K″). Here K″ is the extension of the field K lying in K′ and consisting of all the elements immobile relative to the action of the group D. According to the Broer theory for the finite dimensional representations of a finite group in the vector spaces over the field of rational numbers, two such representations are isomorphous if for any cyclic subgroup the D dimensionalities coincide with the dimensionalities of the respective subspaces of D-invariants (Serre, Sect.12.5 in [51]). Applying this statement to the representations of the group G′ in the spaces V and W, we arrive at their isomorphism. Now, in order to obtain the required statement on isomorphism of the respective infinite dimensional representations of the group G(K) suffice it to consider the enumerable sequence of finite Galois extensions K′ embedded one into another whose union coincides with the algebraic closure K_s of the field K.

If all the endomorphisms of the Abelian variety X are defined over the field K, then the natural mapping

$$\text{End } X \to \text{End}_{G(K)} X(K_s)$$

is bijection (the Faltings theorem [31]). From this statement one can readily obtain that the the natural mapping

$$\text{End } X \to \text{End}_{\text{Aut}(C/K)} X(C) .$$

is also bujection. Note here that all the endomorphisms of the complex Abelian variety X are defined over a certain subfield of the field C finitely generated over Q. If the Abelian variety X can be set by the equations with coefficients from a certain field K, then all the endomorphisms of the variety X will be defined over a certain finite algebraic extension of the field K.

Appendix: l-adic numbers (see [13])

For any simple number l the notation

$$\text{ord}_1 : Z \setminus \{)\varnothing\} \to Z$$

is used for the mapping satisfying the following requirements

$$\text{ord}_l(1) = \text{ord}_l(-1) = \varnothing, \text{ord}_l(1) = 1;$$

$\text{ord}_l(p) = 0$ for any simple number p different from l and $\text{ord}_l(ab) = \text{ord}_l(a) + \text{ord}_l(b)$ for any nonzero integers a and b.These conditions uniquely define the mapping ord_l. Its existence can readily be proven. Namely, $\text{ord}_l(a)$ is the index of the maximum power of a prime l into which the number a can be divided. For example,

$$\text{ord}_l(-1) = 1, \text{ord}_l(1^2) = \text{ord}_l(-1^2) = 2.$$

If l = 2, then $\text{ord}_2 = 0$ for any odd integer a; in the general case

$$\text{ord}_2(2^m) = \text{ord}_2(2^m(2n + 1)) = m \text{ for any integer m and n.}$$

If l = 3, then

$$\text{ord}_3(3^m) = \text{ord}_3(3^m(3n + 1)) = \text{ord}_3(3^m(3m + 2)) = m$$

for any integer m and n.

Let us continue the mapping ord_l up to the mapping

$$ord_1 : Q^* = Q \setminus \{\emptyset\} \to Z$$

by the formula

$$ord_1(a/b) = ord_1(a) - ord_1(b) \quad \text{for} \quad a,b \in Z \setminus \{0\}.$$

It can readily be checked whether the mapping ord_l satisfies the following conditions:

$$ord_1(uv) - ord_1(u) + ord_1(v) \quad \text{for} \quad u, v \in Q^*;$$
$$ord_1(u + v) \geq \max\{ord_1(u), ord_1(v)\} \quad \text{for} \quad u,v \in Q^*,$$

such that u + v is nonzero. (The second condition is a reformulation of the simple fact that if two integers are divisible into a certain power of prime l, then their sum is also divisible into this power).

Let us take a certain positive real number D > 1 and define the funtion $| |_l : Q \to R$ with the values in the field of real numbers R by the following formulae:

$$|u|_l = D^{ord_l(u)} \text{ if u is nonzero and } |\emptyset|_l = \emptyset.$$

It is seen that the following conditions are met.

The condition of positiveness

$$|u|_l \geq \emptyset \text{ and } |u|_l = \emptyset \Leftrightarrow u = \emptyset,$$

multiplicativity

$$|uv|_l = |u|_l |v|_l,$$

and the triangle inequality

$$|u + v|_l \leq |u|_l + |v|_l .$$

(In fact, even the strengthened triangle inequality is valid,

$$|u + v|_l \leq \max \{|u|_l, |v|_l\}).$$

Positiveness, multiplicativity, and the triangle inequality are the properties well known for a conventional absolute value

$$|\,| : Q \dashrightarrow R, \quad |u| = u \text{ if } u \geq \varnothing$$

and $$|u| = - u \text{ if } u \leq \varnothing.$$

The Ostrovskii| theorem {13] states that for any mapping $||\ \ ||$: Q → R satisfying the conditions of positiveness, multiplicativity, and the triangle inequality there exists a real positive D > 1 such that either

$$||\ u\ || = |u|^D \qquad \text{for all} \quad u \in Q$$

or for a certain simple l

$$||\ u\ || = |u|_l = D^{\mathrm{ord}_l(u)} \qquad \text{for all nonzero u.}$$

It is well known that the field of real numbers R is obtained from the field of rational numbers Q by the complement to the metrics

$$\mathrm{dist}(u, v) = |u - v| .$$

The fields of l-adic numbers Q_l is obtained from the field of rational numbers Q by the complement to the metrics

$$\text{dist}(u, v) = |u - v|_l .$$

(The result is independent of the choice of the real number D). The ring of integer l-adic numbers $Z_l \in Q_l$ is obtained by the complement to the same metrics from the ring of integer numbers Z. In conventional real topology, the ring Z is discrete, in particular, any convergent sequence of integer numbers becomes constant upon the reject of a finite number of the first terms. Not so in l-adic topology. For example:

$$1, l, l^2, l^3, \dots , l^n, \dots$$

is a sequence of integer numbers that tends to zero in the l-adic topology and which does not become constant upon the reject of any finite number of the first terms. The field Q is local, compact, and open in Q_l. We have

$$Q_l = \cup\, l^{-n} Z_l \quad (n = \emptyset, 1, 2, \dots).$$

The sets $l^{-n}Z_l$ $(n = 0, 1, 2, \dots)$ form a fundamental system of the neighborhoods of zero in the field Q_l.

The natural mappings

$$Z \to Z/\, l^nZ, \text{ and } | \dashrightarrow a \bmod l^n$$

are continued to the continuous surjective homomorphism $Z \to Z/l^nZ_l$ with the kernel l^nZ_l. We arrive at the natural isomorphism $Z_l/\, l^n\, Z_l \approx Z/\, l^n\, Z$.

Any rational number lies in the ring Z_l for all simple l except of a finite set of simple divisors of its (number's) denominator. Therefore one can define natural surjective homomorphism

$$Q \to\oplus\, Q_l/\, Z_l \quad (l = 2, 3, 5, 7, 11, 13, \ ),$$

whose kernel coincides with the ring Z which is the "intersection" of all Z_l. This statement provides the natural isomorphism

$$Q/Z = \oplus\, Q_l/\, Z_l,$$

identifying the l-component of the periodic group Q/Z with Q_l/Z_l.

When working with l-adic numbers, one should realize the analogies with polynomials, rational functions, and formal series. Here is a small dictionary of some of them.

Z	ring of polynomials C[t]
Q	field of rational functions C(t)
prime l	complex number z
ord_l	$ord_2: C[t] \setminus \{\emptyset\} \to Z$, $ord_z(f)$ is the order of vanishing of the polynomial f at point z
log | |	degree deg of a polynomial
Z_l	C[[t - z]] is the ring of formal power series of t - z
Q_l	C((t - z)) is the field of formal Laurent series of t - z, for which only the finite number of coefficients before the negative degrees of t - z have nonzero values
R	C((1/t))

Despite the fact that the field of l-adic numbers is dully unconnected, it is possible to construct on it quite an interesting analysis, a functional analysis, a theory of Lie groups, and many other theories relating to "real" mathematics [13, 5, 19]. In this article we are interested mainly in l-adic groups and Lie algebras. Let us emphasize the following results. If the Lie algebra of a compact l-adic Lie group is commutative, then this Lie group contains an open commutative normal divisor of a finite index isomorphous to the direct sum of a certain number of rings of integer l-adic numbers considered as an additive group. Any closed subgroup of the l-adic Lie group is also a

Lie group. A set of regular elements of l-adic Lie algebra is open and dense everywhere. If the Abelian variety X is set by equations with integer or rational coefficients, then the group $X(Q_l)$ of its points with the coordinates from the field Q_l is defined. This group is naturally ascribed the structure of the commutative compact l-adic Lie group. Everywhere on the compact l-adic Lie group the logarithmic mapping (l-adic logarithm) is defined which is inverse to the exponential mapping in the vicinity of unity [5]. The l-adic logarithm log of the product of the commuting group elements is equal to the sum of the logarithms of cofactors. J.-P. Serre drew author's attention to the fact that far from unity of the Lie group the l-adic logarithm is not necessarily local diffeomorphism. More precisely, the l-adic logarithm is the local diffeomorphism in the vicinity of the element s of the l-adic Lie group then and only then when under the joint action Ad of the Lie group on its Lie algebra the operator Ad(s) has no eigenvalues which are nontrivial roots of 1. This is associated with the fact that, by definition,

$$\log(s) = \frac{1}{N} \log(s^N),$$

where the natural number N is chosen in such a way that the Nth powers of all the elements of the compact l-adic Lie group would lie in a sufficiently small neighborhood of the group unity.

2. LIE ALGEBRA

For an Abelian variety as for any other algebraic group, its Lie algebra is defined that nevertheless is turned out to be also commutative in virtue of the commutativity of the group law without bringing any additional information. Nevertheless, Serre [18] suggested to use a construction that establishes the correspondence between any Abelian variety x and a finite dimensional Lie algebra $g_l = g_{l,x}$ over the field of l-adic numbers Q_l for all primes l (l = p should be excluded from the p characteristic). For the complex Abelian variety X the finite dimensional Lie algebra $mt = mt_x$ is defined over the field of rational numbers which is the Lie algebra of a certain reductive algebraic Q-group, MT, called the Mumford-Tate group of the Abelian variety X [29. 37]. Any of these Lie algebras turns out to be commutative then and only then when X is a CM-type variety. The Serre hypothesis [18] states that the g_l algebras "are independent" of l, i.e., there exists the Q Lie algebra,

g_0, such that $g \approx g_0 x\ _Q Q_l$ for all l. (It is well known that the g_l algebras in the p characteristic start being independent of l after the closure of the field of constants, in other words, there exists a finite dimensional C Lie algebra g′ such that

$$g_l \otimes _Q C_l \approx g'$$

for all $l \neq p$ (the Klyachko theore|m).
For complex Abelian varieties the Mumford-Tate hypothesis [37] states that the mt algebra can be taken as g_0, i.e., $g \approx mt\ x\ _Q Q_l$ for all l.

Finally let us define the Lie algebras g_l and mt. We begin with the description of the mt algebra based on the existence of a complex structure on the real (!) space

$$L_R := L_Q \otimes _Q R = V,$$

where R is the field of real numbers. Let us consider multiplication by complex numbers in the space $L_R = V$ as R-linear operators in a real vector space L_R. These operators form a two-dimentional ($\dim_R C = 2$) subspace in the algebra $End_R L_R$ of all R-linear operators in L_R hereinafter denoted by t. By definition, the subspace t lies in
$End_R L_R = End_R(L_Q \otimes _Q R) = End_Q L_Q \otimes _Q R$
and contains homotheties, i.e., operators of the form a id, where $a \in R$ and id is an identity operator.

2.1. Lie Algebra of the Mumford-Tate Group

The Lie algebra of the Mumford-Tate group of the complex Abelian variety X is called the smallest Lie Q-subalgebra $mt = mt_X$ in the End $_Q L_Q$ algebra of all the linear mappings of the Q space L_Q into itself such that

$$t \in mt_R := mt \otimes _Q R \in End_Q L_Q \otimes _Q R = End_R L_R$$

i.e., the minimal algebra of all the Lie Q-subalgebras that upon the extension of scalars up to R, contain the subspace t.

It is clear that the Lie algebra mt contains homotheties Q id, $a \in Q$. It is reductive, and its natural representation in the Q-space L_Q is quite reducible (the proof of this fact is based on the existence of polarization). From the minimality condition in the definition of the Lie algebra mt and the explicit description of the End°X algebra it follows that the centralizer of the Lie algebra mt is

$$\mathrm{End}_{\mathrm{mt}}\, L_Q = \mathrm{End}^\circ X$$

In particular, the complex Abelian variety X is simple then and only then when the natural representation of the Lie algebra $\mathrm{mt} = \mathrm{mt}_x$ in the L_Q space is irreducible. This representation is absolutely irreducible, i.e., remains to be irreducible upon the extension of the field of scalars up to C then and only then when End°X = Q.

Example. If X is an elliptical curve, i.e.,

$$n = \dim X = 1,\ \dim_Q L_Q = 2,$$

then

$$\mathrm{mt} = \mathrm{End}_{\mathrm{End}^\circ X}\, L_Q \in \mathrm{End}_Q L_Q.$$

Thus, if X is a CM-type curve, then mt = End°X. If X is the curve without the complex multiplication, then $\mathrm{mt} = \mathrm{End}_Q L_Q$.

The Lie algebra mt is an algebraic Lie subalgebra in $\mathrm{End}\,_Q L_Q$: there exists a connected reductive algebraic Q-subgroup MT in $GL(L_Q)$, called the Mumford-Tate group of the Abelian variety X, whose Lie algebra coincides with mt.

Let us expand the reductive Lie algebra mt into a direct sum

$$\mathrm{mt} = \underline{c} \oplus \mathrm{mt}'$$

of its center $\underline{c}$ and semivoid mt′ algebra. Since mt coincides with the centralizer of the End° X algebra in $\mathrm{End}_Q L_Q$, the center $\underline{c}$ lies in the algebra End°X (and even in its center). It can readily be seen that Q id $\in c \in$ End°X.

Therefore if End° X = Q (or if End° X is the central simple algebra over Q), then $\underline{c}$ = Q id. Under the same assumptions, mt′ is a simple Lie Q-algebra (the Borovoĭ theorem [4]).

Simple linear-algebra considerations (using equality $\dim_Q l_Q = 2n = 2 \dim X$) show that $mt' = \{0\}$. i.e. $mt = \underline{c}$ is the commutative Lie algebra then and only then when the Q-algebra $End^\circ = End_{mt} L_Q$ contains a semivoid commutative Q-subalgebra of dimensionality 2n, i.e., when X is a CM-type Abelian variety.

Any polarization H defines nondegenerate asymmetrical bilinear forms

$$\text{Im } H: L_R \times L_R \to R \quad (\text{we remind here that } L_R = V),$$
$$\text{Im } H: L_Q \times L_Q \to Q;$$

now let

$$sp(L_R) \in End_R(L_R),\ sp(L_Q) \in End\ (L_Q)$$

be the Lie algebras (Q- and R-algebras) of the respective symplectic groups. We have

$$sp(L_R) = sp(L_Q) \otimes {}_Q R.$$

In virtue of Hermicity of the H form the Lie algebra t (of multiplications by complex numbers) preserves the form Im H c within the accuracy of proportionality, i.e.,

$$R\ id \in t \in R\ id \oplus sp(L_R) \in End_R(L_R).$$

Now from the minimality condition in the definition of the Lie algebra mt it follows that

$$Q\ id \in mt \in Q\ id \oplus sp(L_Q) \in End_Q(L_Q).$$

Let us put

$$hdg := mt \cap sp\ (L_Q) \in sp(L_Q).$$

The Lie algebra hgd is the reductive algebraic Lie algebra which is the Lie algebra of the so-called Hodge group, Hdg, of the Abelian variety X that was introduced by Mumford [35]. We have

$$\text{mt} = \text{Q id} \oplus \text{hdg} ;$$

in particular, the center $\mathfrak{c}$ = Q id <=> Lie algebra of the hdg semispace, i.e., hdg = mt′.

It is clear that

$$\text{End}_{\text{hdg}} L_Q = \text{End}_{\text{mt}} L_Q = \text{End}^{\circ} X,$$

and that the following embedding takes place

$$\text{hdg} \in \text{sp}(L_Q) \cap \text{End}_{\text{End}^{\circ} X}\, L_Q.$$

If the Abelian variety X is simple and n = dim X is a simple number, then this inclusion is transformed into an equality (the Tankeev-Riber theorem) [21, 36]. Mumford [35] constructed an example of a four--dimensional complex Abelian variety X c End°X = Q for which the hdg Lie algebra is strictly less than $\text{sp}(L_Q)$, i.e., the above given inclusion is rigorous.

Consider a semivoid complex Lie algebra $\text{mt}'_C = \text{mt}' \times {}_Q C$ acting naturally in the complex vector space

$$L_C = L_Q \otimes {}_Q C = L_R \otimes {}_R C = L \otimes C$$

The semisimple Lie algebra mt'_C is expanded into a direct sum of complex simple Lie algebras called its simple factors. All these simple factors are, in fact, simple classical Lie algebras (A_r, B_r, C_r, D_r) and their weights in L_C are the microweights in the Burbaki sense [5] (the Deligne theorem) [28]. (The highest weight of an irreducible representation of the simple Lie algebra is a microweight then and only then when all the weights of this representation are obtained from the heighest weight due to the action of the Weil group. Here is the list of such representations for the classical Lie algebras: (A) for the Lie algebra sL(r + 1) of operators with the zero spur it is the ith external degree of the standard representation of the dimensionality r + i where $1 \le i \le r$; (B) for the Lie algebra so(2r + 1) of the orthogonal group it is a spinor representation of dimensionality 2^r; (C) for the Lie algebra sp(2r) it is a standard representation of dimensionality 2r; (D) for the Lie algebra so(2r) it is either a standard representation of dimensionality 2r or one of two semispinor representations of dimensionality 2^{r-1}. Everywgere r is the rank of the respective simple Lie algebra.

Example. If X is an RM-type Abelian variety, then it can readily be checked whether all the simple factors of the Lie algebra mt'_C are the Lie algebras sl(2), and their irreducible representations arising in L_C are standard two-dimensional representations. Thus if X is an elliptical curve without complex multiplication, then (see above) $mt = End_Q L_Q$ and, therefore $mt' = sl(L_Q)$ is the Lie algebra of all the operators with the zero spur and

$$mt'_C = sl(L_C) \approx sl(2,C).$$

If $End^o X = Q$ and n = dim X is either an odd number or n = 2 or n = 6, then, using the Deligne and Borovoi| theorems together with the equality $\underline{c} = Q$ id, it can readily be shown that

$$mt' = hdg = sp(L_Q)$$
$$\text{and } mt = Q \text{ id} \oplus sp(L_Q).$$

For an arbitrary complex Abelian variety the following estimate is valid

$$n = \dim X = \frac{1}{2} \dim L_Q \leq 2^{r-1} \dim_Q End^o X$$

where r is the rank of the Lie algebra mt'. This statement is also a corollary of the Deligne theorem.

2.2. Tate Modules

Now we can finally define l-adic Lie algebras g_l. Let X be a complex Abelian variety of dimensionality n and X(C) = V/L. Let us embed it into a projective space and let K be a subfield of the field of complex numbers obtained by the adjunction to the field of rational numbers of all the coefficients of the equations setting X and the group law. Let us put $T_l(X) = L \times Z_l$. This group, the so-called Tate l-module, is a free module of rank 2n over the ring of integer l-adic numbers Z_l. On the Tate module, the Galois group G(K) of the field K acts naturally and continuously. This action is glued of the above defined consistent actions of the Galois group onto the factor-groups

$$T_l(X)/l^i T_l(X) = (L \otimes Z_l)/l^i(L \otimes Z_l) = L/l^i L = X_i ,$$

i.e. of the actions at the finite-order points of th|e Abelian variety X that are annihilated by the multiplication by the power of prime l. Here is a more explicit(?) definition that can readily be transferred to the p characteristics. It is easy to check the existence of natural isomorphism

$$T_l(X) \approx \mathrm{Hom}(Q_l/Z_l, \mathrm{TORS}\ X(C)).$$

Here the natural action of the Galois group on torsion induces the action of the Galois group on the Tate module. Thus, we have defined continuous homomorphism

$$G(K) \to \mathrm{Aut}_Z\, T_l(X) \approx GL(2n, Z_l)$$

of the Galois group into the gr|oup of automorphism of the Tate module which, being isomorphous to the group of nondegenerate matrices of the order 2n over the integer l-adic numbers, is a compact l-adic Lie group. Let us denote by G_l the image of this homomorphism. (The group G_l is the Galois group of an infinite expansion of the field K obtained by the adjunction of the coordinates of all those points of the Abelian variety X which are annihilated by multiplication by a certain power of the prime l). In virtue of the compactness of the Galois group, the G_l group is a closed subgroup of the l-adic Lie group $\mathrm{Aut}_{Z_l} T_l\ (X) \approx GL(2n, Z_l)$.

As in the case of the real Lie group, any closed subgroup of the l-adic Lie group is also a l-adic Lie group ([5]. [19]). Therefore, G_l is an l-adic Lie group, and its Lie algebra is defined. It is just the searched finite--dimensional Lie algebra g_l over the field of l-adic numbers Q_l. When considering the properties of the Lie algebra g_l, it is convenient to represent it as a linear Lie algebra, i.e., the subalgebra of the Lie algebra of all the linear operators acting in a certain finite dimensional vector space over Q_l. Such a space denoted by $V_l(X)$ is obtained from the Tate module by extension of scalars up to Q_l, i.e.,

$$V_l(X) := T_l(X) \otimes_Z Q_l = L \otimes Q_l = L_Q \otimes_Q Q_l.$$

The space $V_l(X)$ is a vecto|r space over the field Q_l of dimensionality 2n containing the Tate module $T_l(X)$ as a Z_l lattice. The following natural embeddings take place

$$T_l(X) \in V_l(X),\ G_l \in \mathrm{Aut}_Z\, T_l(X) \in \mathrm{Aut}_Q\, V_l(X).$$

In particular, the group G_l acts nat|urally and Q_l^1 linearly on the vector space $V_l(X)$ and its Lie algebra g_l turns out to be the Q_l subalgebra of the Lie algebra $\mathrm{End}Q_lV_l(x)$ of all Q_l-linear operators acting in the space $V_l(X)$. The Lie algebra g_l is an algebraic Lie algebra, i.e., is the Lie algebra of a certain algebraic Q_l subgroup in $GL(V_l(X))$ (the Bogomolov theorem [1, 27]). In the p characteristic the Lie algebra g_l is not necessarily algebraic. Thus for any nonsupersingular elliptic curve defined over a finite field, g_l is a one-dimensional nonalgebraic Lie algebra).

The action of the ring End X onto the points of the finite order of the Abelian variety X is naturally continued up to the Z_l linear action onto the Tate module $T_l(X)$ and the following embedding takes place
$\mathrm{End}\, X \otimes Z_l \to \mathrm{End}_g\, T_l(X)$;
(if all the endomorphisms of the Abelian| variety X are defined over K, then End X x Z_l is embedded into $\mathrm{End}_G T_l(X)$. Passing from Z_l modules to the vector spaces, we arrive at the embedding
$\mathrm{End}^\circ X \otimes_Q Q_l = \mathrm{End}\, X \otimes Q_l \to \mathrm{End}_g\, V_l(X)$.
Thus, a finite dimensional semisimple Q_l Lie a|lgebra $\mathrm{End}^\circ X\, x_Q Q_l$ lies in the centalizer of the Lie algebra g_l. What can be said of the Lie algebra g_l and its natural representation in the $V_l(X)$ space? The Lie algebra g_l is reductive; its representation in $V_l(X)$ is quite reducible, and the centralizer

$$\mathrm{End}_g\, V_l(X) = \mathrm{End}^\circ X \otimes_Q Q_l \quad \text{(Faltings theorem [30, 31]).}$$

(In the char|acteristic $p > 2$, this statement was proved by the author [9, 10], the case of characteristic 2 was considered by Mori [34]. (The Galois group for the finite field K of q elements has a canonic generator which is the Frobenius automorphism Fr that acts as raising to the power q. Thus, the problem of the complete reducibility of the action of the Galoi group reduces to the problem of diagolinazability of the action of the Frobenius automorphism on $V_l(x)$. A. Weil proved that the operator Fr is semisimple and its characteristic polynomial has integer coefficients independent of l, and that its eigenvalues are integer algebraic numbers that, being considered as complex numbers, have absolute values equal to a squared root of q [16]. The theorem of centralizer for finite fields was proved by Tate [23] who suggested [22] that it is correct also for the general case).

If X is an Abelian variety without nontrivial endomorphisms, i.e., $End^oX = Q$, then $V_l(x)$ is an irreducible (and even absolutely irreducible) representation of the Lie algebra g_l since its centralizer (according to the Faltings theorem) consists of scalars.

The above cited result obtained by Faltings plays a key role in the Faltings proof of the Mordell hypothesis [30]. We refer the reader to the review [11] describing in detail the relation between the Abelian varieties and the Mordell and Shafarevich hypotheses, but here we limit ourselves to the discussion of the Mordell hypothesis proved by Faltings. (Recently P. Vojta proved the Mordell hypothesis on the basis of the Daison lemma).

2.3. Digression: on the Mordell hypothesis

Let us remind that the Fermat Last theorem not proven up to now states that the equation $x^n + y^n = z^n$ has no natural (positive integer) solutions x, y, z. Here n is a natural number ≥ 3. (At present, the Fermat theorem is proven for all the powers $n \leq 125000$). From the Mordell hypothesis proved by Faltings it follows that if the Fermat equation really has solutions, their number is not large in the following sense. First of all it should be noted that the Fermat equation is homogeneous. In other words, using any of its solutions (x,y,z), it is possible to construct a series of new solutions by multiplying x, y, and z by the same number. Therefore, in what follows we shall consider only the primitive solutions for which the natural numbers x, y, and z do not have nontrivial (different from unity) common divisors. Any natural solution (if at all) of the Fermat equation can be obtained from a certain primitive solution by multiplication by a natural number. From the Faltings theorem it follows that for $n \geq 4$ the set of primitive solutions of the Fermat equation is finite. A statement on the finiteness of the number of solutions which is formally weaker than the Fermat theorem, turns out to be correct for a larger class of equations. Thus, if to ascribe arbitrary (integer) nonzero coefficients to terms x^n, y^n, z^n in the Fermat equation (at $n \geq 4$), then the newly obtained equation can have solutions but their number will be finite all the same. Thus, although the equation $x^4 + y^4 = 2z^4$ has the solution, $x = y = z = 1$, the number of all its (primitive) solutions is finite.

In order to give the rigorous formulation of the Faltings theorem it is convenient to rewrite the Fermat theorem in the form $u^n + v^n - 1 = 0$, $u = x/z$, $v = y/z$. The statement about the finiteness of the number of primitive solutions of the Fermat equation would follow from the finiteness of the number of rational solutions (solutions in rational numbers) (u, v) for the equation $u^n + v^n - = 0$. It is just the finiteness of the number of rational (u, v) solutions that was proven by Faltings for a large class of algebraic equations $f(u,v) = 0$, where f is a polynomial of u,v with rational coefficients. Let us describe a class of equations for which the Faltings theorem is valid. First of all, let us require that the polynomial f would be absolutely irreducible, i.e., could not be represented as the product $f = f_1 f_2$, with f_1 and f_2 being polynomials of u,v (generally speaking, with complex coefficients) but not constants. Thus the polynomials $u^n + v^n - 1$ and $u^2 - v^n - a$ (for any nonzero rational a) are absolutely irreducible, whereas polynomials the $u^2 - v^4 = (u - v^2)(u + v^2)$, $u^2 - 3v^2 = (u + \sqrt{3}\ v)(u - \sqrt{3}\ v)$ are not. We have to exclude also all the homogeneous polynomials of degrees higher than unity, since they can be expanded into the product of linear multipliers (the corollary of the main algebraic theorem).

The problem of a finite number of rational solutions of the equation $f = 0$ is solved using a very crude characteristic of the set of its complex points, a complex graph. In fact, the only topological invariant of the complex graph is its genus. Roughly speaking, a set of complex solution is (topologically) constructed as a two-dimensional sphere with several glued handles. (More precise: let us carve up from a complex graph a (finite) set of singular points (u, v) at which both partial derivatives, df/du and df/dv, go to zero. Topologically, what is remained is constructed as a two-dimensional sphere with a certain number of glued handles, then a certain finite set of points is carved up). It is this number that is called the genus of the equation $f = 0$ (or, which is the same, the genus of the respective Riemann surface). If the genus is zero, the complex graph is a two-dimensional sphere, if it is 1, then it is a tore. In these cases, the set of rational solutions can be infinite. The genus of the equation $u^n + v^n - 1 = 0$ is equal to $(n - 1)(n - 2)/2$; in particular, it exceeds unity at $n \geq 4$.3. The same genus is also possessed by the equation $u^n = bv^n + c = 0$, where a, b, c are arbitrary nonzero rational numbers. In general, the genus of the equation $f = 0$ for polynomial f of degree n does not exceed $(n - 1)(n - 2)/2$. In particular, for polynomials of the first and second degrees the genus is always zero. The genus of the equation $u^2 - (v^3 + bv + c) = 0$ (b and c are arbitrary rational numbers) is equal to unity if the polynomial $v^3 + bv + c$ has no multiple roots and to zero otherwise. The genus of the equation $u^2 - v^n - a = 0$ is equal to $(n - 1)/2$ for odd n and $(n - 2)/2$ for even. Thus $u^2 - v^{2m+1} - a = 0$ is an example of the equation of genus m for any nonnegative integer m (here $a \neq 0$). The genus of the equation $u^2 - v^{2m+1} = 0$ is equal to zero.

The Mordell hypothesis proved by Faltings states that a set of rational solutions of the equation of a genus exceeding 1 is finite. The fact that the genus of the Fermat equation exceeds 1 for $n \geq 4$ leads to the statement on the finite number of the solutions of the Fermat equations that began our consideration. Faltings also proved the finiteness of the number of solutions (u, v) lying in an arbitrarily given field of algebraic numbers K of the finite degree over the field Q for the equations whose genus exceeds 1. (In this case, the coefficients of the polynomial f are not necessarily rational numbers but they should belong to the field K). As an example, consider the field K of all the numbers of the form a + b i where $i = \sqrt{-1}$ \and a and b are rational numbers. The equation $u^4 + v^4 + 4 = 0$ has no solutions in rational numbers but has obvious solution $u = 0$, $v = 1 + i$ in field K. According to the Faltings theorem, a set of all the solutions of the above equation in the field K is finite.

2.4. l-adic Representations and Lie Algebras

Revenons à nos moutons, i.e., come back to the l-adic Lie algebras g_l and their natural representations in $V_l(X)$. Note also that the Faltings theorem of centralizer applied to the product of the complex Abelian varieties X and Y leads to the following statement (the theorem of isogeny) [31].

Let K be a subfield of finite type of complex numbers over which the Abelian varieties X and Y and all the endomorphisms of their product X x Y are defined (such a field always exists). The Abelian varieties X and Y are isogeneous to one another then and only then when the representations $V_l(X)$ and $V_l(Y)$ of the Galois group of the field K are isomorphous to one another for a certain prime l. (For isogeneous Abelian varieties these representations are isomorphous to one another for all primes l). (A similar statement is also valid in the p characteristic).

The Lie algebra g_l contains the homotheties Q_l id (the Bogomolov theorem) [1, 27]. (In the p charracteristic, an ordinary elliptic curve is a counter-example for this statement).

For complex Abelian varieties X a natural isomorphism

$$V_l\,(x) = L_Q \oplus {}_Q Q_l$$

provides the determination of the Q_l-Lie algebra

$$mt_l = mt \oplus {}_Q Q_l \in End_Q L_Q \oplus {}_Q Q_l = = End_Q\ V_l(x).$$

The Mumford-Tate hypothesis [37] states that [1]

$$g_l = mt_l.$$

This hypothesis was verified for the CM-type Abelian varieties (Shimura-Taniyama, Weil [37]). Simple linear algebraic considerations show that it is correct also for RM-type Abelian varieties. It is well known that the following inclusion always takes place

$$g_l \in mt_l$$

(Pyatetskii -Shapiro-Deligne-Borovoi [17, 4, 24]). The equality

$$g_l = mt_l.$$

has been verified for the Abelian varieties X without nontrivial endomorphisms for which n = dim X is an odd number or n = 2 or n = 6 (the Serre theorem [39]). Thus if X is an elliptic curve without complex multiplication, then

$$g_l = mt_l = \mathrm{End}_{Q_l} V_l\ (x).$$

Thus, in this example G_l is an open subgroup of the finite-index group $\mathrm{Aut}_{Z_l} T_l\ (X) \approx GL(2,Z_l)$. In fact, $G_l\ \ \mathrm{Aut}_{Z_l} T_l\ (X) \approx GL(2,\ Z_l)$ for all primes l except of finite numbers (this statement is a particular case of another Serre theorem valid for the same X: in its formulation GL(2) is substituted by a group of symplectic similarities [40, 38]).

Let us expand the reductive Lie algebra g'_1 into a direct sum

$$g_l = \underline{c}_l \oplus g_l$$

of its center $\underline{c}_l$ and a semisimple algebra h_l. Since g_l coincides with the centralizer of the algebra $\mathrm{End}^\circ X \times_Q Q_l$ in $\mathrm{End}_{Q_l} V_l\ (X)$, the center $\underline{c}_l$ lies in the algebra $\mathrm{End}^\circ X \times {}_Q Q_l$ (and even in its center). In particular,

$$\dim_{Q_l} \underline{c}_l \leq \dim_Q \mathrm{End}^0 X.$$

It can readily be seen that

$$Q_l id \in \underline{c}_l \in \underline{c} \oplus {}_Q Q_l \in \mathrm{End}^0 X \oplus {}_Q Q_l,$$

where $\underline{c}$ is the center of the Lie algebra mt. If $\mathrm{End}^\circ X = Q$ (or if $\mathrm{End}^\circ X$ is the central simple algebra over Q), then all these inclusions are transformed into equality).

As in the case of the Lie algebra mt, the simple linear algebraic considerations (using the equality $\dim_{Q_l} V_l\ (X) = 2n = 2 \dim X$) show that $g'_l = \{0\}$, i.e., $g_l = \underline{c}_l$ is a commutative Lie algebra then and only then when the Q_l-algebra $End^{\circ}X\ x_Q\ Q_l = End_{g_l}\ V_l(X)$ contains a semisimple commutative Q_l-subalgebra of dimensionality 2n. It can readily be shown that the latter condition is satisfied then and only then when X is an CM-type Abelian variety [40].

Let Q_l be an algebraic closure of the field Q_l. Extending the field of scalars up to Q_l, we arrive at the vector space

$$\bar{V}_l(X) = V_l(X)\ x_{Q_l} Q_l$$

of dimensionality 2 dim X in which the semisimple $\bar{Q}_l$ Lie algebra $g_{l,X} = g_l\ x_{Q_l} \bar{Q}_l$ acts naturally . It can readily be seen that $\bar{V}_l(X)$ is expanded into a direct sum of not more than $\dim_Q End^{\circ}X$ irreducible representations of the Lie algebra $g_{l,X}$. In turn, the semisimple Lie algebra $g_{l,X}$ is expanded into a direct sum of simple Q_l Lie algebras called its simple factors.

The hypothesis of microweights put forward by the author [7,11] states that all these simple factors are classical Lie algebras (A_r, B_r, C_r, D_r) and their weights in $\bar{V}_l(X)$ are microweights in the Burbaki sense. For complex Abelian varieties, this hypothesis follows from the Mumford-Tate hypothesis and the Deligne theorem. (In the p characteristic the hypothesis of microweights was proven by the author for ordinary Abelian varieties [7]). From the Serre results [41] it follows that the hypothesis of microweights is valid for the complex Abelian variety X and the prime l if

$$\underline{c}_l = Q_l\ \text{id},\ g'_l \text{ is a simple } Q_l \text{ Lie algebra.}$$

The following result obtained by the author confirms the hypothesis of microweights for the low-rank Lie algebras g'_1.

Theorem 1. Let X be a complex Abelian variety. If g'_1 is a Lie algebra of rank 1, i.e., $g_{1,X} \approx sl(2,\bar{Q}_1)$, then all the nonunidimensional irreducible representations W of the algebra $g_{1,X}$ entering the space $V_1(X)$ are of dimensionality 2 and the image of the Lie algebra $g_{1,X}$ in End(W) coincides with the Lie algebra sl(W) of all the operators with the zero spur.

If g'_1 is a Lie algebra of rank 2, then the following situations are possible.

a1) $g_{1,X} = g_1 \oplus g_2$ where $g_i \approx sl(2,\bar{Q}_1)(i = 1, 2)$. All unidimensional irreducible representations **W** of the algebra g_i entering the space $V_1(X)$) are of dimensionality 2. Then the image of the Lie algebra g_i in End(**W**) coincides with the Lie algebra sl(**W**).

a2) $g_{1,X} \approx sl(3,\bar{Q}_1)$ is the A_2-type Lie algebra. All nonunidimensional irreducible representations **W** of the $g_{1,X}$ algebra entering the $V_1(X)$ space have dimensionality 3, and the image of the Lie algebra $g_{1,X}$ in End(**W**) coincides with the Lie algebra sl(**W**).

b2) $g_{1,X} \approx sp(4,\bar{Q}_1)$ is the B_2-type Lie algebra. All the nonunidimensional irreducible representations **W** of the algebra $g_{1,X}$ enetering the space $V_1(X)$ have dimensionality 4, and the image of the Lie algebra $g_{1,X}$ in End(**W**) coincides with the Lie algebra sp(**W**) of a certain symplectic group of the space **W**.

The proof of Theorem 1 will be published elsewhere.

The hypothesis of microweights leads to the following estimate

$$n = \dim X = \frac{1}{2} \dim_{Q_1} V_1(X) \leq 2^{r-1} \dim_Q \mathrm{End}^\circ X$$

where r is the rank of the Lie algebra g'_1. (From the results of Ref.[42] it follows that this rank is l-independent). The following result (also obtained by the author) has the same "qualitative" structure as the above considered inequality.

Theorem 2. For any natural numbers r and e there exists a natural number $D \doteq D(r,e)$ depending on r and e alone that satisfies the following conditions. If X is an Abelian variety such that the rank of the semisimple algebra g'_1 is equal to r and the dimensionality $\dim_Q End^{\circ}X$ is equal to e, then $n = \dim X \leq D(r,e)$. (This inequality is also valid in the p characteristic).

Note to the Proof of Theorem 2. Since $V_l(X)$ is a direct sum of not more than e irreducible representations of the Lie algebra $g_{l,X}$, then suffice it to prove that for all simple factors g of the Lie algebra $g_{l,X}$ any of their irreducible representations W entering $V_l(X)$ is of dimensionality bounded from above by the constant depending on r and e alone.

The method used for proving Theorems 1 and 2 will be developed in the next Section. Here we should like to note that it is sufficient to prove these theorems for Abelian varieties defined over the number field using the modified Hilbert theorem of irreducibility [14] for infinite extensions formulated by Serre. For the case of the number field, Theorem 2 will be proven in the next Section.

3. ABELIAN VARIETIES OVER NUMBER FIELDS

3.1. Good Reduction and Newton Diagrams

Now let us assume that K is a field of algebraic numbers of a finite degree over Q and O is the ring of all integer algebraic numbers of the field K (e.g., K = Q, O = Z). It can be taken that the Abelian variety X is given by homogeneous algebraic equations all the coefficients of which are in the ring O. Let v be the maximum ideal of the ring O and $k(v) = O/v$ be the respective finite field of residues of the characteristic p consisting of q elements, where q is a certain degree of the prime p.

(If K = Q and O = Z, then v = pz for a certain prime p, i.e., v is a set of all integers divisible into a given prime p, k(v) = Z/pZ is a (simple) field of residue of the characteristic p of q = p elements). Assuming that v does not belong to a certain finite exclusive set of simple ideals, it can be taken that there exists an Abelian variety X(v) defined over the finite residue field k(v) by the equations obtained from the equations for X by the substitution of the coefficients by their residues (mod v) [20, 26]. In this case it is said that X has good reduction in v. The Faltings theorem [30] refined by the author [45] states that for a given number n and a finite set M of simple ideals of the ring O the set of n-dimensional Abelian varieties over K with good reduction everywhere outside M considered within the accuracy of isomorphism (over K) is finite. A similar hypothesis was also put forward by Parshin [46]. The recent result of Fontain [47] obtained independently also by Abrashkin [48-50] indicates that there exist no Abelian varieties over the field of rational numbers Q having good reduction everywhere. Neither exist they over the fields K = Q(-1), Q(-2), Q(-3). Q(-7), Q(2), Q(1) (Ibid).

Thus, let X have a good reduction in v. The mapping of reduction, mod v, induces homomorphism $X(K) \to X(v)(k(v))$ whose kernel does not contain nonzero points of the finite order which is mutually simple with p [20, 26]. Whence, due to the coincidence of the orders of groups X_m and $X(v)_m$ for m mutually simple with p, it follows that if all the points of the order m on the Abelian variety X are defined over the field K, then the reduction mapping induces the isomorphism $X_m \approx X(v)_m$. In order to determine this isomorphism in the general case one should take the finite Galois extension K' of the field K, which is a field where all the points of the order m on X are defined, and choose in the ring O' of all the integer algebraic numbers of the field K' a simple ideal v' lying over v. Then the reduction mapping mod v′ induces the isomorphism $X_m \approx X(v)_m$, defined not uniquely but within the accuracy of the choice of the ideal v'. Since the Galois group of extension K'/K acts transitively on the set of all simple ideals v' lying over v, we obtain that the isomorphism $X_m \approx X(v)_m$ is uniquely defined within the accuracy of the action of the Galois group of the field K. Passing to the Tate modules, we obtain that for $l \neq p$ there exists a natural isomorphism of the spaces $V_l(X(v))$ and $V_l(X)$ (defined within the multiplication by automorphism from $G_l \in \mathrm{Aut}_{Q_l} V_l(X)$), for which the Frobenius automorphism of the field k(v) acting naturally on $V_l(X(v))$ passes into a certain element of the group G_l (defined within the accuracy of conjugacy) is called the Frobenius element and denoted by Fr_v. An element Fr_v considered as an operator in $V_l(X)$ is semisimple. Its characteristic polynomial

$$P_v(t) = \det(t\ \mathrm{id} - Fr_v, V_l(X)) = t^{2n} = + a_1 t^{2n-1} + \ldots + a_{2n}$$

lies in Z[t] and is independent of l, i.e., all a_i are conventional integers and are the same for all l. Thus, $a_{2n} = q^n$.

All the complex roots of a polynomial P_v are integer algebraic numbers with the absolute values equal to the square root of q (the Weil theorem). It is well known that the sum

$$1 + a_1 + \ldots + a_{2n} = P_v(1)$$

is equal to the number of points on the Abelian variety X(v) with the coordinates in the field k(v) (the Weil theorem). In a similar way, the number of points of the variety X(v) with the coordinates in the field $k(v)_m$ from q^m, which is the (only) extension of the field k(v) of degree m, is equal to $P_{v,m}(1)$, where $P_{v,m}$ is the reduced polynomial of degree 2n with the roots which are the mth degrees of the roots of a polynomial P_v [16]. The following equalities are valid

$$P_v(t) = t^{2n}\, P_v(q/t), \quad P_{v,m}(t) = t^{2n}\, P_{v,m}(q^m/t)$$

If v runs over a certain set of simple ideals of density 1 (e.g. a set of all simple ideals except of a finite number), then the elements Fr_v (or,more exactly, their conjugacy classes) form a subset dense everywhere in the group G_l (the corollary of the Chebotarev density theorem). As another example of the set of simple ideals v for which the Frobenius elements form a subset dense everywhere one can take the set of all simple ideals of degree 1 (i.e., with q = p) except of a finite number.

In order to obtain additional information about the set R(v) of all the roots of the polynomial P_v let us consider its field of expansion L(v), i.e., the subfield in Q_l obtained by the adjunction to Q of a set R(v). The field L(v) is a finite Galois extension of the field Q, the set R(v) lies in the ring of all integer algebraic numbers of the field L(v); for any embedding $L(v) \to C$ and the root $\tau \in R(v)$ we have $|\tau| = q^{1/2}$, i.e., $\tau\bar{\tau} = q$, where $\bar{\tau} \in R(v)$ is the number which is complex conjugate with τ [33]. Whence it follows that for any simple ideal @ in L(v) not lying over p, τ is a @-adic unit. If @ is a simple ideal in L(v) lying over p and $ord_@: L(v)^* \to Q$ is a discrete normalization of the field L(v) corresponding to @ [6] and normalized by the condition $ord_@(q) = 1$, then

$$1 = ord_@(q) = ord_@(\bar{\tau}) + ord_@(\tau).$$

Since τ and $\bar{\tau}$ are integer algebraic numbers, the rational numbers $ord_@(\tau)$ and $ord_@(\tau)$ are nonnegative, and we arrive at the inequality

$$0 \le ord_@(\tau) \le 1 \text{ for } \tau \in R(v).$$

Let us put $A_v = ord_1(R(v)) \in Q$ and define on the finite set A_v the function mult: $A_v \to Z_+$ with the values in the set Z_+ of all the positive integer numbers by the formula mult(c) = card{$\tau \in R(v)$ (with multiplicities | $ord_@(\tau) = c$}. Hereinafter for any finite set A we denote the number of its elements by card(A) and the multiplicity of the number $\tau \in R(v)$ by its multiplicity as a root of the popynomial P_v. From the above considerations it is clear that

$$0 \le c \le 1,\ 1 - c \in R(v),\ \text{mult}(c) = \text{mult}(1 - c) \quad \text{for } c \in A_v.$$

Since L(v) is the Galois extension of the field Q, its Galois group G(L(v)/Q) acts transitively on the set S of all simple ideals @ lying over p [6] and transforms into itself the set of roots R(v) of the polynomial P_v. Therefore the set A_v and the function mult are independent of the choice of a simple ideal @ $\in$ S. The pair (A_v, mult) allows the following explicit description. For the sake of simplicity, we assume that q = p, i.e., that v is an ideal of degree 1. In this case the mapping ord_p coincides on rational numbers with a conventional p-adic normalization

$$ord_p: Q^* \to Z,\quad ord_p(p) = 1,$$

that puts each nonzero integer number in correspondence with the maximum degree $ord_p(a)$ of the prime p the division by which comes out even. In particular, $ord_@(a_i)$ $ord_p(a_i) \in Z$ for all nonzero coefficients a_i of the polynomial P_v. For example,

$$ord_@(a_{2n}) = n, ord_@(a_o) = 0,$$

where the free term is $a = q^n$, and $a_o = 1$ is the highest coefficient of the polynomial P. As usual [13], the normalization ord_p provides the determination of the Newton diagram M(v) of the polynomial P_v that is constructed by a convex shell of points (i, ord a_i) $\in Z \times Z$ on the coordinate plane. Then the set of polygon M(v) tilts coincides with A_v and the function mult establishes the correspondence between the tilt and its length. Under our assumption that q = p it is clear that all the vertices of a polygon M(v) lie at the integer points. Thus, the end vertices are always the points (0,0) and (2n,n). The property of vertex integrity signifies that for any rational number $c \in A_v$ the product with mult (c) is an integer. In particular, if for a certain rational number c there exists exactly one root $\tau \in R(v)$ of the multiplicity unity with $ord_@(\tau) = c$, then c is an integer, i.e., c = 1. If $1/2 \in T(v)$, then there exist either at least two roots or one multiple root τ with $ord_@(\tau) = c$, etc. (All these statements are valid for an arbitrary q). In order to get rid of the words "root multiplicity", let us consider the reduced polynomial $P_1(t)$ without multiple roots whose set of roots coincides with R(v). It is clear that $P_1(t) \in Z[t\}$, i.e., all its coefficients are conventional integers. After similar considerations as above, we arrive at the integrity of the vertices corresponding to ord_p of the Newton diagram of the polynomial P_1. This leads to the following statements. If for any rational number c, N(c) denotes the number of different (!) roots $\tau \in R(v)$ with $ord_@(\tau) = c$, then its product with N(c) is an integer. Thus, if there is only one $\tau \in R(v)$ with $ord_@(\tau) = c$, then c = 0 or c = 1. If $1/2 \in R(v)$, then the number of different roots $\tau \in R(v)$ with $ord_@$ = 1/2 is even and therefore not less than two. If $1/3 \in A_v$, then $2.3 \in A_v$, and the number of different roots $\tau \in R(v)$ with $ord_@(\tau) = 1.3$ is divisible by 3, and, hence, is not less than tree. Of course, under the same assumptions, the number of different roots $\tau \in R(v)$ with ord (τ) = 2/3 is also divisible by 3, and hence, is not less than three. Note also that these statements remain valid also if to substitute the set R(v) by a certain G(l(v)/Q)-invariant subset, i.e., substitute the polynomial P_v (or, if necessary, P_l) by any reduced polynomial with integer coefficients and without multiple roots that divide it. It is just this generalization together with Theorem 3 given below that are used for proving Theorem 1 in the spirit of the author's studies [8, 12, 53] (see also [55]).

Here are some important examples of Newton diagrams. X is said to have an ordinary reduction in v if $A_v = \{0,1\}$, i.e., the set A_v consists of two integer tilts 0 and 1; it is clear that in this case mult (0) = mult (1) = dim X = n. The Abelian variety X has an ordinary reduction in v then and only then when X(v) is an ordinary Abelian variety.

Another limiting case is that where $A_v = \{0, 1/2, 1\}$ and mult (0) = mult (1) = 1; in this case, X is said to have a K3-type reduction in v.

X is said to have a supersingular reduction in v if $A_v = \{1/2\}$. The elliptic curve X has supersingular reduction in v then and only then when the X(v) is a supersingular elliptic curve. For an arbitrary dimensionality, the reduction in v is supersingular then and only then when X(v) is isogeneous to the product of supersingular curves. Any elliptical curve X over the field of rational numbers K = Q has supersingular reduction in an infinite set of simple v = p (the Elkies theorem [52]).

Note that in the general case the main ideal in the field L(v) is

$$p = (n_{@ \in S} @)^{[L(v):Q]/\mathrm{card}(S)}$$

and therefore for q = p

$$\mathrm{ord}_@(L(v)^*) = \frac{\mathrm{card}(S)}{[L(v):Q]}\,\mathrm{ord}_@(Q^*) = \frac{\mathrm{card}(S)}{[L(v):Q]}Z \in \frac{1}{[L(v):Q]}Z$$

In particular, since A_v lies on the unit segment, $A_v \in \frac{1}{[L(v):Q]}Z$, i e $[L(v):Q]A_v \in Z$. Thus, $\mathrm{card}(A_v) \le [L(v):Q] + 1$.

Now let us try to consider nomalizations $ord_{@}$ for all $@ \in S$. It is easy to show that if $ord_{@}(\tau) = ord_{@}(\tau)$ for all $@ \in S$, then τ_1/τ_2 is a root of 1. The formula of the product [6] for the field L(v) provides the formulation of a more precise result (see [42]).

Let $\Gamma(v)$ be a multiplicative subgroup in L(v) generated by a set R(v). By definition, $\Gamma(v)$ is a commutative group with the finite number of generators whose rank is denoted by r(v). The Galois group F(L(v)/Q) acts naturally on $\Gamma(v)$. This action sets the embedding $G(L(v)/Q) \to \mathrm{Aut}\,\Gamma(v)$. Then the homomorphism kernel,
$ord{:}g(v) \to Q^s,\ \tau \to \{ord_{@}(\tau)\}_{@ \in s}$
consists of roots of 1.

Now let us assume that $\Gamma(v)$ does not contain nontrivial roots of 1. This condition is met, e.g., if the Frobenius element $Fr_v \in 1 + 1^2$

$End_{Z_1}T_1$ (X) since in this case all its eigenvalues are comparable with 1

modulo 1^2. Then Aut $r(v) \approx GL(r(v), Z)$ and, therefore, the Galois group G((L(v)/Q) is isomorphous to a certain finite subgroup of the complete linear group GL(r(v),Z). As is well known [19], the order of any finite subgroup in GL(r(v), Z) does not exceed the order of the group GL(r(V), Z/3 Z). Therefore, under our assumptions that $q = p$ and that nontrivial roots of 1 are absent in the group r(v), we have

$$\begin{aligned} card(A_v) &\le [L(v){:}Q] + 1 = card(G(L(v)/Q)) + 1 \le \\ &\le card\ (GL(\tau(v),\ 1, Z/3\ Z)) + 1. \end{aligned}$$

In particular, the number $card(A_v)$ is limited from above by the constant dependent on r(v) alone.

The previous estimate can be refined if we take into account that the multiplicative group $\Gamma(v)$ contains an infinite cyclic subgroup generated by the natural number q which, obviously, is immobile with respect to the action of the Galois group G(L(v)/Q). Therefore, if the group $\Gamma(v)$ does not have nontrivial roots of 1, it is isomorphous to the finite subgroup of the group GL(r(v) - 1, Z). If, in addition, $q = p$, then

$$\begin{aligned} card(A_v) &\le [L(v){:}Q] + 1 = card(G(L(v)/Q)) + 1 \le \\ &\le card\ (GL(r(v) - 1, Z/3\ Z)) + 1. \end{aligned}$$

Now, let us return to the mapping ord. Continuing it over the Q-linearity on Γ(v) x Q, we arrive at the embedding of the vector Q-spaces

$$g(v) \times Q \to Q^s,$$

which are also denoted by ord. Therefore, the vector Q-space Hom (r(v) x Q, Q) is generated by the mappings $ord_@: r(v) \times Q \to Q$ obtained by continuing the above defined homomorphisms $ord_@$: r(v) $\to$Q over the Q-linearity. Thus any homomorphism $\Gamma(v) \times Q \to Q_l$ is a linear combination of the mappings $ord_@$ with coefficients from Q_l. This fact will be used for studying the replicas of the l-adic logarithm [5] of the Frobenius element

$$fr_v := \log Fr_v \in g_l \in End_{Q_l} V_l(X) \in End_{\bar{Q}_l} V_l(X).$$

The operator vr_v is semisimple, and the set $spec(fr_v)$ coincides with log $R(v) \in Q_l$. Hereinafter, if not otherwise, for any semisimple linear operator f in $V_l(x)$ the term spec (f) denotes the set of its eigenvalues. Then Q(f) denotes the vector Q-subspace in Q_l generated by the set spec(f). Thus, the l-adic logarithmic mapping maps R(v) into $spec(fr_v)$ and, hence, r(v) into $Q(fr_v)$; the latter mapping is continued over Q-linearity up to the isomorphism of the vector Q-spaces

$$g(v) \times Q \longrightarrow Q(fr_v),$$

which is also denoted by log. (The group Γ(v) consists of l-adic unities, and the l-adic logarithm log τ of the l-adic unity τ is zero then and only then when τ is a root of 1). Therefore

$$r(v) = \dim_Q g(v) \times Q = \dim_Q Q(fr_v).$$

Example. If X has a supersingular reduction in v, then it is easy to see that

$$fr_v = -\frac{1}{2}(\log q)\ id \text{ is a scalar operator.}$$

If the reduction is not supersingular, then the operator fr_v is not a scalar.

From the theory of representations of semisimple Lie algebras with the heighest weight [5] it follows that for any semisimple element

$$f \in g_1 \times Q_1{}^{Q_1} = g_1' \times Q_1{}^{Q_1} \times \underline{c}_1 \times Q_1{}^{Q_1} \in \text{End}_{Q_1} V_1(X)$$

the dimensionality $\dim_Q Q(f)$ does not exceed the rank of the reductive Lie algebra $g_1 \times {}_{Q_1}Q_1$ equal to the rank of the reductive Lie algebra g_1 which, in turn, is equal to the sum of the rank r of the semisimple Lie algebra g_1' and the dimensionality $\dim_{q_1} \underline{c}_1$ of the center $\underline{c}_1$. Since $\dim_{Q_1} c_1 \leq \dim \text{End}^o X$ (see above), we arrive at

$$r(v) \leq r + e.$$

If to assume as earlier that q = p and the group $\Gamma(v)$ does not contain nontrivial roots of 1, then G(L(v)/Q) is isomorphous to the finite subgroup of the group GL(r + e - 1, Z) and

$$\text{card}(A_v) \leq \text{card}\ (GL(r + e - 1, Z/3\ Z)) + 1$$

In particular, the number card (A_v) is bounded from above by the constant depending on r and e alone. Thus, if r = e = 1, the group G(L(v)/Q) is embedded into the group GL(1,Z) and, hence, is either trivial (equal to {1)) or is a cyclic group of order 2. (If the Galois group G(L(v)/Q) is trivial, then L(v) = Q, the reduction of X in v is supersingular, and q is the even degree of number p. In particular $q \neq p$).

3.2. Some l-adic Lie Subalgebras

Let us denote by t_v the minimal algebraic Lie subalgebra in $\mathrm{End}_{Q_l}V_l$ (X) containing the operator fr_v. The Lie algebra t_v is the Lie algebra of the so-called Frobenius tore introduced by Serre [39] and defined as the minimal algebraic subgroup in $GL(V_l(X))$, the group of Q_l points of which contains the Frobenius element Fr_v. (In the actual fact, the Frobenius tore introduced by Serre is constructed over the Abelian variety X(v) and defined as the algebraic Q-group; our tore is obtained from the Serre construction by extending the scalar field up to Q_l).

In virtue of the algebraicity of the Lie algebra g_l, the inclusions $t_v \in g_l$ take place

$$\bar{t}_v: = t_v \times Q_l{}^{Q_l} \in \bar{g}_l: = g_l \times Q_l{}^{Q_l} \in \mathrm{End}_{\bar{Q}_l} V_l(X).$$

From the well known description of the algebraic shell, i.e., the minimal algebraic Lie subalgebra containing the given semisimple operator [2, 25], it follows that the Q_l Lie algebra $\bar{t}_V$ consists of $fr_v{}^{(\pi)}$ replicas of the operator fr_v of semisimple operators in $V_1(X)$ parametrized by homomorphisms π: $Q(fr_v) \to Q$ and defined by the following condition. Any eigenvector $x \in V_1(X)$ of the operator fr_v is the eigenvector of all the $fr_v{}^{(\pi)}$ replicas and $fr_v{}^{(\pi)} = \pi(a)x$ if $fr_v x = ax(a \in \mathrm{spec}(fr_v) \in Q(fr_v))$. For any simple ideal @ in S, $fr_@ \in t_v$ denotes the replica $fr_v{}^{(\pi)}$ corresponding to homomorphism π which is the composition

$$\pi = \mathrm{ord}_@ \log^{-1}:Q(fr_v) \to g(v) \times Q \to Q \in \bar{Q}|_l .$$

It is clear that the $fr_@$ operators generate the Lie algebra $\bar{t}_v$ as the vector $\bar{Q}|_l$ space. By definition, we have

$$\mathrm{spec}(f_@) = \mathrm{ord}_@\log^{-1}(\mathrm{spec}(fr_v)) = \mathrm{opd}_@(R(v)) = A_v \in Q,$$

$Q(fr_@) = Q$, and the multiplicity of any eigen- value a of the $fr_@$ operator is equal to the number mult(a).

Let us denote by

$$mt_v \in g_l \in End_{Q_l} V_l (X)$$

the Q_l-subspace in the Lie algebra g_l generated by subspaces $Ad(\sigma)\, t_v = \sigma t_v \sigma^{-1}$, where σ runs over the group $G_l \in Aut\, V_l(X)$, Ad: $G_l \to Aut(g_l)$ is the joint representation of the l-adic Lie group G_l. (We remind here that the Frobenius element Fr_v is defined within the conjugacy in the group G_l, and, hence, the element fr_v and the Lie subalgebra t_v are defined within the joint action of the group G_l. Thus defined subspace mt_v depends on v alone and is independent of the nonuniqueness of the choice of the Frobenius element). By definition, the subspace mt_v is invariant with respect to the joint action of the group G_l and, hence, is an ideal in the Lie algebra g_l. In particular, mt_v is the reductive Lie algebra. If $fr_v = \log Fr_v$ is a regular element of the Lie algebra g_l (in virtue of the above considered corollary from the Chebotarev density theorem, there are a lot of such Frobenius elements), then, in an obvious way, mt_v contains a semisimple component g'_l of the reductive Lie algebra g_l. It is clear that the Q_l Lie algebra $mt \times_{Q_l} Q_l$ is generated as the Q_l vector space by the operators

$$\sigma\, fr_@ \sigma^{-1} \quad (\sigma \in G_l,\ @ \in S).$$

For each such operator

$$spec(\sigma\, fr_p \sigma^{-1}) = spec(fr_@) = A_v \in Q,$$

and the multiplicity of any eigen number a is equal to mult(a).

Using the above cited Weil theorem, it is easy to see [1, 37] that t_v contains the homotheties $Q_l id$. Therefore, the Lie algebra mt_v also contains $Q_l id$. (The t_v and Q_l id algebras coincide then and only then when X has a supersingular reduction in v. If the reduction is supersingular, then

$$mt_v = t_v = Q_l\, id.)$$

Let us denote by

$$hdg_v \in g_l \in End_{Q_l} V_l\ (X)$$

the Q_l subspace in the Lie algebra g_l generated by the operators

$$Ad(\sigma)\, fr_v = \sigma\, fr_v \sigma^{-1} \qquad (\sigma \in G_l \in Aut\, V_l(X)\,)$$

Note that

$$spec(\sigma\, fr_v\, \sigma^{-1}) = spec(fr_v) = \log R(v) \in \bar{Q}|_l$$

By definition, the subspace hdg_v is invariant with respect to the joint action of the group G_l and, hence is an ideal in the Lie algebra g_l. It is also clear that $hdg_v \in mt_v$, and the Lie algebra mt_v is an algebraic shell of the Lie algebra hdg_v, i.e., the minimal algebraic Lie subalgebra in $End_{Q_l} V_l\ (X)$ containing hdg_v. Since the commutant of the Lie algebra experiences no changes in the transition to its algebraic shell [2, 25], then the semisimple components of the Lie algebras hdg_v and mt_v coincide. This signifies that there exists a semisimple Q_l Lie algebra which is an ideal in the g'_l algebra and coincides with it if fr_v is a regular element in g_l satisfying the following conditions:

(1) $g_v \in hdg_v \in mt_v$;

(2) The Lie algebra hdg_v is the direct sum of its center and the Lie algebra g_v;

(3) The Lie algebra mt_v is the direct sum of its center and the Lie algebra g_v.

Let us put

$$g_{v,X} = g_v \times {}_{Q_1}\mathbb{Q}_1 \in \mathrm{End}_{\mathbb{Q}}\, \mathbb{V}_1(X);$$

it is a semisimple Q_l Lie algebra which is an ideal in the Lie algebra $g_{l,X}$ and coincides with it if the fr_v element is regular in the Lie algebra g_l. Let us consider the natural representation of the $g_{v,X}$ algebra in the space $\mathbb{V}_1(X)$. Let g be a simple factor of the semisimple Lie algebra $g_{v,X}$. It is clear that then g is a simple factor also of the Lie algebara $g_{l,X}$. What can be said about the nonunidimensional irreducible representations **W** of the simple Lie algebra g entering $\mathbb{V}_1(X)$? The answer will be given in terms of the set A_v, the function mult, and the number of polynomial irreducible over Q by which the characteristic polynomial P_v can be divided.

Theorem 3. (a) The highest weight of the irreducible representation **W** of the simle Lie algebra g is the sum of not more than [card(A_v) - 1] fundamental weights;

(b) If X has an ordinary reduction in v, then g is a simple classical Lie algebra, **W** is the fundamental representation and its highest weight is a microweight in the sense of Burbaki;

(c) If X has a K3-type reduction in v, then the image of the Lie algebra g in End (**W**) coincides either with sl(**W**) or with the Lie algebra so(**W**) of a certain orthogonal group or else with the Lie algebra sp(**W**) of a certain symplectic group of the space **W**.

3.3. Newton Diagrams and Represetations of Simple Lie Algebras

Let us remind that if X has an ordinary reduction in v, then $A_v = \{0,1\}$, in particular, $\mathrm{card}_v(A) = 2$. If X has a K3-type reduction in v, then A_v is the arithmetic progression $\{0, 1/2, 1\}$ and mult(0) = mult(1) = 1.

Now let us derive Theorem 2 from Theorem 3.

The proof of Theorem 2 (by the module of Theorem 3). Suffice it to prove that the dimensionality of any irreducible representation V of the Lie algebra g entering V(x) is bounded from above by the constant depending on r and e alone. Using the Chebotarev density theorem, we choose a simple ideal v such that the following condition is fulfilled:

(1) v is an ideal of degree 1, i.e., its residue field k(v) is a simple field of p elements; (2) p $\neq$ 1; (3) X has a good reduction in v; (4) the Frobenius element is

$$Fr_v \in 1 + 1^2 \ End_{Z_1}T_1 \ (X) \in \ Aut_{Z_1}T_1 \ (X);$$

in particular, all its eigenvalues are comparable with 1 modulo 1^2, and, hence, the mutiplicative group G(v) generated by them contains no nontrivial roots of 1; (5) the element fr_v = log Fr_v is a regular element of the Lie algebra g_l.

Then from condition 5 it follows that

$$g_l' = g_v, g_{l,x} = g_{v,x},$$

i.e., any simple factor g of a semisimple Lie algebra $g_{l,X}$ is at the same time a simple factor of a semisimple Lie algebra $g_{v,X}$. Applying statement (a) of Theorem 3, we obtain that the dimensionality of any irreducible representation W of the Lie algebra g entering $V_1(X)$ is bounded from above by the constant dependent on card(A) and the rank of the simple Lie algebra g alone. Since the rank r of a semisimple Lie algebra $g_{l,X}$ is equal to the sum of the ranks of its simple factors g, and any irreducible representation V of the Lie algebra g entering $V_1(X)$ is expanded into a tensor product of irreducible represnetation W of simple factors that also enter $V_1(X)$, we obtain that the dimensionality dim(V) is bounded from above by the constant dependent on r and card(A_v) alone because the number of simple factors of the algebra $g_{l,X}$ does not exceed its rank r. Now we have to note that, as has already been shown earlier, the number card(A_v) is bounded from above by the constant dependent on r and e alone.

In order to prove Theorem 3 we need the following elementary statement of the theory of representations.

Lemma 1. Let H be a finite dimensional vector space over the algebraically closed field k of the zeroth characteristic,

$$g^r \in End(H)$$

is the linear reductive Lie algebra whose natural representation in the space H is expanded into a direct sum of irreducible representations. Now let us expand the Lie algebra g^r into a direct sum

$$g^r = \underline{c}^r + g^s$$

of its center $\underline{c}^r$ and a semisimple Lie algebra g^s. Now let us expand g^s into a direct sum of simple algebras g called its factors. For any simple factor g there will be found an ideal g' in the Lie algebra g^s such that $g^s = g + g'$; thus, $g^r = \underline{c}^r + g + g'$. Let f_0:H →H be a semisimple operator acting in the space H and lying in the Lie algebra g^r: spec(f_0) ∈ k is the set of its eigenvalues, mult: spec(f_0) →Z_+ is the function that establishes the correspondence between each eigenvalue and its multiplicity, i.e., the dimensionality of the respective characteristic subspace in H, Q(f_o) ∈ k is the vector Q-subspace in k spanned over the set spec(f).

Let f be a projection of the element f_0 onto the factor g, i.e., f ∈ g and $f_0 - f \in \underline{c}^r + g'$.

Then for any irreducible representation **W** of the Lie algebra g entering H there will be found a number b ∈ Q(f_0) satisfying the following conditions: a set spec(f) of eigenvalues of the element f considered as an operator in space **W** lies in the set b + spec(f_0), i.e., spec(f) ∈ b + spec(f_o) = $\{b + a \mid a \in \text{spec}(f_o)\}$ ∈ Q(f_o) ∈ k.
The multiplicity mult_W (b + a) of any eigen value b + a of the operator f in the space **W** does not exceed the number mult(a). In particular: (1) card(spec(f)) ≤ card(spec(f_0)); (2) if f = 0, then the operator f_0 has an eigenvalue of multiplicity not less than the dimensionality dim(**W**) of the space **W**.

The proof of Theorem 3 (by the module of Lemma 1). Let us put $H = \bar{V}_1(X)$, $k = \bar{Q}_1$, $g^r = mt_1 \times_{Q_1} Q_1$, $g^S = g_{V,X}$. Since the elements $f_@$ generate the Lie algebra g^r, there will be found an element $f_0 = f_@$ such that its projection f onto a simple factor g is nonzero. Applying Lemma 1 we arrive at

$$\mathrm{spec}(f) \in b + A_v \in Q(f_o) \in Q(f_@) = Q,$$
$$\mathrm{card}(\mathrm{spec}(f)) \le \mathrm{card}(A_v),$$

and the multiplicity $\mathrm{mult}_W (b + a) \le \mathrm{mult}(a)$ for any eigenvalue b + a of the operator f in the space W, $a \in A_v$. Now statement 1 of Theorem 3 follows immediately from the following result obtained by the author ([43], Theorem 2.2a).

Lemma 2. Let W be a finite dimensional vector space over the algebraically closed field k of the zeroth characteristic and $g \in \mathrm{End}(W)$ be a linearly irreducible simple Lie algebra, in other words, a natural representation of the simple k Lie algebra g in the space W is irreducible. Let $f \in g$ be a nonzero simple elements of the Lie algebra g, spec(f) be a set of eigenvalues of the element f considered as an operator in the space W. Then the highest weight of the representation W is a sum of not more than [card(spec(f) - 1] fundamental weights.

A few words about the proof of Lemma 2. In [7, 43], the following invariant of the irreducible representation W of the simple Lie algebra g called a representation length and denoted as length(W) was introduced. It is defined in such a way. Let us represent the difference between the highest and lowest weights as a linear combination of simple roots. All the coefficients of this expansion are positive integers (if W is not onedimensional). The smallest of them is called the representation length. The representation length is a positive integer (for nononedimensional W; if W is onedimensional, then length (W) = 0). It is proven that the representation length is always not smaller than the sum of its heights, i.e., if the highest weight is a sum m of its fundamental weights, then $m \le \mathrm{length}(W)$. In particular, a set of pairs of nonisomorphous irreducible representations of the simple Lie algebra g with the given length is finite. Thus, if g is a simple Lie algebra of rank 1, i.e., $g \approx sl(2)$, then length(W) = dim(W) - 1, and for any natural number m there exists the unique (within the isomorphism) irreducible representation of the Lie algebra g of the length m. The dimensionality of this representation is equal to m + 1. If the representation W has length 1, then g is a simple classical Lie algebra and the highest weight of the representation is a microweight (Serre [41]). It was proved by the author elsewhere [43] that always

$$\text{length}(W) \leq \text{card}(\text{spec}(f)) - 1$$

(and there will be found an element f for which such an estimate is exact). Whence it follows that statement 1 of Theorem 3 allows the following strengthening:

Theorem 4. For any simple factor g of the semisimple Lie algebra g and its irreducible representation **W** entering $\nabla_1(X)$ we have length(**W**) $\leq$ card(A_v) - 1.

Now statement 2 of Theorem 3 turns out to be a particular case of Theorem 4 for card(A_v) = 2.

In order to prove statement 3 of Theorem 3 suffice it to note that it follows directly from the following Lemma which is an obvious combination of Theorem A.2.1 from Appendix of Ref.[7] and Note 1.1.1 of Ref.[43].

Lemma 3. Under the assumptions and notations of Lemma 2 let the arithmetic progression $(a_1,a_2,a)_3 \in Q$ exist such that spec(f) $\in (a_1, a_2, a_3)$. Let m_1, m_2, m_3 be a triad of natural numbers satisfying the following conditions. First, $m_1 = m_3 = 1$. Second, if a_i is an eigenvalue of the operator f, then its multiplicity does not exceed m_i. Then the image of the Lie algebra g coincides either with sl(**W**), so (**W**), or sp(**W**).

Now, we have to prove only Lemma 1. Our approach is based on the constructions of Ref.([7], p.280-281).

The proof of Lemma 1. The subspace g^r**W** $\in$ H is the g^r submodule in H which, if considered as g-module, is isomorphous to the direct sum of a certain number of the copies of a simple g-module of **W**. Let **V** be a simple g^r-submodule in g^rH. In virtue of the Schur lemma, the center c^r of the Lie algebra g^r acts on the space **V** by scalars; in particular, **V** is as simple as the g^s-module. From the expansion $g^s = g + g'$ the existence of simple g-module of **W**$'$ and g'-module of **V**$'$ follows such that the g^s-module of **V** is isomorphous to the tensor product **W**$'$ x **V**$'$. Since the space **V** considered as a g-module is isomorphous to the direct sum dim(**V**$'$) of the copies of the g-module of **W**$'$, then the g-modules of **W** and **W**$'$ are isomorphous and, therefore, the g^s-module of **V** is isomorphous to the tensor product **W** x **V**$'$.

Let us represent the element $f_0 \in g^r = \underline{c}^r + g + g'$ as the sum

$$f_0 = f + f' + f_e;\ f \in g,\ f' \in g'_, f_e \in \underline{c}^r$$

The element f_e belongs to the center $\underline{c}^r$ and, hence, acts on the space V as multiplication by a certain scalar denoted by j. Thus,

$$f_{oz} = (f + f')z + jz \quad \text{for} \quad z \in V \approx W \times V'$$

Let spec(f′) ∈ k be a set of eigenvalues of the element f′ considered as an operator in the space V′, mult′: spec(f′) | Z_+ is the function that establishes the correspondence between the eigenvalue of the operator f′ and its multiplicity. Let us remind that the element f + f′ of the Lie algebra g + g′ acts on the tensor product W x V' according to the formula

$$(f + f')\ (y \times y') = fy + f'y' \text{ for } y \in W,\ y' \in V'.$$

Therefore, the set spec(f + f′) of eigenvalues of the element f + f′ considered as an operator in the space V ≈ W x V′ coincides with
spect(f) + spec(f′) = {a + a′ | a ∈ spec(f),a′ ∈ spec(f′)} ∈ k,
and the multiplicity $\text{mult}_W(a + a')$ of the eigenvalue a + a′ of the operator f + f' in the space V is not less than the product
mult(a) mult′(a′) ≤ mult(a) for a ∈ spec(f),a′ ∈ spec(f′)
Taking a certain fixed a′ ∈ spec(f′), we obtain that for any a ∈ spec(f) the sum a + a′ is the eigenvalue of the operator f + f′ in the space V of multiplicity not less than mult(a). Hence, a + a′ + j is the eigenvalue of the operator $f_0 = (f + f') + f_e$ in the space V of multiplicity not less than mult(a). Since V is the g^r-submodule and, hence, f_0 is an invariant subspace in H, we obtain that for any a ∈ spec(f) the sum a + (a′ + j) is the eigenvalue of the operator f_0 in the space H of multiplicity not less than mult(a). Now let us put b + -(a′ + j) and check whether b ∈ $Q(f_0)$. This is the only nontrivial point of our consideration based on the semisimplicity of the Lie algebra g.

The spur

$$\sum \text{mult}_W(a)\ a \qquad (a \in \text{spec}(f))$$

of the operator $f \in g$ in the space **W** is zero since the image of a simple Lie algebra g in End(**W**) lies in the Lie algebra sl(**W**) of operators with the zero spur. Whence it follows that

$$a' + j = \frac{1}{\dim(W)} \sum \text{mult}_W(a)(a + a' + j) \in Q(f_0)$$

(the sum of the multiplicities of all the eigenvalues of the operator f in the space **W** is equal to the dimensionality dim(**W**)). Therefore

$$b = -(a' + j) \in Q(f_0).$$

Another method for estimating irreducible representations of simple factors of the semisimple Lie algebra $g_{l,X}$ entering into $\widetilde{V}_l(X)$, is based on the expansion of the characteristic polynomial P_v into the product of the degrees of polynomials irreducible over Q.

<u>Theorem 5.</u> Let us assume that the characteristic polynomial P_v satisfies the so-called purity condition: if τ_1 and τ_2 are the roots of the polynomial P_v and τ_1/τ_2 is the root of 1, then $\tau_1 = \tau_2$. Now expand the polynomial P_v into the product of the degrees of polynomials P_i irreducible over Q:

$$P_v(t) = p\ P_i(t)^{n_i} \qquad (1 \le i \le d).$$

Then for any simple factor g of the semisimple Lie algebra $g_{l,X}$ and any of its irreducible representations entering $V_l(X)$ one of the following conditions is fulfilled:

(1) g is a simple factor of the semisimple Lie algebra $g_{l,X}$, and the highest weight of the irreducible representation **W** is the sum of not more than (2d - 1) fundamental weights;

(2) g is not a factor of the Lie algebra $g_{v,x}$ and $\dim(W) \le \max \{ n_i \mid 1 \le i \le d \}$.

Note. From the results obtained by Tate [23] it follows that the purity condition is equivalent to the requirement that all the endomorphisms of the Abelian variety X(v) would be defined over the finite field k(v). It is clear that they all are defined over the finite extension

$k(v)_m$ of the field k for a certain natural m, and the respective polynomial $P_{v,m}$ satisfies the purity condition, i.e., if the ratio of its two roots is the root of 1, then these roots of the polynomial $P_{v,m}$ coincide. We remind that $P_{v,m}$ is a characteristic polynomial of the element Fr_v^m and its roots are the mth powers of the roots of the polynomial P_v. The following statement generalizes Theorem 5.

Theorem 5a. Let us choose natural m such that the polynomial $P_{v,m}$ would satisfy the purity condition and expand it into a product of the degrees of polynomial P_i irreducible over Q:

$$P_{v,m}(t) = p\ P_i(t)^{n_i} \quad (1 \leq i \leq d).$$

Then for any simple factor g of the semisimple Lie algebra $g_{1,X}$ and any of its irreducible representation entering $V_1(X)$ one of the following conditions is fulfilled:

(1) g is a simple factor of the semisimple Lie algebra $g_{v,X}$ and the highest weight of the irreducible representation W is the sum of not more than (2d - 1) fundamental weights;

(2) g is not a simple factor of the Lie algebra $g_{v,X}$ and $\dim(w) \leq \max \{ n_i | \ 1 \leq i \leq d \}$.

Note. Numbers d and n_i are independent of the choice of m and, in accordance with the Tate results [23], can be invariantly described in the following way. The End° algebra is a finite dimensional semisimple Q-algebra and, hence, can be expanded into a direct sum of a certan number of simple Q-algebras: this number is just d. The numbers n_i are the indices (square roots of dimensionality) of these simple algebras over their centers.

Note. Under the assumptions of Theorem 5 (or 5a) if P_v (or $P_{v,m}$) is a polynomial without multiple roots, then $g_{1,X} = g_{v,X}$. Indeed, if g is a simple factor of the algebra $g_{1,X}$ not entering $g_{v,X}$, then all its irreducible representations entering $V_1(X)$ are onedimensional, i.e. g W = {0} and, therefore, gV(x) = 0, i.e., g = {0}.

When proving Theorem 5 and 5a, we shall use the following statement.

Lemma of progression. Let $\tau_0, \tau_1, \ldots, \tau_s \in R(v)$ be a sequence consisting of (s + 1) roots of the polynomial P_v satisfying the following condition: for a certain natural number N the sequence of the Nth degree τ_0^N, τ_1^N,..., τ_s^N is the geometric progression whose denominator $(\tau_1/\tau_0)^N$ is not a root of 1. Then:

(1) If the polynomial P_v can be expanded into the product

$$P_v(t) = p\ P_i\ (t)^{n_i} \qquad (1 \leq i \leq d)$$

of d P_i-polynomials irreducible over Q, then $s \leq 2d - 1$;

(2) If a polynomial $P_{v,m}$ can be expanded into the product

$$P_{v,m}(t) = p\ P_i\ (t)^{n_i} \qquad (1 \leq i \leq d)$$

of d P_i-polynomials irreducible over Q, then $s \leq 2d - 1$.

Note. For any homomorphism of the group $\Gamma(v)$ in the vector space over the field of the zeroth characteristic the sequence τ_0, τ_1,...,τ_g is transformed into an arithmetic progression. For example, ord(τ_0), ord(τ_1), ..., ord(τ_s) is the arithmetic progression in the space Q^s, and its denominator is not zero, since the homomorphism ord annihilate only the roots of 1. In a similar way, the l-adic logarithm transforms the sequence τ_0, τ_1, ..., τ_s into the arithmetic progression log(τ_0), log(τ_1), ..., log(τ_s), lying in the field $\bar{Q}_1$.

The proof of the Lemma of progression is based on the following elementary observation: on any line in the real Hilbertian space not more than two vectors have one and the same preliminary prescribed length. Therefore, if

$$X_0, X_1, \dots, X_s$$

is the arithmetic progression with the nonzero difference consisting of the (s + 1) vectors of the real Hilbertian space and the function that establishes the correspondence between the vector and its squared length takes on the set

$$\{ X_0, X_1, \dots, X_s \}$$

not more than d different values, then $s + 1 \leq 2d$ and, hence, $s \leq 2d - 1$. As a real Hilbertian space we take the coordinate space

$$R^s = Q^s \times_Q R$$

with the standard scalar product. We remind here that the following homomorphism is defined

$$\text{ord}: g(v) \to Q^s \in R^s,$$

whose kernel consists of roots of 1. From the explicit description of the mapping ord it follows that if the elements $\tau, \tau' \in \Gamma(v)$ are conjugate to one another with respect to the action of the Galois group $G(L(v)/Q)$ of the field $L(v)$, then the vectors

$$\text{ord}(\tau), \text{ord}(\tau') \in Q^s \in R^s$$

are obtained from one another by the permutation of coordinates corresponding to the action of the Galois group on the set of ideals S; in particular, the lengths of vectors ord(τ) and ord(τ') are equal. Therefore the vectors ord(τ) have the same lengths for all the τ running over the set of all the roots of any given polynomial P_i (in the notation of (1) or (2)).

Now, in the situation described in (1), the lemma of progression yields that the function that establishes the correspondence between the vector and its squared length takes on the set ord(R(v)) not more than d different values. (The set of the roots R(v) of the polynomial P_v is the union of the sets of roots of polynomials P_i). Therefore any arithmetic progression lying in ord(R(v)) consists of not more than (2d - 1) terms and, hence, any geometric progression in R(v) whose denominator does not lie at the kernel of the homomorphism ord, i.e., is not a root of 1, also consists of not more than (2d - 1) terms.

Now let us assume that we are in the situation (2) of the Lemma of progression. Then we immediately obtain that the function establishing the correspondence between the vector and its squared length takes not more than d different values on the set

$$\text{ord}(R(v)^m) = m\ \text{ord}(R(v)), \text{ where } R(v)^m = \{ \tau^m \mid \tau \in R(v) \}$$

is the set of all the roots of the polynomial $P_{v,m}$ which is the union of the set of the roots of polynomials P_i. Therefore, also on the set

$$\mathrm{ord}(R(v)) = \frac{1}{m}\,\mathrm{ord}(R(v)^m)$$

the function that establishes the correspondence between the vector and its squared length takes not more than d different values. The remaining part of the proof is similar to that of (1).

Corollary of the Lemma of progression. In the notation and the assumptions of Theorem 5 (or Theorem 5a), the set

$$\mathrm{spec}(fr_v) = \log R(v) \in \bar{Q}_l$$

of all the eigenvalues of the operator

$$fr_v = \log Fr_v \in \mathrm{End}_{Q_l} V_l(X) \in \mathrm{End}_{\bar{Q}} \bar{V}_l(X)$$

satisfies the following conditions:

(1) The multiplicity of any eigenvalue of the operator fr_v does not exceed max $\{n_i \mid 1 \leq i\ d:$

(2) if $\{a_0, a_1, \ldots, a_s\} \in (fr_v) \in \bar{Q}_l$ is the arithmetic progression with the nonzero difference consisting of (s + 1) terms, then $s \leq 2d - 1$.

Proof. We know that the multiplicity of any eigenvalue of the operator Fr_v (or Fr_v^m) does not exceed

$$\max \{n_i \mid 1 \leq i \leq l|\}$$

and that the ratio of two different eigenvalues is not the root of 1, i.e., l-adic logarithms of two different eigenvalues are different. Now we have to note that

$$\mathrm{spec}(fr_v) = \log (R(v)) = \frac{1}{m} \log (R(v)^m),$$

where R(v) is the set of all the eigenvalues of the operator Fr_v, and $R(v)^m$ is the set of all the eigenvalues of the operator $Fr_v^{\,m}$.
(2) Let

$$\{ a_0, a_1, \ldots, a_s \} \in \mathrm{spec}(fr_v) \in \bar{\mathbb{Q}}_l$$

be an arithmetic progression with the zeroth difference consisting of (s + 1) terms. Then there is a sequence $\tau_0, \tau_1, \ldots, \tau_s \in R(v)$ such that $a_0 = \log(\tau_0)$, $a_1 \log(\tau_1), \ldots, a_s = \log(\tau_s)$.

It is clear that

$$\tau_0^{\,N}, \tau_1^{\,N}, \ldots, \tau_s^{\,N}$$

is the geometric progression where N is the number of roots of 1 in the group Γ(v); it is clear that the denominator of this progression is not a root from 1. Now the last step is the application of the Lemma of progression.

Proof of Theorems 5 and 5a. Let us apply Lemma 1 to the vector space $H = V_l(X)$ over the field $k = \bar{\mathbb{Q}}_l$, of the reductive Lie algebra $g^r = g_l \times_{\mathbb{Q}_l} \mathbb{Q}_l$, its semisimple component $g^s = g_{l,X}$, and the element
$f_0 = \sigma\, fr_v \sigma^{-1} \in hdg_v \in hdg_v \times_{\mathbb{Q}_l} \bar{\mathbb{Q}}_l \in g_l \times_{\mathbb{Q}_l} \bar{\mathbb{Q}}_l = g^r$,
$\sigma \in G$. Since, by definition, the Lie al|gebra hdg_v i|s the minimal ideal in g_l containing all the elements $f_0 = \sigma\, fr_v \sigma^{-1}$, then for a simple factor g of the Lie algebra g_l there exists an element f_0 of the given form whose projection f on g is nonzero then and only then when g is a simple factor of the ideal $g_{v,X}$ of the Lie algebra $g_l \times_{Q_l} Ql$. (Let us remind here that $g_{v,X}$ is a semisimple component of the Lie algebra $hdg_v \times_{Q_l} Q_l$).

If the projection f of the element $f_0 = fr_v$ is zero, then, according to Lemma 1, for any irreducible representation W of the Lie algebra g entering into $H = V_l(X)$, the operator f_0 has the eigenvalue of multiplicity not less than dim(W). Applying the corollary of the Lemma of progression, we obtain that

$$\dim(W) \le \max \{ n_i \mid 1 \le i \le d \}.$$

Now let us assume that for a certain element f_0 of the indicated form the projection f onto the factor g is not zero. According to Lemma 1, for any irreducible representation W of the Lie algebra g entering H = $V_1(X)$ there is a number $b \in k = \bar{Q}_1$ such that

$$b + \mathrm{spec}(f) \in \mathrm{spec}(f_0) = \mathrm{spec}(fr_v).$$

Here $\mathrm{spec}(f) \in Q_1$ is understood as a set of eigenvalues of the element f considered as an operator in the space W. If $b_0, b_1, \ldots, b_s \in \mathrm{spec}(f) \in \bar{Q}_1$ is an arithmetic progression with the nonzero difference consisting of (s + 1) terms and lying in spec(f), then

$$b_0 + b, b_1 + b, \ldots, b_s + b \in \mathrm{spec}(fr_v) \in \bar{Q}_1 -$$

is the arithmetic progression with the nonzero difference consisting of (s + 1) terms and lying in spec(fr). Applying the corollary of the Lemma of progression, we arrive at the inequality $s \leq 2d - 1$.

Now theorems 5 and 5a follow immediately from the statement proved by author elsewhere (Theorem 2.2b, [43]).

Lemma 4. Let k, W, g, f, and spec(f) be the same as in the formulation of Lemma 2. If the highest weight of the irreducible representation W of the simple Lie algebra g is the sum M of the fundamental weights, then the set spec(f) of eigenvalues of any nonzero semisimple element f contains the nonconstant arithmetic progression

$$a_0, a_1, \ldots, a_m,$$

consisting of (M + 1) terms.

A few words about the proof of Lemma 4. In Refs.[7, 43] the following invariant of the irreducible representation W of the simple Lie algebra g was introduced. It is called the depth of the representation and denoted by depth(W). The number depth(W) is determined as the value of the highest weight on the maximal dual root. The depth has the value depth(W) = 1 then and only then when the highest weight of the simple g-module of W is the microweight [5]. It is proven [7, 43] that if the highest weight is the sum M of the fundamental weights, then $M \leq$ depth(W). It can be proven for any nonzero semisimple element f of the

Lie algebra g that its spectrum spec(f) in the space **W** contains nonconstant arithmetic progression

$$a_0, a_1, \ldots, a_{\text{depth}(W)}$$

consisting of [depth(W) + 1] terms [7, 43]. In particular, if spec(f) does not contain nontrivial arithmetic progression, i.e., if for any triad a, a′, a″ ∈ spec(f) such that 2a = a′ + a″ the equality a = a′ = a″ is valid, the highest weight of the irreducible representation **W** should be a microweight. Whence the following strengthening of Theorem 5 and the corrolary.

Theorem 6. Let us assume that the characteristic polynomial P_v satisfies the so-called purity condition: if τ_1 and τ_2 are the roots of the polynomial P_v and τ_1/τ_2 is the root of 1, then $\tau_1 = \tau_2$. Let us expand the polynomial P_v into the product of the degrees of the polynomials P_i irreducible over Q:

$$P_v(t) = p\, P_i\ (t)^{n_i} \qquad (\ 1 \le i \le d\ ).$$

Then for any simple factor g of the semisimple Lie algebra $g_{1,X}$ and any its irreducible representation **W** entering $V_1(X)$ one of the following conditions is met:

(1) g is a simple factor of the semisimple Lie algebra $g_{v,X}$ and depth(W) ≤ 2d - 1.

(2) g is not a factor of the Lie algebra $g_{v,X}$ and dim(W) ≤ max { n_i | 1 ≤ i ≤ d }.

Corollary of Theorem 6. Let P be an irreducible polynomial over Q the roots of which satisfy the following condition. If τ, τ' are two different roots, then their ratio τ/τ' is not a root of 1. Then for any simple factor g of the semisimple Lie algebra $g_{1,X}$ the highest weight of any irreducible **W** entering $V_1(X)$ is a microweight. In particular, **W** is always a fundamental representation.

Note. Squared length of the vector ord(τ), where τ is the root of the polynomial P_i irreducible over Q, depends only on the p-adic Newton diagram of this polynomial. More exactly, let τ be a root of polynomial P_i and $A_i \in A_v$ be a set of inclinations of polynomial P_i, mult: A $\to Z_+$ be the function of the inclination length. Then the squared length of the vector ord(τ) is

$$\mathrm{card}(S)\frac{\sum \mathrm{mult}_i(a)\ a^2}{\deg(P_i)} = \mathrm{card}(S)\frac{\sum \mathrm{mult}_i(a)\ a^2}{\sum \mathrm{mult}_i(a)}$$

where the sum is taken over all the elements of the set of inclinations A_i.

Therefore, the formulations of Theorem 5, 5a, and 6 can be strengthened by substituting the number d of the irreducible factors by the number of differemt Newton diagrams corresponding to these multipliers and even by the number of different values of the expression

$$\frac{\sum \mathrm{mult}_i(a)\ a^2}{\sum \mathrm{mult}_i(a)}$$

which is called the index of the Newton diagram of the polynomial P_i. More exactly, taking that the Newton diagram consists of the finite set A of nonnegative rational numbers and the function of length mult: A $\to Z_+$, we define the index ind(A, mult_A) of the diagram (A, mult_A) as the ratio

$$\frac{\sum \mathrm{mult}_A(a)\, a\, 2}{\sum \mathrm{mult}_A(a)}$$

The above given explicit form for the squared length of the vector ord(τ) yields the following strengthening of Therem 6.

<u>Theorem 7.</u> Let us assume that the charateristic polynomial P_v satisfies the so-called purity condition: if τ_1, τ_2 are the roots of polynomial P_v and τ_1/τ_2 is the root of 1, then $\tau_1 = \tau_2$. Let us expand the

polynomial P_v into the product of the degrees of polynomials P_i irreducible over Q:

$$P_v(t) = p\, P_i\ (t)^{n_i} \qquad (1 \le i \le d).$$

Let (A, mult) be a Newton diagram of the polynomial Pi and ind(Ai, $mult_i$) be its index. We denote by Ind_v the finite set of rational numbers formed by indices

$$Ind(A, mult) \quad (1 \le i \le d).$$

$$\text{Then } card(Ind_v) \le d$$

and for any simple factor g of a semisimple Lie algebra $g_{l,X}$ and any its irreducible representation W entering $V_1(X)$ one of the following conditions is met:

(1) g is a semiple factor of the semisimple Lie algebra $g_{v,X}$ and depth(W) $\le$ 2 card(I_v) - 1 $\ge$ 2d - 1.

(2) g is not a factor of the Lie algebra $g_{v,X}$ and dim(W) $\ge$ max $\{ n_i |$ $1 \le i \le d \}$.

Corollary of Theorem 7. Let P_v be a polynomial over Q for whose roots the following condition is fulfilled. If τ and τ' are two different roots, then their ratio τ/τ' is not a root of 1. Let us assume that the Newton diagrams (A_i, $mult_i$) have one and the same index for all the irreducible representations of polynomials P_i over Q that divide P_v (e.g., all the Newton diagrams (A,mult) coincide).

Then for any simple factor g of semisimple Lie algebra $g_{l,X}$ the highest weight of any its irreducible W entering $V_1(X)$ is a microweight. In particular, W is always a fundamental representation.

Examples. If X has a conventional reduction in v, then all ind_i are equal to 1/2 and if it has a supersingular reduction, then ind_i are equal to 1/4.

The expansion of the polynomial P_v into the product of the degrees of polynomials P_v irreducible over Q yields, in accordance with the Tate theorem [23], the existence of the Abelian varieties X_i over the field k(v) such that X is isogeneous to the product of varieties X_i, and for each i the characteristic polynomial of the Frobenius endomorphism of the variety X_i coincides with $P_i^{n_i}$. If X_i is an ordinary Abelian variety, then $ind_i = 1/2$, if it is supersingular, then ind = 1/4.

Note. Considering the norm of the space $L^r(r > 1)$ on the space R^s instead of L^2 norm it is possible to generalize Theorem 7 for the case of "r-index"

$$\frac{\sum \text{mult}_A(a)\ a^r}{\sum \text{mult}_A(a)}$$

of the Newton diagram (A, mult_A).

REFERENCES

1. F.A. Bogomolov. "Points of finite order on the Abelian varieties." Izv. AN SSSR, Ser. matem., 1980, 44, 782-803 (in Russian).
2. A. Borel'. Linear algebraic groups. Moscow: Mir, 1972 (in Russian).
3. M.V. Borovoi|. "On the action of the Galois group on the rational classes of type (p,p) cogomologies of the Abelian varieties." Matem. sbornik, 1974, 94, 649-652 (in Russian).
4. M.V. Borovoi|. "Hodge group and the algebra of endomophisms of the Abelian variety." In: Problems of the group theory and homological algebra. Jaroslavl', 1981, 124-128 (in Russian).
5. N. Burbaki. Groups and Lie algebras. Chapters 1-8. Moscow: Mir, 1972, 1976, 1978 (in Russian).
6. A. Weil, Fundamentals of the theory of numbers. Moscow: Mir, 1972 (in Russian).
7. Yu.G. Zarkhin. "Weights of simple Lie algebras in cogomologies of the algebraic varieties." Izv. AN SSSR, Ser. matem. 1984, 48, 264-304.
8. Yu.G. Zarkhin . "Abelian varieties, l-adic representations and SL_2." Izv. AN SSST Ser. matem., 1979, 43, 2, 294-308.
9. Yu.G. Zarkhin . "Endomorphisms of the Abelian varieties over the fields of the finite characteristic." Izv. AN SSSR, Ser. matem. 1975, 39, 272-277.

10. Yu.G. Zarkhin , "Abelian varieties in the P characteristic." Matem. zametki, 1976, 19, 393-400.
11. Yu.G. Zarkhin and A.N. Parshin. Problems of finitness in the Diophantine geometry. Supplement to book [3], 369-438 (the English translation to be published by the AMS).
12. Yu.G. Zarkhin . "The representations of the Lie algebra sl(2) in l-adic cogomologies," 1985, 19, 3, 65-66.
13. N. Koblits. p-adic numbers, p-adic analysis and zeta-functions. Moscow: Mir, 1982 (in Russian).
14. S. Lang. Fundamental of the Diophantine geometry. Moscow: Mir, 1986 (in Russian).
15. Yu.I. Manin. The Mordell-Weil theorem. Supplement to book [16].
16. D.Mumford. Abelian varieties. Moscow: Mir, 1971 (in Russian).
17. I.I. Pyatetskii-Shapiro. "Relationships between the Tate and Hodge hypotheses for Abelian varieties." Matemat.sb. 1971, 85, 610-620.
18. J.-P.Serre. Abelian l-adic representations and elliptical curves. Moscow:Mir, 1973 (in Russian).
19. J.-P. Serre. Lie groups and algebras. Moscow: Mir, 1969.
20. J.-P. Serre and J.Tate. "Good reduction of the Abelian varieties." Matemetika, 1971, 15:5, 140-165 (in Russian).
21. S.G. Tankeev. "Cycles on the simple Abelian varieties of simple dimensionalities." Izv. AN SSSR, Ser. matem., 1982, 46, 155-170.
22. J.Tate. "Algebraic classes of cohomologies." Uspekhi matem. nauk, 1965, 20, 6, 27-40.
23. J. Tate. "Endomorphisms of the Abelian varieties over the finite fields." Matematika, 1968, 12:6, 31-40.
24. Hodge cycles and motives. Moscow: Mir, 1985 (in Russian).
25. K. Chevalley. Theory of Lie groups, v.2. Algebraic groups. Moscow: IL, 1958 (in Russian).
26. G. Shimura. Introduction into the arithmetic theory of automorphic functions. Mowcow: Mir, 1973 (in Russian).
27. F.A. Bogomolov. Sur l'algébraicite des representations l-adique. C.R. Acad. Sci. Paris. 1980, 290
28. P. Deligne. Varietes de Shimura: interpretations modulaire et techniques de construction de modeles canonique. Proc. Symp. Pure. Math., 1979, 33, pt.2, 247-290.
29. P. Deligne. "La conjecture de Weil pour les surfaces K3." Invent. Math. 1972, 15, 206=226.

30. G. Faltings. "Endlichkeitssätze für Abelsche Varietäten über Zahlkorpern." Invent. Math., 1983, 73, 349-366; Erratum, 1984, 75, 381.
31. G. Faltings. "Complements to Mordell." In: G. Faltings, G. Wustholtz, et al. Rational points, Vieveg, 1984.
32. S. Lang, Abelian varieties, Springer Verlag, 1983.
33. B. Mazur, "Eigenvalues of Frobenius acting on algebraic varieties over finite fields." Proc. Symp. Pure Math. 1975, 29, 231-261.
34. L. Moret-Bally. "Pinceaux des varietes abeliennes." Astérisque, 1985, v.129.
35. D. Mumford. A note of Shimura's paper "discontinuous groups and Abelian varieties". Math. Ann., 1969, 181, 345-351.
36. K.A. Ribet. "Hodge classes on certain types of Abelian varieties." Amer. J. Math., 1983, 105, 523-238.
37. J.-P. Serre. "Representations l-adiques." In: Kyoto Symposium on Algebraic Number Theory, Kyoto, 1976. Japan Sopciety for the Promotion of Science, Tokyo, 1977, 177-193.
38. J.-P. Serre. "Propriétés galoisiennes des points d'ordre finis des courbes elliptiques." Invent. Math., 1972, 15, 259-331.
39. J.-P. Serre. Resumé des cours de 1984-1985. Annuaire du College de France, Paris, 1985.
40. J.-P. Serre. Resumé du College de France, Paris, 1986.
41. J.-P. Serre. "Groupes algébraiques associes aux modules de Hodge-Tate." Astérisque, 1979, 65, 155-188.
42. Yu.G. Zarkhin. "Abelian varieties, l-adic representations and Lie algebras. Rank independence on l." Invent. Math. 1979, 55, 165-176.
43. Yu.G. Zarkhin. "Linear semisimple Lie algebras containing an operator with small number of eigenvalues." Arch.Math. Basel, 1986, 46, 522-532.
44. Arithmetic geometry (Eds.:G. Cornell, J.H. Silverman) Springer-Verlag, 1986.
45. Yu.G. Zarkhin. "A finiteness theorem for unpolirized Abelian varieties with prescribed places of bad reduction." Invent. Math., 1985, 79, 309-321.
46. A.N. Parshin. "Quelques conjestures de finitude en Geometrie Diophantiennes." Actes Congress Intern. Math., Nice, 1970, 1, 467-471.

47. J.-M. Fontaine. "Il n'y pas varietes abeliennes sur Z." Invent. Math., 1985, 81, 515-583.

48. V.A. Abrashkin. Group schemes of period p over the ring of the Witt vectors. DAN SSSR, 1985, 283, 1269-1294.

49. V.A. Abrashkin. "Khonda systems of the group scheme of period p." Izv. AN SSSR, Ser. matem., 1987, 51, 451-484.

50. V.A. Abrashkin. "Galois modules of the group schemes of period p over the ring of the Witt vectors." Izv. AN SSSR, Ser. matem., 1987, 51, 691-736.

51. J.-P. Serre. Linear representations of the finite groups. Moscow: Mir, 1968 (in Russian).

52. N.D. Elkies. "The existence of infinitely many supersingular primes for every elliptic curve over Q." Invent. Math. 1987, 89, 3, 561-567.

53. Yu.G. Zarkhin , "Torsion of the Abelian varieties over GL(2)-extensions of number fields." Submitted to Mathematische Annalen.

54. D. Bertrand, "Minimal heights and polarizations on Abelian varieties." In: Problèmes Diophantiennes 1986/87. Publications de l'Université de Paris VI, No.84 (1988), No.3.

55. Yu.G. Zarkhin . "Torsionpunkte abelscher Varietäten über algebraischen Zahlkörpern." In: Mathematisches Forschunginstitut Oberwolfach, Tagungsbericht 31/1988 "Arithmetische Algebraische Geometrie" (17.7 bis 23.7. 1988).

Mathematics
and Modelling

ISBN 1-56072-104-9

...AND THE EFFECT OF ERROR AUTOCORRECTION

G.L. Litvinov

> When he has spare money, he goes to a book store and buys some useful book. Once he bought a book entitled Inverse Trigonometric Functions and Chebyshev polynomials.
>
> N.N. Nosov, "Merry Little Family" (Moscow, 1975).

1. INTRODUCTION

The author encountered the phenomenon of error autocorrection at the end of the seventies when developing nonstandard algorithms for computing elementary functions on small computers. It was necessary to construct rational approximations of the form

$$R(x) = \frac{a_i + a_1x + a_2x^2 + \dots + a_nx^n}{b_0 + b_1x + b_2x^2 + \dots + b_mx^m} \qquad (1)$$

to certain functions of one variable x defined on finite segments of a real axis. With this aim a simple method (described in [1] and considered below) that allowed one to determine a set of coefficients a_i, b_i of approximation (1) as a solution of a certain set of algebraic linear equations. It turned out that thus arising sets of linear equations can be ill-conditioned, in other words, the determination of the approximation coefficients is, generally speaking, ill-posed in the sense of [2]. Nevertheless, the method provides paradoxically high quality approximations with errors close to best [1].

Thus the approximation of form (1) to the function cos x on the segment $[-\pi/4, \pi/4]$ obtained by the above method has a relative error of $0.55 \cdot 10^{-13}$ at m = 4 and m = 6 whereas the best possible relative error is $0.46 \cdot 10^{-13}$ [3]. The respective set of algebraic linear equations has the conditioning number 10^9 which threatens us with the loss of nine reliable decimal places in the solution because of the computational error. The computer experiment confirms such a threat. The above mentioned algorithm (described below) was implemented in a program written in FORTRAN. The computations were performed with a double accuracy (sixteen decimal places) on ICL-4-50 and ES-1045 computers of similar architecture. But the transition from one computer type to another perturbs a set of linear equations and the computational process owing to the computational error including rounding-off error. As a result, the coefficients of the above mentioned approximation to the function cos x experience a perturbation already in the sixth-ninth decimal place. But the error of the most rational approximation remains constant and is $0.4 \cdot 10^{-13}$ for the absolute error and $0.55 \cdot 10^{-13}$ for the relative error. In the same way, for the type (1) approximation to the function $\tan^{-1}$ obtained by this method at the segment [-1,1] for m = 8, n = 9 the relative error is $0.5 \cdot 10^{-11}$ and remains unchanged in the transition from ICL-4-5 to ES-1-45 although the respective set of linear equations has the conditioning number of the order of 10^{11} and the approximation coefficients are perturbed with a relative error of the order of 10^{-4}.

Thus the errors of the numerator and the denominator of a rational approximation compensate one another. The effect of error autocorrection is associated with the fact that the errors in the coefficients of a rational approximation are distributed not arbitrarily but form the coefficients of a new approximation to the approximated function. It can readily be understood that the methods of interval arithmetics (see, e.g., [54]) are insufficient for taking into account this effect and correct evaluation of the error of a rational approximation.

Note also that the use of standard procedures known from the theory of ill-posed problems would result in substantial loss of accuracy. For example, the application of the regularization method would lead to the loss of two thirds of correct digit places [4]; also the volume of necessary computations would drastically increase. The point is that the exact solution of a set of equations in this case is not our ultimate aim. In fact we are interested in the construction of a sufficiently accurate approximation. Such an approximation allows one to "justify" and simplify some algorithms for the aooriximation construction and also obtain, without any additional transformations, the required approximations in the form convenient for applications.

Professor Y.L. Luke kindly drew author's attention to his works [5,6] where the effect of the error autocorrection was first revealed and interpreted at an heuristic level for the classical Padé approximations. The above mentioned method leads to the generalized Padé approximations (linear Padé-Chebyshev approximations) if the computational error is neglected. In the present article the mechanism of error autocorrection has been considered for rather a general case with the invocation of heuristic considerations and computer experiments (the linear methods for constructing rational approximation, nonlinear generalized Padé approximations). It seems that the efficiency of the widely used algorithms for constructing rational approximations is determined by the error autocorrection (at least in the case where the number of coefficients is sufficiently large).

The nowaday understanding of the mechanism of the error autocorrection provides a certain control for the computational error by varying the procedure of constructing the approximation depending on its form.

We also consider the algorithm for constructing Padé-Chebyshev approximations and briefly describe the respective program (see [7]). It has been demonstrated that in some cases the introduction of a controlling parameter taking account of the mechanism of the error autocorrection decreases the computational error. The computational results illustrate the program potentialities and the quality of the obtained approximations in comparison with the best approximations. In brief, some other (linear and nonlinear) methods for constructing rational approximations are also considered as well as the methods for constructing linear and nonlinear Padé-Chebyshev approximations with the use of the REDUCE system of analytical computations (see [8]). The characteristics of the relative accuracy of the above indicated methods are considered. The effect associated with the fact that a small variation of the approximated function can drastically deteriorate the quality of the Padé-Chebyshev approximation is also considered with due regard for the error autocorrection.

The article can be considered, to a large extent, as a review; the statement is as elementary as possible. The reviewing parts of the article were meant to make the further statement clear, consistent and "closed in itself" but not necessarily complete. The point is that the literature devoted to the theory of rational approximations and its applications in the numerical analysis (including the computer calculations of respective functions, the numerical solution of equations, acceleration of series convergence, quadratures), in theoretical and experimental physics (including the quantum theory of field, theory of scattering, nuclear and neutron physics), in the theory and practice of experimental data processing, in mechanics, in the theory of control, and in many other fields is, in fact, immense (see, in particular, the reviews and reference books [3, 10-16]).

The author is indebted to E.E. Shnol and Yu.G. Zarhin for their support and to Y.L. Luke for fruitful discussions and important references. The author is also grateful to I.A. Andreeva, A.Ya. Rodionova, and V.N. Fridman who participated in writing programs and organizating computer experiments. Without their valuable help this article would not have been written.

2. BEST APPROXIMATIONS

In our considerations we shall need some information and results closely related to the ideas put forward by P.L. Chebyshev (see [17]). Let [A,B] be a segment of a numerical axis, in other words, let [A,B] consist of such real numbers x that $A \leq x \leq B$) and let f(x) be a continuous function defined on [A,B]. Let us consider a function of absolute errors of approximation (1) to the function f(x), i.e., the quantity

$$\Delta(x) = f(x) - R(x), \tag{2}$$

and the absolute error of such an approximation, i.e.,

$$\Delta = \max_{A \le x \ge B} | \Delta(x) | = \max_{A \le x \ge B} | f(x) - R(x) | \tag{3}$$

The classical problem of the theory of approximations consists in that it is necessary to determine the values of the coefficients in the numerator and denominator of (1) at given values of powers m and n in (1) such that (3) would be minimal. The respective approximation is called best (with respect to the absolute error). Of great importance here is the following result.

The Generalized Valle-Poussin Theorem [17]. If polynomials

$$P(x) = a_0 + a_1 x + \ldots + a_{n-v} x^{n-v}, \; Q(x) = b_0 + b_1 x + \ldots b_{m-\mu} x^{m-\mu}, \tag{4}$$

where $o \le \mu \le m$, $0 \le v \le n$, and $b_{m-\mu} \ne 0$ have no common divisor (i.e., the fraction $R(x) = P(x)/Q(x)$ is irreducible) and the expression $R(x) = P(x)/Q(x)$ is finite on the sector [A, B] and if the error function $\Delta(x) = f(x) - R(x)$ of the approximation R takes at the sequential points

$$x_1 < x_a < \ldots < x_N$$

of the segment [A, B] nonzero values

$$\lambda_1, -\lambda_2, \ldots, (-1)^N \lambda_N$$

with alternating signs (so that numbers λ_i are either all positive or all negative), and $N = m + n + 2 - d$, where d is the least of numbers μ and v, then the error Δ of any approximation of form (1) satisfies the inequality

$$\Delta \ge \lambda = \min \{ |\lambda_1|, |\lambda_2|, \ldots, |\lambda_N| \} \tag{5}$$

Proof. Let us assume that there exists an approximation R(x) of form (1) for which inequality (5) is invalid. Let us consider the difference $\varepsilon(x) = R(x) - R(x) = [f(x) - R(x)] - [f(x) - R(x)] = \Delta(x) - \Delta(x)$ From our assumption it follows that the numbers $\varepsilon(x_1)$, $\varepsilon(x_2),\ldots,\varepsilon(x)_N$ have nonzero values and alternating signs. Since the function $\varepsilon(x)$ is continuous on the segment [A, B], then the function $\varepsilon(x)$ has not less than $N - 1 = m + n + - d$ zeroes inside the segment [A, B]. On the other hand, from the definition of the function $\varepsilon(x)$ it is seen that $\varepsilon(x) = U(x)/V(x)$ where U(x) and V(x) are polynomials and that the degree of polynomial U(x) does not exceed $m + n - d$, so that the function $\varepsilon(x)$ cannot have more than $N - 2 = m + n - d$ zeroes. The obtained contradiction proves the theorem.

The quantity d used in the theorem is called the approximation defect R(x); in practice, d usually is $d = \mu = \nu = 0$. The generalized Valle-Poussin theorem provides a sufficient condition under which the approximation $R(x) = P(x)/Q(x)$ is the best (P(x) and Q(x) are polynomials of type (4)). The points $x_1 < x_2 < \ldots < x_N$ on the segment [A, B] are called the points of the Chebyshev alternant for the approximation R(x) if the absolute values of the error function $\Delta(x) = f(x) - R(x)$ at these points coincide with the absolute error Δ of the approximation R whereas the signs are alternating. In other words, the quantity $\Delta(x)$ at points $x_1, \ldots, x_N$ has the extrema with equal absolute values and alternating signs. From the generalized Valle-Poussin theorem it follows that the existence of the points of Chebyshev alternant is sufficient to make approximation R(x) best.

Chebyshev's Theorem. The existence of the points of Chebyshev alternant is a necessary and sufficient condition for the approximation to be best. Such an approximation exists and is unique if not to distinguish between two fractions that coincide upon their reduction.

A relatively simple proof is given in [17]. Let us note that both Chebyshev and Valle-Poussin considered the case of polynomial approximations. The general case was first considered by N.I. Akhiezer, his results being valid also in the case where the error function has the form $\Delta_\rho = -\dfrac{f(x) - R(x)}{\rho(x)}$ where the weight $\rho(x)$ does not go to zero; if the weight satisfies certain additional conditions, the segment [A, B] would not necessarily be taken finite [17]. Note also that for $\rho(x) \equiv 1$ we arrive at the absolute error in form (2), and if f(x) does not go to zero on [A, B], then for $\rho(x) = f(x)$ we arrive at the function of relative errors

$$\delta(x) = \frac{\Delta(x)}{f(x)} = \frac{f(x) - R(x)}{f(x)} = 1 - \frac{R(x)}{f(x)}, \tag{6}$$

Therefore, the relative error is

$$\delta = \max_{A \leq x \geq B} |\delta(x)|, \tag{7}$$

and one can consider best approximations in the sense of relative error.

Now let us assume that the segment [A, B] is symmetric with respect to zero, i.e., that A = - B. If the function f(x) is even, then it is easy to see that all its best rational approximations on this segment (in the sense of an absolute or relative error) are also even functions so that they can be sought directly in the form $R(x^2) = P(x^2)/Q(x^2)$, where P and Q are polynomials. If the function f(x) is odd, then its best approximations are also odd functions and can be sought directly in the form $xR(x^2) = x\dfrac{P(x^2)}{Q(x^2)}$, where P and Q are polynomials. Then it is possible to consider the best approximation in the sense of the relative error if the odd function f(x) goes to zero only at x = 0 and is continuously differentiable and if $f'(x) \neq 0$. In this case f(x) can be represented in the form $x\varphi(x)$, where $\varphi(x)$ is a continuous even function not going to zero. Then the description of rational approximations to f(x) with the best relative error reduces to the solution of the same problem for an even function $\varphi(x)$; indeed,

$$\frac{f(x) - xR(x^2)}{f(x)} = \frac{x\varphi(x) - xR(x^2)}{x\varphi(x)} = \frac{\varphi(x) - R(x^2)}{\varphi(x)}$$

3. METHODS FOR CONSTRUCTING BEST APPROXIMATIONS

Let us assume that the rational approximation of form (1) is the best approximation to a continuous function f(x) on a sector [A, B]. For the sake of simplicity, we take that we deal with the zero defect. Let x_1, x_2, ..., x_{m+n+2} be the points of a Chebyshev alternant. Then the error function $\Delta_\rho(x)$ corresponding to the weight ρ (see above) satisfies the following set of equations

$$\Delta_\rho(x_k) = (-1)^k \lambda, \tag{8}$$

where $|\lambda| = \Delta_\rho = \max_{A \leq x \geq B} |\Delta_\rho(x)|$; $k = 1, \ldots, m + n + 2$.

For fixed values of $x_1, \ldots, x_{m+n+2}$, relationships (8) can be considered as a set of m + n + 2 equations in unknowns a_i, b_j, λ, where i = 0, 1, ..., n and j = 0, 1,...,m. Since the numerator and denominator of the fraction R(x) can be multiplied by one and the same number, set (8) can be complemented with one more condition, e.g., $b_0 = 1$, so that the number of equations would coincide with the number of unknowns. This idea lies in the basis of the iterative methods for calculating the coefficients of the approximation R(x) suggested by A.Ya. Remez (see [18]) for polynomial approximations and extended to the general case. Different variants of the generalized Remez method were considered in many studies, see, e.g. [3, 12, 20-27].

The approximation is constructed by the following scheme. At the first stage, the initial approximations $x_1 < x_2 < \ldots x_{m+n+2}$ to the points of the Chebyshev alternant are chosen on the sector [A, B], and the set of equations (8) is solved. As a result, a certain rational approximation $R_1(x)$ with the error function $\Delta_1(x) = \dfrac{f - R_1(x)}{\rho(x)}$ is obtained. Then the extreme points for this function are sought, and the obtained information is used for modifying the set $\{x_1, \ldots, x_{m+n+2}\}$. Then the procedure is repeated again to yield a new approximation $R_2(x)$ etc.

Taking into account that $\Delta_\rho(x) = ((f(x) - R(x))/\rho(x)$ and R(x) has form (1), the set of equations (8) can be modified to the form

$$f(x_k) - \frac{\sum_{j=0}^{n} a_i (x_k)^i}{\sum_{j=0}^{m} b_j (x_k)^j} = (-1)^k \rho(x)\lambda$$

whence, upon some elementary transformations, we arrive at the set of

equations $\sum_{i=0}^{n} a_i \ (x_k)^i + \mu_k(\lambda) \sum_{j=0}^{m} b_j \ (x_k)^j = 0$

where $\mu_k(\lambda) = (-1)^k \rho(x_k)\lambda - f(x_k)$, $k = 1, 2, \ldots, m + n + 2$.

Note that for the fixed value of λ (and also for points $x_1 \ldots, x_{m+n+2}$) the coefficients a_i, b_j of the approximation satisfy the set of homogeneous linear algebraic equations (9). But λ is also to be determined, and therefore (9) is a nonlinear set of equatitons whose solution is rather a difficult problem. An exception here is the case where it is necessary to find a polynomial approximation, i.e., the case $m = 0$; then the set of equations (9) becomes linear.

As a rule, the solution of nonlinear set of equations (9) reduces to multiple solution of a set of linear equations. The below described method is rather popular on the West (see, e.g., [3,12, 21, 22, 25, 28]), it was used for tabulating rational approximations to elementary and special functions [3]. Let us put $b_o = 1$ (normalization); then (9) acquire the form

$$\sum_{i=0}^{n} a_i \ (x_k)^i + \mu_k(\lambda) \sum_{j=1}^{m} b_j \ (x_k)^j + (-1)^k \rho(x_k)\lambda = f(x_k) \qquad (9^1)$$

Substituting λ in the nonlinear terms of system (9^1) by a given number λ_0, we arrive at the linear system

$$\sum_{i=0}^{n} a_i \ (x_k)^i + \mu_k(\lambda_o) \ \sum_{j=1}^{m} b_j (x_k)^j + \ (-1)^k \rho(x_k)\lambda = f(x_k) \qquad (10)$$

The iteration process is applied to the initial set of values for critical points $\{x_k\}$, i.e., the initial approximations to the points of the Chebyshev alternant and the given value λ_0. First we determined from (10) a new value of λ and substitute it as λ_0 into the nonlinear terms of equations (10), then the set of equations (10) is solved again and a new value of λ is determined, etc. As a result, a new value of λ and the set of coefficients a_i, b_j are obtained. The next step consists in the determination of still new set of critical points $\{x_k\}$ as extreme points of the error function for the approximation determined at the previous step. Both steps form one cycle of the iteration process. The computation is terminated when the absolute value of λ coincides, with the a priori set accuracy, with the maximum value of the error function. The complete text of the respective program in ALGOL is given elsewhere [22].

Unfortunately, the above described iterative process does not necessarily converge even if the initial approximation differs infinitesimally from the problem solution (see [28]). It has been proved that for some variants of the Remez method the iteration process converges if the initial approximation is sufficiently good (see [20, 23-25, 29, 12). And nevertheless in some concrete cases it is difficult to indicate the initial approximation that will provide the convergence of the iteration process prior to computations; also, for the initial approximation it is rather difficult to check the validity of the conditions that guarantee convergence. One of the practical methods here reduces to the construction of the best polynomial approximation of the degree m + n (in this case no difficulties are met). Then, using the points of the Chebyshev alternant of this approximation as the initial set of critical points for the iteration process, the best approximation is constructed in the form of a ratio of a polynomial of degree m + n - 1 to a linear function. Then, in a similar way, the degree of the numerator is gradually reduced, and the degree of the denominator is gradually increased until the attainment of a necessary approximation of form (1) (see [3,12]).

Along with the iteration methods, best rational approximations are constructed by the methods based on the apparatus of linear and convex programming (see [18,30]). As a rule, the iterative methods are more efficient [27] but do not allow the direct generalization for functions of several variables.

4. THE ROLE OF APPROXIMATE METHODS AND THE ESTIMATE OF THE APPROXIMATION QUALITY

Since the algorithms for constructing best rational approximations are rather cumbersome, of wide use are now more simple methods of approximate solution of the problem (see, e.g., [1, 5, 7-9, 11-15, 24, 25, 31-37]). Below we shall describe the methods that can readily be used on computers and yield approximations close to best at relatively low expenditure of computer resouces. Such an approximation can also be used as initial for the interation algorithm providing an exact result. An approximation best in the sense of the absolute error, generally speaking, is not best in the sense of the relative error. In practice, it is very important to see that both absolute and relative errors were small. Therefore, very often it is easier to use not best approximations but rather approximations constructed with the aid of an approximate method that have acceptable absolute and relative errors. Finally, the method for solving the problem of a rational approximation can be used also in cases where the information on the function to be approximated is not full (e.g., the values of the function are known only for a finite number of argument values, or only the first terms of the function expansion into a series are known, or the initial information has an error, etc.).

The generalized Valle-Poussin theorem (see above, Sect.2) provides the estimate of the closeness of the approximate solution to the best approximation even if the latter is unknown.

Let us assume that we should like to estimate the closeness of the given approximation of form (1) to a certain approximation of the same form to the given function f(x) with the best absolute error. For the sake of simplicity, let us also assume that the defect of the best approximation is zero (in practice this cond ition is usually fulfilled). Then, in virtue of the Chebyshev theorem of the alternant, the following is true: if the given approximation R(x) is sufficiently close to the best one, the function of absolute errors $\Delta(x)$ takes at the sequential points $x_1 < x_2 < \ldots x_{m+n+2}$ of the variation range of the argument x the non zero values $\lambda_1, -\lambda_2, \ldots, (-1)\lambda_{m+n+2}$ with alternating signs. In this case, it is said that an alternant appears. If $|\lambda_1| = |\lambda_2| = \ldots = |\lambda_{m+n+2}|$, then it is a Chebyshev alternant. Let us denote by Δ_{min} the best possible absolute error of approximation (1) to f(x) (numbers m and n are given). Let us put $\lambda = \min \{ |\lambda_1| , \ldots, |\lambda_{m+n+2}| \}$. Then, in virtue of the generalized Valle-Poussin theorem, the inequality $\Delta_{min} \geq \lambda$ is valid; thus

$$\Delta \geq \Delta_{min} \geq \lambda \tag{11}$$

It is clear that Δ coincides with the maximum absolute value of the extremum of $\Delta(x)$, and it is possible to take as λ the smallest absolute value of this extremum (within the accuracy of a sign). The quantity

$$g = \lambda / \Delta \tag{12}$$

characterizes the degree of closeness of the error of the given approximation to the error of the best approximation. It is clear that if this approximation is the best, then $0 < q \leq 1$ and $q = 1$. The closer the quantity q to unity, the higher is the approximation quality. From (11) and (12) it follows that

$$\Delta_{min} \geq g \bullet \Delta \tag{13}$$

As a rule, estimate (13) is rather rough. The appearance of the alternant indicates the closeness of the error of the given approximation to best, and the quantity Δ_{min}/Δ is usually substantially exceeds q.

In a similar way, the quality of the approximation is estimated with respect to the best relative error.

If it is possible to compute the values of the function to be approximated for all the points of the segment [A,B] (or, at least, of a sufficiently " dense" set of such points) and if the coefficients of the rational approximation R(x) are already known, then it is easy to determine the points of the local extremum of the error function and the quantities λ_1, ..., $\lambda_{m+n+2}$and also λ and q using a special standard subroutine. The same subroutine is also necessary for constructing the best approximations by the iterative method. A good application package for constructing rational approximations should also include such a subroutine as well as a good soubroutine for solving sets of linear algebraic equations. Another subroutines using the algorithms for the approximate solution of the approximation problem and constructing best approximations should also be included in the application package.

5. CHEBYSHEV POLYNOMIALS AND POLYNOMIAL APPROXIMATIONS

An important role in the approximation theory and the computational practice is played by Chebyshev polynomials (see, e.g., [12, 13, 17, 18, 24, 33, 38]). We shall consider Chebyshev polynomials of the first kind. They were initially defined by the author in the form

$$T_n(x) = \cos(n \cos^{-1}x) \tag{14}$$

where n = 0,1,2,... . Let us put $\varphi = \cos^{-1}x$; expressing $\cos n\varphi$ in terms of $\sin \varphi$ and $\cos \varphi$, we can see that the right-hand side of (4) coincides with a certain polynomial. In particular, $T_0(x) = \cos 0 = 1$, $T_1(x) = \cos\varphi = \cos(\cos^{-1}x) = x$, $T_2(x) = \cos 2\varphi = \cos^2\varphi - \sin^2\varphi = T_1^2(x) - (1 - T_1^2(x)) = 2x^2 - 1$, etc. For practical computation of $T_n(x)$ the use is made of a recurrence formula

$$T_n(x) = 2xT_{n-1}(x) - T_{n-2}(x) \tag{15}$$

Sometimes it is more convenient to consider polynomials $T(x) = 2^{-n+1}T_n(x)$, since the coefficient of x^n in the polynomials $T_n(x)$ is unity. The above indicated polynomials satisfy the recurrence relation

$$T_n(x) = xT_{n-1}(x) - \frac{1}{4} T_{n-2}(x \tag{16}$$

Let us consider the particular case of the problem of the best approximation when an approximation in the form of a polynomial $\bar{P}(x)$ of degree n-1 to the function $f(x) = x^n$ is sought on the segment [-1,1]. From the Valle-Poussin theorem it follows that the unknown approximation has the form $P(x) = x^n - T_n(x)$. In this case the error function $\Delta(x)$ coincides with $T(x)$ and the points of the Chebyshev alternant can be indicated explicitly: $x = -\cos\frac{k\pi}{n}$ where $k = 0,1,\ldots,n$. Indeed, $T(x) = 2^{-n+1}$

$\cos n(\pi - k\pi/n) = 2^{-n+1}\cos(n - k)\pi = \frac{(-1)^{n-k}}{2^{n-1}}$, i.e., at the indicated points, $T_n(x)$ ta- kes its maximum value, $1/2^{n-1}$, with alternating signs. Whence an important corollary follows: the best polynomial approximation of degree n - 1 to the polynomial $a_0 + a_1x + \ldots a_nx^n$ on the segment [-1,1] has the form $a_0 + ax_1 + \ldots + a_nx^n - a_nT_n(x)$ This result allows one to reduce the polynomial degree (e.g., of a certain polynomial approximation) with the minimum loss of accuracy. The reduction of the polynomial degree by a sequential application of the above procedure introduced by K. Lanzosh is called economization [38].

The monomials $x^0_, x^1_, \ldots, x^m$ can be expressed via Chebyshev polynomials $T_0, T_1, \ldots, T_m$. At m > 0 the following formula is valid

$$x^m = 2^{1-m}\sum_{k=0}^{[m/2]} a_k \binom{m}{k} T_{m-2k}(x) \tag{17}$$

where $\binom{m}{k}$ is the binomial coefficient, [m/2] is the integer part of the number m/2, a = 1/2 at k = m/2 and a = 1 at k ≠ m/2. The expansion of the polynomial T in degrees of x at m > 0 is given by the formula

$$T_m(x) = \frac{1}{2} m \sum_{k=0}^{[m/2]} \frac{(-1)^k (m - k - 1)!}{(m - 2k)!} (2x)^{m-2k} \tag{18}$$

Finally, $x^0 = T_0 = 1$. It is clear that the set of polynomials $\sum_{i=0}^{n} c_i T_i$, (where c_i are the numerical coefficients), coincides w|i|t|h the set of all polynomials of degree n.

The above considered economization procedure can be described also in another way. The initial polynomial $\sum_{i=0}^{n} a_i x^i$ can be represented with the aid of formula (17) in the form $\sum_{i=0}^{n} c_i T_i$. A polynomial of degree k obtained upon economization coincides with $\sum_{i=0}^{k} c_i T_i$. For functions represented in the form of a power series $f(x) = \sum_{i=0}^{\infty} a_i x^i$ one can readily obtain with the aid of ecomoniza|t|i|on procedure the polynomial approximations close to best on the segment [-1,1]. With this aim it is necessary to substitute f(x) by a segments of its Taylor series at the point x = 0, i.e., by the polynomial $\sum_{i=0}^{n} a_i x^i$ that approximates this function with a high degree of precision and then, economizing this polynomial, to obtain the polynomial $\sum_{i=0}^{k} c_i T_i$ of the given degree k. For $n \to \infty$ the quantity $\sum_{i=0}^{k} c_i T_i$ tends to the sum of the first k+1 terms of the expansion of f(x) into a Fourier series in Chebyshev polynomials.

Let us denote by L_w^2 a Hilbertian space of functions with an integrable square by measure w(x)dx on the segment [-1,1]. Let us put $w(x) = 1/\sqrt{1-x^2}$. It can readily be checked that the Chebyshev polynomials T_n form an orthogonal (but not orthonormalized) basis in L_w^2. The expansion

$$f(x) = \sum_{i=0}^{\infty} c_i T_i$$

of the function f(x) into a series of Chebyshev polynomials (the Fourier-Chebyshev series) can readily be reduced to the expansion of the function f(cos x) into a conventional Fourier series in cosines.

Among polynomials of degree n the best approximation to the f(x) function in L_w^2 shows the polynomial $P_n(x) = \sum_{i=0}^{n} c_i T_i$. The following result shows that on the segment [- 1,1] this approximation is close to best in the sense of the absolute error.

<u>The Cheney Theorem</u>. Let φ(x) be a function integrable on the segment [-1,1]. If at i = 0, 1, 2, ..., n the equality

$$\int_{-1}^{1} \varphi(x)\, T_i(x)\, w(x)\, dx = 0 \,, \tag{20}$$

is valid, then f(x) either changes the sign on [-1,1] not less than k+1 times or vanishes almost everywhere.

<u>Proof</u>. Let us assume that f(x) changes the sign exactly m times at points $x_1, \ldots, x_m$ where $0 \le m \le k$. Let us put $P(x) = \prod_{i=0}^{m} (x - x_i)$; since P(x) is a polynomial of degree m, it can be represented as a linear combination of polynomials T_0, T_1, ...T_m where $m \le k$. Therefore the following equality follows from (20): $\int_{-1}^{1} \varphi(x)P(x)w(x)dx = 0$. Since it is clear that the function f(x)P(x) preserves its sign, this equality signifies that φ(x) goes to zero almost everywhere, which proves the theorem.

For a more general case this result has been proved in Ref.[39, p.110]. The above proof can be extended to the case of a system of orthogonal polynomials of rather a general form and an arbitrary integration interval (even infinite) [40].

Now let us return to function (19) and its approximation $P_n(x) = \sum_{i=0}^{n} c_i T_i$. The function of absolute errors $\Delta(x) = f(x) - P(x) = \sum_{i=n+1}^{\infty} c_i T_i$ orthogonal to polynomials T_0, T_1, ..., T_n, i.e., $\int_{-1}^{1} \Delta(x) T_i(x) w(x) dx = 0$

for all i = 0, 1,..., n. If the function f(x) i|s| continuous, then the error function $\Delta(x)$ is also continuous. From the Cheney theorem it follows in this case that either $\Delta(x)$ is identical zero, or it changes the sign n+1 times. This indicates the existence of an alternant, i.e., that the approximation $P_n(x)$ is close to best and that the degree of this closeness can be estimated by relationships (11)-(13).

Whereas the segment of the Taylor series

$$\sum_{i=0}^{n} \frac{1}{n!} f^{(n)}(0)\, x^n$$

provides the best approximation only in the neighborhood of zero, the segment of the Fourier-Chebyshev series $\sum_{i=0}^{n} c_i T_i$ for the function f(x) with the same number of terms provides the approximation close to best over the whole segment [-1,1].

The change of the variable $x \to \frac{1}{2}[(B-A)x + A + B]$ reduces the approximation problem on an arbitrary finite segment [A,B] to the case [-1,1]. As a rule, it is just this case that will be considered in what follows.

6. ILL-POSED PROBLEMS AND RATIONAL APPROXIMATIONS

Let $\{\varphi_0, \varphi_1,\ldots,\varphi_n\}$ and $\{\psi_0, \psi_1,\ldots,\psi_n\}$ be sets consisting of linearly independent functions of the argument x running over a certain (probably multidimensional) set X. Let us consider a problem of constructing an approximation in the form

$$R(x) = \frac{a_0\varphi_0 + a_1\varphi_1 + \ldots + a_n\varphi_n}{b_0\psi_0 + b_1\psi_1 + \ldots + b_m\psi_m} \tag{21}$$

to the given function f(x) defined on X. If X coincides with the segment [A,B] of the real axis, $\varphi_k = x^k$ and $\psi_k = x^k$ for all k, then expression (21) becomes a rational function (1) (see Introduction). It is clear that expression (21) yields a conventional rational function also in the case where φ_k and ψ_k are Chebyshev polynomials $T_k(x)$ or else Legendre, Laguerre, or Hermite polynomials, etc.

Let us choose a certain theoretical method for construction of approximations of form (21) and consider the problem of computating coefficients a, b. Such a problem is often ill-conditioned, i.e., small perturbations of the function f(x) to be approximated or computational errors result in substantial errors in the coefficient values. Thus, the problem of computing coefficients of best rational approximations (including polynomial approximations) at large degrees of the numerator or denominator is ill-conditioned.

The instability with respect to the caomputational error can be associated with both the choice of the theoretical method of the approximation construction (i.e. with the problem formulation) and with the concrete algorithm implementing this method. Ill-conditioning of the problem of computing the coefficients of best approximations is associated with the formulation of this problem. The said is true also for other methods for constructing rational approximations with large number of coefficients. But an inappropriate choice of the algorithm implementing the method can aggravate the negative consequences caused by ill-conditioning.

Some methods for constructing approximants of the form (21) is associated with the solution of sets of linear algebraic equations. This procedure may give rise to quite a substantial error, if the respective matrix is ill-conditiond. Let us consider an arbitrary set of linear algebraic equations

$$Ay = h, \tag{22}$$

where A is the given square matrix of order N with the components a_{ij} (i,j = 1,..., N), h is the given vector-column with the components h_i $\overset{,}{\mathrm{i}}$, and y is an unknown vector-column with components y_i. Let us define the vector norm by the equality

$$\| y \| = \sum_{i=1}^{N} | x_i | \tag{23}$$

(this norm is more convenient for calculations than $\sqrt{x^2+...x^{2-}}$). Then the matrix norm is defined by the equality

$$\| A \| = \max_{\| y \| = 1} \| A \| = \max_{1 \le i \ge N} \sum_{j=1}^{N} |a_{ij}|. \tag{24}$$

If the matrix A is nondegenerate, then the quantity

$$\mathrm{cond}(A) = \| A \| \; \| A^{-1} \| \tag{25}$$

is called the conditioning number of the matrix A (see, e.g., [41]). Since $y = A^{-1}h$, the absolute error Δy in the vector y is related to the absolute error Δh of the vector h as $\Delta y = A^{-1} \Delta h$, whence

$$\|\Delta y\| \le \|A^{-1}\| \cdot \|\Delta h\| \text{ and } \|\Delta y\|/\|y\| \le \|A^{-1}\| \left(\|h\|/\|y\|\right) \cdot \left(\|\Delta h\|/\|h\|\right).$$

Taking into account that $\|h\| \le \|A\| \cdot \|y\|$, , we finally arrive at

$$\|\Delta y\|/\|y\| \le \|A^{-1}\| \cdot \|A\| \cdot \|\Delta h\|/\|h\|, \tag{26}$$

i.e., the relative error of the solution y is estimated in terms of the relative error of the vector h with the aid of the conditioning number. It is clear that equality in (26) can be attained. Therefore if the conditioning number is of the order of 10^k, then, due to the error caused by rounding-off in h, we can loose k decimal places in y. In a similar way the contribution of the error into matrix A is estimated. Finally, cond(A) is only slightly dependent on the norm choice. The method for fast estimation of a conditioning number see in Sect.3.2 of Ref.[41]. The analysis of the cases where the conditioning number yields too pessimistic estimate of the error is made elsewhere [42].

As an example we note that the coefficients of polynomials P(x) that yield the best approximation to the function f(x) in the metrics of the Hilbertian space L^2_w (see earlier in Sect.5) can be determined from the set of equations

$$\int_{-1}^{1}(f(x) - P_n(x))x^k w(x)dx = 0, \tag{27}$$

where k = 0,1,... ,n. The above equations are linear and algebraic with respect to the coefficients of polynomials $P_n(x)$ (in powers of x or in Chebyshev polynomials). But since the monomials x^k are "almost linearly dependent", set (27) is ill-conditioned. The equivalent system

$$\int_{-1}^{1}(f(x) - P_n(x))T_k(x)w(x)dx = 0, \tag{27'}$$

is better conditi|o|ned, but in this case as well it is preferable to apply the economization procedure or to determine the coefficients c_i from (19) by the formulae

$$c_0 = \frac{1}{\pi}\int_{-1}^{1}f(x)w(x)dx; \quad c_i = \frac{2}{\pi}\int_{-1}^{1}f(x)T_i(x)w(x)dx, \quad i\geq 1, \tag{28}$$

where $w(x) = 1/\sqrt{1-x^2}$.

7. EFFECT OF ERROR AUTOCORRECTION

Let us consider the theoretical method (problem) of constructing approximation (21) to the function f(x). Let coefficients a_i, b_i yield the exact or approximate solution of the problem and let a_i,b_i be another approximate solution obtained by the same method. Let us denote by Δa_i, Δb_i the absolute errors in coefficients, i.e., $\Delta a_i = a_i - a_i$, $\Delta b_j = b_j - b_j$; these errors arise due to perturbation of the function f(x) to be approximated or due to the computational error. Let us put

It can readily be seen that the following exact equality is valid

$$P(x) = \sum_{i=0}^{n} a_i\varphi_i, \quad Q(x) = \sum_{j=0}^{m} b_j\psi_j, \quad \Delta P(x) = \sum_{i=0}^{n} \Delta a_i\varphi_i,$$

$$\Delta Q(x) = \sum_{j=0}^{m} \Delta b_j\psi_j, \quad P = P + \Delta P, \; Q = Q + \Delta Q.$$

It can readily be seen that the following exact equality is valid

$$\frac{P + \Delta P}{Q + \Delta Q} - \frac{P}{Q} = \frac{\Delta Q}{Q}\left(\frac{\Delta P}{\Delta Q} - \frac{P}{Q}\right). \tag{29}$$

As was indicated in Introduction, the ill-conditioning of the problem of calculating coefficients does not exclude the attainment of a high accuracy of the obtained approximations. This signifies that the approximations P/Q and P/Q are close to function to be approximated and, hence, are close to one another, although the coefficients of these approximations are rather different. In this case, the absolute value of the relative error $\Delta Q/Q = \Delta Q/D+\Delta Q$ of the denominator substantially exceeds that of the left-hand side of equation (29). This is possible only if the difference $\Delta P/\Delta Q$ - P/Q is small, i.e., the function $\Delta P/\Delta Q$ is close to P/Q and, hence, to the function to be approximated. Therefore the function $\Delta P/\Delta Q$ is called an erroneous approximation. In fact, for a special case this approximation was introduced in [5]. Later on, we shall see that in many instances an erroneous approximation, in fact, approximate the function quite well, and, hence, P/Q and P/Q differ by the product of small quantities in the right-hand side of equation (29). The point is that the errors Δa_i, Δb_j are not arbitrary but related by certain relationships.

Let the theoretical method for constructing the approximation of form (21) be linear in the sense that the coefficients of the approximation can be determined from the set of linear algebraic equations. The condition of their homogeneity is associated with the fact that the multiplication of the numerator and denominator of fraction (21) by one and the same nonzero number does not change approximation (21). Let us denote by y a vector whose components are the coefficients a_0, a_1, ..., a_n, b_0, b_1, ..., b_m. Let us assume that the coefficients can be determined from the homogeneous set of equations

$$H_Y = 0, \tag{30}$$

where H is the matrix of dimensionality (m + n + 2) x (m + n + 1).

The vector y is an approximate solution of (30) if the quan tity $\|Hy\|$ is small. If y and $\tilde{y}$ are approximate solutions of the set (30), then the vector $\Delta y = \tilde{y} - y$ is also an approximate solu tion of this set, since $\|H\Delta y\| = \|H\tilde{y} - Hy\| \le \|H\tilde{y}\| + \|Hy\|$. Therefore it is natural to take that the function **ΔP/ΔQ** corrrresponding to the solution **Δy** is an approximation to f(x). Obviously, that the order of the error of the approximate solution **Δy** of (30) (i.e., quantities $\|H\Delta y\|$, coincides with the order of the larger of the difference error of the approximate solution y and $\tilde{y}$. For the given order of the error, an increase in the error **Δy** is compensa ted by the fact that **Δy** satisfies with large "relative" accuracy the set of equations (30). Generally speaking, this increases the accuracy of the erroneous approximation.

In order to obtain a definite solution of set (30) it should be complemented with the normalization condition:

$$\sum_{i=0}^{n} \lambda_i a_i + \sum_{j=0}^{m} \mu_j b_j = 1, \tag{31}$$

where λ_i, μ_j are numerical coefficients. As a rule, the equality $b_o = 1$ is chosen as such a normalization condition, but this is not always satisfactory with respect to the minimization of the computational error. Complementing set (30) with equation (31), we arrive at a homogeneous set of m + n + 2 linear algebraic equations of type (22). If the approximate solutions y and y of set (30) satisfy condition (31), then the vector **Δy** satisfies the condition

$$\sum_{i=0}^{n} \lambda_i \ \Delta a_i + \sum_{j=0}^{m} \mu_j \ \Delta b_j = 0 \tag{31^1}$$

Of course, the above considerations are not rigorous; for each concrete method for constructing approximations an additional analysis is necessary. More accurate considerations for the classical Padé approximations are given below in Sect.8 and for linear and nonlinear

Padé-Chebyshev approximation in Sect.14. The existence of the above described mechanism of error autocorrection can also be checked by numerical experiments (see below).

The effect of the error autocorrection was also revealed for some nonlinear methods of the construction of rational approximations. One of such methods is considered below in Sect.12-14 (nonlinear Padé-Chebyshev approximation).

It should be emphasized that, as was noted in Sect.3, the coefficients of the best Chebyshev approximations satisfy the set of linear algebraic equations (9) and are calculated as approximate solutions of this set at the concluding stage of the iteration process using the Remez-type algorithms. Thus, the methods for obtaining best rational approximations can be considered as linear. At least for some functions (e.g., for cos at $-1 \le x \le 1$) both linear and nonlinear Padé-Chebyshev approximations are very close to best approximations in the sense of the relative and absolute errors, respectively. The results of the application of the algorithms for the calculation of the Padé-Chebyshev approximations can be considered as the approximate solutions of set (9) determining the best approximations. Therefore the existence of error autocorrection for Padé-Chebyshev approximations is an additional argument in favor of the assumption that this effect takes place also for best approximations.

Finally, let us note that the main relationship (29) looses sense if the approximation is sought in the form $a_0\varphi_0 + a_1\varphi_1 + \dots + a_n\varphi_n$, i.e., if the denominator in (21) reduces to unity. Nevertheless, even in this case the effect (although much weaker) of error autocorrection is possible, which is associated with the fact that the errors in Δa_i satisfy approximately certain relationships. Such a situation can occur when applying the least squares procedure.

8. PADÉ APPROXIMATION

Let the expansion of a function f(x) into a power series (Taylor series at zero) be given as

$$f(x) = \sum_{i=0}^{\infty} c_i x^i \ . \tag{32}$$

The classical Padé approximation for f(x) is a rational function of the form

$$R(x) = P_n(x)/Q_m(x), \tag{33}$$

where P(x) and Q(x) are the polynomials of degrees n and m, respectively, for which the following relationship is valid

$$Q_m(x)f(x) - P_n(x) = O(x^{m+n+1}) \tag{34}$$

Let

$$P_n(x) = a_0 + a_1 x + \ldots + a_n x^n,\ Q_m(x) = b_0 + b_1 x + \ldots + b_m x^m \tag{35}$$

If b ≠ 0, then (34) signifies that

$$f(x) - P_n(x)/\ Q_m(x) = O(x^{m+n+1}) \tag{34^1}$$

i.e., the first m + n + 1 terms of the Taylor expansion in powers of x (up to x^{m+n}) for f(x) and R(x) coincide. The Padé approximation yields the best approximation in a small neighborhood of zero and is a natural generalization of the function expanded into a Taylor series and tightly related to the function expansion into a continuous fraction. Padé approximations are considered in enormous number of studies (see e.g. [11-16, 5, 6]).

The coefficients b_j in the fraction denominator in (33) can be calculated by solving the homogeneous set of linear equations:

$$\sum_{j=1}^{m} c_{n+k-j} b_j = -b_j c_{n+k} \ , \tag{36}$$

where $k = 1, 2, \ldots, m$ and $c_l = 0$ at $l < 0$. Any nonzero constant can be taken as b_0. The coefficients a_i are calculated by the formulae

$$a_i = \sum_{k=0}^{i} b_k c_{i-k} = \sum_{k=0}^{i} b_{i-k} c_k . \tag{37}$$

The text of the respective program in FORTRAN can be found elsewhere [11].

For large m, system (36) is ill-conditioned. Moreover the problem of calculating the coefficients of the Padé approximations is also ill-conditioned irrespective of a concrete algorithm used for the solution of this problem (6, 43, 44]. Y.L.Luke made the following consideration. Let Δb_j be errors in coefficients b_j arising in the numerical solution of set of equations (36). We neglect the errors in x and c_j and take that, in virtue of (27), the errors in coefficients a_i have the form

$$\Delta a_i = \sum_{k=0}^{i} \Delta b_{i-k} c_k . \tag{37^1}$$

From (37^I) it follows that

$$\Delta Qf - \Delta P = \sum_{j=0}^{m} \Delta b_j x^j \sum_{i=0}^{\infty} c_i x^i - \sum_{i=0}^{n} \Delta a_i x^i =$$

$$= \sum_{j=0}^{m} \sum_{i=0}^{\infty} \Delta b_j c_i x^{i-j} - \sum_{i=0}^{n} \sum_{k=0}^{i} \Delta b_{i-k} c_k x^i ,$$

whence upon the change of the subscripts the following relationship is obtained

$$\Delta Q(x) f(x) - \Delta P(x) = O(x^{n+1}),$$

which is analogous to relationship (34). Thus it can be expected that the erroneous approximation really approximates the function f(x). Thus the effect of the error autocorrection really takes place. The experimental data for the function e^{-x} at m = n = 6, 7, ..., 14 and x = 2.5 and the experiments with the functitons $x^{-1}\ln(1 + x)$, $(1 + x)^{\pm 1/2}$, and $xe^{x}\int_{x}^{\infty} t^{-1}e^{-t}dt$ are described in [5]; see also [6].

The natural generalization of the classical Padé approximation is the multipoint Padé approximation (or second-order Padé approximation), i.e., the rational function of the form (33) with the values at certain points $x_i (i = 1, 2, ..., m + n + 1)$ coinciding with the values of the approximated function f(x). This definition is extended also to the case of multiple points; at $x_i = 0$ it leads for all i to the classical Padé approximation [11, 14, 15]. The calculation of the coefficients of multipoint Padé approximations can be reduced to the solution of a set of linear equation. There are grounds to take that in this case as well the effect of the error autocorrection takes place.

9. LINEAR PADÉ-CHEBYSHEV APPROXIMATIONS AND THE PADE PROGRAM

Let us consider the approximation of form (33) to the function f(x) on the sector [-1,1]. The function of absolute errors of this approximation has the form

$$\Delta(x) = \Phi(x)/ Q_m(x),$$

where

$$\Phi(x) = f(x)Q_m(x) - P_n(x). \tag{38}$$

The funtion $R(x) = P_n(x)/Q_m(x)$ is called a <u>linear Padé-Chebyshev approximation</u> to the function f(x) if

$$\int_{-1}^{1} \Phi(x)T_k(x)w(x)dx = 0, \quad k = 0, 1, \dots, m+n, \tag{39}$$

where $T_k(x)$ are the Chebyshev polynomials, $w(x) = 1/\sqrt{1-x^2}$. This notion (in another form) was introduced in [45]. It can be extended to the case of other orthogonal polynomials [11, 33, 34, 39, 40]. Such an approximation always exists [39]. Using the considerations similar to those in Sect.5 and applying the Cheney theorem, one can realize why linear Padé-Chebyshev approximations are close to best. Let $P_n(x)$ and $Q_m(x)$ have the form (35). Then the set of equations (39) is equivalent to the following set of linear algebraic equations in coefficients a_i, b_j

$$\sum_{j=0}^{m} b_j \int_{-1}^{1} \frac{x^j T_k(x) f(x)}{\sqrt{1-x^2}} dx - \sum_{i=0}^{n} a_i \int_{-1}^{1} \frac{x^i T_k(x)}{\sqrt{1-x^2}} dx. \tag{40}$$

Homogeneous system (40) can be transformed into inhomogeneous system by adding a normalization condition; in particular, such a condition can be any of the below given ones:

$$b_0 = 1, \tag{41}$$

$$b_m = 1, \tag{42}$$

$$a_m = 1. \tag{43}$$

A program Padé approximationDE written in FORTRAN (with double accuracy) provides the construction of rational approximations by solving a set of equations of type (40). The complete text of one version of this program and its detailed description can be obtained from the Foudation of Algorithms and Programs of the Computation Center of the USSR Academy of Sciences [7]. For even functions the approximation is sought in the form

$$R(x) = \frac{a_0 + a_1x^2 + \dots + a_n(x^2)^n}{b_0 + b_1x^2 + \dots + b_m(x^2)^m}, \tag{44}$$

for odd ones in the form

$$R(x) = x \frac{a_0 + a_1x^2 + \dots + a_n(x^2)^n}{b_0 + b_1x^2 + \dots + b_m(x^2)^m}. \tag{45}$$

The program computes the values of the approximation coefficients, the absolute and relative errors and provides the information that allows the estimation of the approximation quality (see Sect.4). In particular, a variant of the PADE program for a SM-4 minicomputer outputs the graphs of the error functions, establishes the existence of alternants, and yields the estimate of the approximation quality with the aid of quantity (12). With the help of a special subroutine, the user can set any function determined with the aid of any algorithm on an arbitrary segment [A, B], the boundaries of this segment, numbers m and n, and some controlling parameters. In particular, one can choose the normalization condition of type (41)-(43), seek for the approximation in form (44) or (45), etc. Since the change of the variable reduces the approximation on any segment [A,B] to the approximation on the segment [-1,1], in what folllows we shall limit ourselves to the case A = -1, B = 1, if not indicated otherwise.

The integrals are computed with the use of the Hauss-Hermite-Chebyshev quadrature formula:

$$\int_{-1}^{1} \frac{\varphi(x)}{\sqrt{1-x^2}}\, dx = \frac{\pi}{s} \sum_{i=1}^{s} \varphi\left(\cos \frac{2i-1}{2s}\pi\right) \tag{46}$$

where s is the number of nodes. For polynomials of degree 2s-1 this formula is exact, so that the accuracy of (46) rapidly increases with an increase of the number of nodes. It depends on how well the functiton $\varphi(x)$ is approximated by polynomials. In order to calculate the values of the Chebyshev polynomials, recurrence relationship (15) is used.

If the function f(x) is even and the approximation is sought in form (44), then set (40) is transformed into the following set of equations

$$\sum_{i=0}^{n} a_i \int_{-1}^{1} \frac{x^{2i}\, T_{2k}(x)}{\sqrt{1-x^2}}\, dx - \sum_{j=0}^{m} b_j \int_{-1}^{1} \frac{x^{2j}\, T_{2k}(x) f(x)}{\sqrt{1-x^2}}\, dx \qquad (47)$$

where k = 0, 1, 2, ..., m + n. If f(x) is an odd function and the approximation is sought in form (45), then first the approximation of form (44) to the even function f(x)/x is sought by solving the set of equations (47) complemented with one normalization condition. Then the obtained approximation is multiplied by x. This method allows us to avoid a large relative error at x = 0.

The possibilities provided by the PADE program are illustrated by Table 1 which lists the errors of some approximations obtained by this program. Table 1 gives for each approximation the absolute error Δ and relative error δ, and, for comparison, the best possible relative error δ_{min} from [3]. The function $\sqrt{x}$ on the segment [1/2, 1] is approximated by the expression of form (1), the function $\cos \frac{\pi}{4} x$ on the segment [-1,1] by (44), the other functions on the same segment are approximated by expressions of type (45).

The PADE program is relatively simple and compact and includes also the standard subroutine DGELG for the solution of a set of linear algebraic equations taken from [46] and the subroutine for numerical integration, and also several service, test, and auxiliary modules. No additional programs are used. The program sets the minimum requirements to apparatus and can, in fact, be implemented on any computer possessing a translator from FORTRAN, sufficient operative storage, and a printer or a display. The modification of the PADE program described in [7] was implemented on IBM-360/370 computers. It requires 60 K operative storage. The program in FORTRAN, including comments, consists of 581 lines. The time of program execution depends on the computer type, the function to be approximated, and the values of the controlling parameters. Thus the execution time of the processor of an ES-1040 computer used for the determination by PADE program of

Table 1.

Function	m	n	Δ	δ	δ_{min}
$\sqrt{x}$	2	2	$0.8 \cdot 10^{-6}$	$1.13 \cdot 10^{-6}$	$0.6 \cdot 10^{-6}$
$\sqrt{x}$	3	3	$1.9 \cdot 10^{-9}$	$2.7 \cdot 10^{-9}$	$1.12 \cdot 10^{-9}$
$\cos \frac{\pi}{4} x$	0	3	$0.28 \cdot 10^{-7}$	$0.39 \cdot 10^{-7}$	$0.32 \cdot 10^{-7}$
$\cos \frac{\pi}{4} x$	1	2	$0.24 \cdot 10^{-7}$	$0.34 \cdot 10^{-7}$	$0.29 \cdot 10^{-7}$
$\cos \frac{\pi}{4} x$	2	2	$0.69 \cdot 10^{-10}$	$0.94 \cdot 10^{-10}$	$0.79 \cdot 10^{-10}$
$\cos \frac{\pi}{4} x$	0	5	$0.57 \cdot 10^{-13}$	$0.79 \cdot 10^{-13}$	$0.66 \cdot 10^{-13}$
$\cos \frac{\pi}{4} x$	2	3	$0.4 \cdot 10^{-13}$	$0.55 \cdot 10^{-13}$	$0.46 \cdot 10^{-13}$
$\sin \frac{\pi}{4} x$	0	4	$0.34 \cdot 10^{-11}$	$0.48 \cdot 10^{-11}$	$0.47 \cdot 10^{-11}$
$\sin \frac{\pi}{4} x$	2	2	$0.32 \cdot 10^{-11}$	$0.45 \cdot 10^{-11}$	$0.44 \cdot 10^{-11}$
$\sin \frac{\pi}{4} x$	0	5	$0.36 \cdot 10^{-14}$	$0.55 \cdot 10^{-14}$	$0.45 \cdot 10^{-14}$
$\sin \frac{\pi}{2} x$	1	1	$0.14 \cdot 10^{-3}$	$0.14 \cdot 10^{-3}$	$0.12 \cdot 10^{-3}$
$\sin \frac{\pi}{2} x$	0	4	$0.67 \cdot 10^{-8}$	$0.67 \cdot 10^{-8}$	$0.54 \cdot 10^{-8}$
$\sin \frac{\pi}{2} x$	2	2	$0.63 \cdot 10^{-8}$	$0.63 \cdot 10^{-8}$	$0.53 \cdot 10^{-8}$
$\sin \frac{\pi}{2} x$	3	3	$0.63 \cdot 10^{-13}$	$0.63 \cdot 10^{-13}$	$0.5 \cdot 10^{-13}$
$\tan \frac{\pi}{4} x$	1	1	$0.64 \cdot 10^{-5}$	$0.64 \cdot 10^{-5}$	$0.57 \cdot 10^{-5}$
$\tan \frac{\pi}{4} x$	2	1	$0.16 \cdot 10^{-7}$	$0.16 \cdot 10^{-7}$	$0.14 \cdot 10^{-7}$
$\tan \frac{\pi}{4} x$	2	2	$0.25 \cdot 10^{-10}$	$0.25 \cdot 10^{-10}$	$0.22 \cdot 10^{-10}$
$\tan^{-1} x$	0	7	$0.75 \cdot 10^{-7}$	$\cdot 10^{-7}$	$\cdot 10^{-7}$
$\tan^{-1} x$	2	3	$0.16 \cdot 10^{-7}$	$0.51 \cdot 10^{-7}$	$0.27 \cdot 10^{-7}$
$\tan^{-1} x$	0	9	$0.15 \cdot 10^{-8}$	$0.28 \cdot 10^{-8}$	$0.23 \cdot 10^{-8}$
$\tan^{-1} x$	3	3	$0.54 \cdot 10^{-9}$	$1.9 \cdot 10^{-9}$	$0.87 \cdot 10^{-9}$
$\tan^{-1} x$	4	4	$0.12 \cdot 10^{-11}$	$0.48 \cdot 10^{-11}$	$0.17 \cdot 10^{-11}$
$\tan^{-1} x$	5	4	$0.75 \cdot 10^{-13}$	$3.7 \cdot 10^{-13}$	$0.71 \cdot 10^{-13}$

the type (1) approximation to the function $\sqrt{x}$ on the segment [1/2, 1] at m = n = 2 is 4.4 s. In this case normalization (43) is used, the number of the points of control used for the determination of the error is 1200, the time of the program translation being neglected*/. One of the program modifications provides the estimate of the quality of the obtained approximation by formula (12) (see Sect.4). Thus for the function ${}_2^{-1}x$ at m = n = 2 and for the relative error we have q = 0.0625, whence it follows that $\delta_{min} \geq q\delta \approx 0.4\ 10^{-9}$. As is seen from Table 1, this estimate is rather rough and in fact $\delta_{min}/\delta \approx 0.84$. For the absolute error the program yields q = 0.71, which indicates that this error is close to best. The above indicated modification of the program allows computations in the dialogue mode, varying the degrees m and n, the boundaries of the segment [A, B], the number of the points of control in the computation of the errors, the number of nodes in quadrature formula (46) and the choice of the algorithm branch, and also estimation of the quality of the approximation from the character of the error curve.

Note. The subroutine for constructing classical Padé approximations given in [11] is quite different from the above considered program.

10. PADE PROGRAM. ANALYSIS OF ALGORITHM

The quality of the approximation obtained with the aid of the PADE program depends mainly on the behavior of the approximation denominator and on computational error. The most unpleasant source of errors for the method is ill-conditioning of arising sets of linear algebraic equations. Probably it is just this fact that hinders the wide use of the above indicated methods.

The conditioning numbers are also very large for the sets of equations arising in the calculations of approximations computed by the above considered PADE programs. Thus during computation of an approximation of form (5) to the function $\sin \frac{\pi}{2} x$ at m = n = 3 on the segment [-1,1] a set of equations arise with the conditioning number of the order of 10^{13}. As a result, the approximation coefficients are determined with a large error. In particular, small perturbation of the set of linear equations arises if we pass from an ICL-4-50 computer to an ES-1045 one (due to computatinal error). This brings large perturbations in the approximation coefficients. Happily enough, the effect of the error autocorrection (see Sect.7) helps in this case and the approximation errors for such a perturbation vary only slightly. This situation has already been described in the Introduction where it was illustrated by concrete examples.

Let us consider some more examples associated with the transition from ICL-4-50 to ES-1045 computer. We take the algorithm branch corresponding to normalization condition (41), i.e., $b_0 = 1$. For $\tan^{-1}x$ the computation of the approximation of form (45) on the sector [-1,1]] at m = n = 5 on an ICL-4-50 computer yields the approximation with the absolute error $\Delta = 0.35 \bullet 10^{12}$ and the relative error $\delta = 0.16 \bullet 10^{-11}$. The respective system of linear algebraic equations has the conditioning number of the order of 10^{30}. Upon the transition to the ES-1045 computer we have $\Delta = 0.5 \bullet 10^{-14}$, $\delta = 0.16 \bullet 10^{-12}$ and the conditioning number of the order of 10^{14}. In this case the absolute values of the errors, Δa_1 and Δb_1, exceed unity. This example indicates that the problem of computing the conditioning number for an ill-conditioned system is, in turn, ill-conditioned itself. Indeed, roughly speaking, the conditioning number is determined by the values of the inverse matrix coefficients (see eqns.(24) and (25) in Sect.6). Then each column of the inverse matrix is a solution of the set of equations with the initial matrix of coefficients, i.e., of an ill-conditioned system.

Let us consider in more detail the effect of the error autocorrection for the approximation of form (44) to the function cos at m = 2 and n = 3 on the segment [-1,1]. The construction of this approximation both on ICL-4-50 and ES-1045 leads to the approximation close to best with the absolute error $\Delta = 0.4 \bullet 10^{-13}$ and the relative error $\delta = 0.55 \bullet 10^{-13}$. The conditioning number in both cases is of about 10^9. The approximation coefficients obtained on the above indicated computers and the coefficients of the erroneous approximation (see Sect.7) are:

$a_0 = 0{,}9999999999999600;$

$a_0 = 0{,}9999999999999610;$ $\quad \Delta a_0 = -10^{-15};$

$a_1 = -\ 0{,}2925310453579570;$

$a_1 = -\ 0{,}2925311266471621б;$ $\quad \Delta a_1 = 10^{-7} \bullet 0{,}811136646;$

$a_2 = 10^{-1} \bullet 0{,}1105254254716866;$

$a_2 = 10^{-1} \bullet 0{,}1105256585556549;$ $\quad \Delta a_2 = -10^{-7} \bullet 0{,}2330839683;$

$a_3 = 10^{-3} \bullet 0,1049474500904401;$

$a_3 = 10^{-3} \bullet 0,1049482094850086;$ $\Delta a_3 = 10^{-9} \bullet 0,7593947685;$

$b_0 = 1;$

$b_0 = 1;$ $\Delta b_0 = 0;$

$b_1 = 10^{-1} \bullet 0,1589409217324021;$

$b_1 = 10^{-1} \bullet 0,1589401105960337;$ $\Delta b_1 = 10^{-7} \bullet 0,8111363684;$

$b_2 = 10^{-3} \bullet 0,1003359011092697;$

$b_2 = 10^{-3} \bullet 0,10033419180835229;$ $\Delta b_2 = 10^{-8} \bullet 0,17093009168.$

Thus, the erroneous approximation has the form

$$\frac{\Delta P}{\Delta Q} = \frac{\Delta a_0 + \Delta a_1 x^2 + \Delta a_2 x^4 + \Delta a_3 x^6}{\Delta b_1 x^2 + \Delta b_2 x^4} \tag{48}$$

If to reject a relatively small quantity $\Delta = -10^{-15}$ in (48), then, as show the computer tests (with up to 2000 points of control) this expression is an approximation to the function cos on the sector [-1,1] with the absolute and relative errors being $\Delta = \delta = 0.22 \bullet 10^{-6}$.

But the polynomial ΔQ goes to zero at $x = 0$, and the polynomial ΔP at $x = 0$ takes a small but nonzero value. Fortunately, equality (29) can be transformed to the form

$$\frac{P}{Q} - \frac{P}{Q} = \frac{\Delta P}{Q} - \frac{\Delta Q}{Q} \bullet \frac{P}{Q} . \tag{49}$$

Therefore at $\Delta Q \mid 0$, the effect of error autocorrection is provided because of quantity ΔP is close to zero, and the approximation error P/Q is determined by the error in coefficient a_0. The same situation arises also in the case where the polynomial ΔQ goes to zero at an arbitrary point x_0 of the segment [A,B] on which the function is approximated. It is clear that for the standard choice of normalization ($b_0 = 1$) the erroneous approximation has two coefficients less than the initial one. From relationships (38) and (39) it is seen that, in the general case, the normalization condition $b_m = 1$ or $a_n = 1$ leads to the fact that the coefficients of the erroneous approximation form an approximate solution of a homogeneous set of linear algebraic equations whose exact solution is set by the Padé-Chebyshev approximation having one coefficient less than the initial one. A higher accuracy of such an erroneous approximation is promoted by the effect of error autocorrection so that we arrive again at the situation where "a snake catches its own tail". A similar situation arises also if the approximation of form (44) is constructed to an even function by solving the set of equations (47).

Sometimes the approximation error can be reduced by an appropriate choice of the normalization condition. As an example consider the approximation of the function e^x by rational functions of form (1) at $m = 15$, $n = 0$ on the segment $[-1,1]$. For the generally accepted normalization $b_0 = 1$, the PADE program yields the approximation with the absolute error $\Delta = 1.4 \cdot 10^{-14}$ and relative error $\delta = 0.53 \cdot 10^{-14}$. Upon the transition to the normalization condition $b_{15} = 1$ the errors decrease approximately by a factor of two: $\Delta = 0.73 \cdot 10^{-14}$ and $\delta = 0.27 \cdot 10^{-14}$. Note also that the conditioning number increases. In the first case it is $2 \cdot 10^{-6}$ and in the second one $6 \cdot 10^{16}$. Thus, the error decreases despite the drastic deterioration of the conditioning of the set of equations. This example shows that a higher accuracy of the erroneous approximation can be accompanied by an increase in the conditioning number and, as shows the experiment, also an increase in the error in the numerator and denominator of the approximation. The appropriate choice of the normalization condition depends on the concrete situation.

A specific situation arises where the degrees of the numerator and denominator of the approximation are zeroes. In this case, an inappropriate choice of the normalization condition can reduce the erroneous approximation down to zero or make it senseless. At $n = 0$ it is expedient to choose condition (42) in a way it is made in the above given example. At $m = 0$ (the case of the polynomial approximation) it is expedient, as a rule, to choose condition (43). Otherwise, we have to solve the set of equations (27) in the situation described earlier in Sect.6.

Since for ES computers the double accuracy corresponds to 16 decimal places of the mantissa in the computer representation of numbers, it is convenient to vary the normalization conditions working on such computers only if the conditioning number exceeds $\delta \cdot 10^{16}$ (δ is the relative error of the obtained approximation). The values of the conditioning number of the respective set of linear algebraic equations are output within the PADE program along with other computational results.

The theoretical error of the method is determined to a large extent by the behavior of the approximation denominator. For the analysis it is convenient to divide the numerator and denominator of the fraction by b_0 and then to equate b_0 to unity. If the coefficients b, ..., n are small in comparison with $b_0 = 1$ (which is often the case in practical computations), then the absolute error $\Delta(x)$ and its numerator $\Phi(x) = f(x)Q(x) - P(x)$ are of the same order, so that the minimization of $\Phi(x)$ also minimizes the error $\Delta(x)$ (see Sect.9). Note also that the coefficients b_1, ..., b_m of the approximation of form (45) to the function $\tan^{-1}x$ on the segment [-1,1] are not small in comparison with b_0. Thus, for $m = n = 3$ the coefficient b_1 exceeds b_0 by a factor of almost 1.5. Therefore, as is seen from Table 1, the error of the approximation to $\tan^{-1}x$ obtained by the PADE program is several times larger than the errors of best approximations.

Note that sometimes it is possible to "improve" the approximation denominator or to reduce the conditioning number of the respective set of equations by extending the segment [A, B] on which the function is approximated. This effect is observed, e.g., in the computation of the approximants to certain hyperbolic functions.

Note also that the change of the standard subroute DGELG of the solution of a set of linear algebraic equations [46] by another subroutine used for the same purpose, e.g., by DECOMP from [41] has no essential influence on the approximation quality obtained by the PADE programs.

The numerator and denominator of the approximation can also be sought in the form

$$P = \sum_{i=0}^{n} a_i T_i \ , \quad Q = \sum_{j=0}^{m} b_j T_j \ , \tag{50}$$

where T_i are Chebyshev polynomials. In this case the set of linear equations defining the coefficients is better conditioned. But the computation of polynomials (50), e.g., by the Clenshaw method, requires longer computational time, although has a good effect on the computational error (see Ch.4 in [47] and [59]). The transformation of polynomials P and Q from form (50) into conventional form (35) also requires some additional efforts.

In practice, it is more convenient to deal with the approximations (1), (44), or (45), and to compute the numerator and denominator of the fraction by the Gorner scheme. In this case the normalization $b_m = 1$ or $a_n = 1$ decreses the number of multiplications. Therefore PADE yields the approximation coefficients in two forms: with the condition $b_0 = 1$ and with one of the conditions $b_m = 1$ or $a_n = 1$ independently of what of the conditions (41) -(43) is used in the solution of the set of equations (39) or (40).

The PADE program (and the respective algorithm) can readily be modified, e.g., for a case where some a_i,b_j coefficients be set a priori. It is also possible to modify the set of equations under consideration by varying the weight w(x) or the interval on which the functions and the sets of orthogonal polynomials are approximated. Using somewhat more sophisticated set of equations of type (40), it is possible to attain the minimum norm of the numerator $\Phi(x)$ of the error function $\Delta(x)$ in the Hilbertian space L_w^2 (see Sect.5).

The use of the PADE program does not require any preliminary expansion of the approximated function into a series or a continued fraction. From equations (39) or (40) and quadrature formula (46) it is seen that the PADE program uses only the information on the values of the approximated function f(x) at the nodes of the quadrature formula (which are zeroes of a certain Chebyshev polynomial).

On the segment [-1,1], the linear Padé-Chebyshev approximations yield much smaller errors than the classical Padé approximation. Thus the Padé approximation of form (1) to the function e^x at $m = n = 2$ has the absolute error at a point $x = 1$ equal to $\Delta(1) = 4 \cdot 10^{-3}$. The PADE program yields the approximation of the same form with the absolute error on the whole segment equal to $\Delta = 1.9 \cdot 10^{-4}$, i.e., 20 times smaller. The absolute error of the best approximations is $0.87 \cdot 10^{-4}$.

11. CROSS-MULTIPLICATIVE SCHEME OF CONSTRUCTING PADÉ-CHEBYSHEV LINEAR APPROXIMATIONS

As a rule, Padé-Chebyshev linear approximations are constructed by the following scheme [45, 3, 11, 12]. Let the function we are going to approximate be expanded into a series in Chebyshev polynomials

$$f(x) = \sum_{i=0}^{\infty} c_i T_i(x) = \frac{1}{2} c_0 + c_1 T_1(x) + c_2 T_2(x) + \ldots, \tag{51}$$

where the notation $\sum_{i=0}^{m}{}' u_i$ indicates that the first term u_0 in the sum should be subst|i|t|uted by $u_0/2$. A rational approximation is then sought in the form

$$R(x) = \frac{\sum_{i=0}^{n}{}' a_i T_i(x)}{\sum_{j=0}^{m}{}' b_j T_j(x)} \tag{52}$$

the coefficients b_j are determined from the set of linear algebraic equations

$$\sum_{j=0}^{m}{}' b_j (c_{i+j} + c_{|i-j|}) = 0, \quad i = n+1, \ldots, n+m, \tag{53}$$

and the coefficients a_i are given by the equations

$$a_i = \frac{1}{2} \sum_{j=0}^{m}{}' b_j (c_{i+j} + c_{|i-j|}), \quad i = 0, 1, \ldots, n. \tag{54}$$

It can readily be verified whether this algorithm leads to the same result as the algorithm described earlier in Sect.9 if to neglect the computational error.

Expressions (53) and (54) include the coefficients c_k at $k = 0, 1, \ldots, n + 2m$. In other words, it is necessary to know the first $n + 2m + 1$ terms of series (51). As a rule, the coefficients c_i are known only approximately. For their determination, one should take the segment of the expansion of f(x) into a series in powers of x (the Taylor series) and, using the economization procedure, transform it to the form

$$\sum_{i=0}^{n+2m} c_i T_i(x). \tag{55}$$

12. NONLINEAR PADÉ-CHEBYSHEV APPROXIMATIONS

A rational function R(x) of form (1) or (52) is called a <u>nonlinear Pade-Chebyshev approximation</u> to the function f(x) on the segment [-1,1] if

$$\int_{-1}^{1}(f(x) - R(x))T_k(x)w(x)dx = 0, \quad k = 0, 1, \ldots, m + n \tag{56}$$

where T(x) are Chebyshev polynomials, $w(x) = 1/\sqrt{1-x^2}$. The Cheney theorem (see Sect.5) shows that the function of absolute errors $\delta(x) = f(x) - R(x)$ has an alternant, so that there are grounds to believe that nonlinear Padé-Chebyshev approximations are close to best in the sense of the absolute error.

In [32] the following algorithm for the computation of the approximation coefficients is given. Let the function f(x) we need to approximate be expanded into series (51) in Chebyshev polynomials. Let us determine the auxiliary quantities γ_i from the set of linear algebraic equations

$$\sum_{j=0}^{m} \gamma_j \, c|_{k-j}| = 0, \; k = n + 1, n + 2, \ldots n + m, \tag{57}$$

assuming that $\gamma_0 = 1$. The coefficients of the denominator in (52) are defined by equalities

$$b_j = \mu \sum_{j=0}^{m-j} \gamma_i \gamma_{i+j},$$

where $\mu^{-1} = \frac{1}{2} \sum_{i=0}^{n} \gamma_1^2$, which provides $b_o = 2$. Finally, the coefficients of the numerator are determined by (54). System (57) can be explicitly solved. It is possible to indicate the formulae for computing γ_i. It is also possible to evaluate in the explicit form the absolute error of the approximation. The respective algorithm is described in details in [33] (see also [11]). In distinction from linear Padé-Chebyshev approximations, the nonlinear approximations of this type do not always exist, but it is possible to indicate the conditions (that can be explicitly verified) that guarantee the existence of such approximations [33]. Nonlinear Padé-Chebyshev approximations in comparison with linear ones provide somewhat lower absolute error, but may have higher relative error. As an example, consider the approximation of form (1) or (52) at m = n = 3 to the function e on the segment [-1,1]. For a nonlinear Padé-Chebyshev approximation, the absolute error in this case is $\Delta = 0.258 \cdot 10^{-6}$ and the relative error is $\delta = 0.252 \cdot 10^{-6}$; for the linear approximation, $\Delta = 0.33 \cdot 10^{-6}$ and $\delta = 0.20 \cdot 10^{-6}$.

13. THE USE OF REDUCE SYSTEM OF ANALYTICAL COMPUTATIONS FOR CONSTRUCTING RATIONAL APPROXIMATIONS

A set of REDUCE analytical computations [48,49] allows the manipulations with the appropriate formulae at a level of symbols and is convenient tool for implementing the algorithms for computing rational approximations. The use of this system makes it unnecessary to write algorithms for computing the function to be approximated if this function can be represented in an analytical form, if the coefficients of the Taylor series are known, or if they can be analytically determined from the differential equations. It is possible to reject the rounding-off errors in the work with the exact arithmetics of rational numbers represented in the form of the ratios of integers.

Within the REDUCE system, the application package is implemented for computations with high accuracy and the construction of rational approximations (see, e.g. [8]). In particular, the algorithms close to those described in Sects. 11 and 12 were developed. The function to be approximated is first expanded into a power series (Taylor series)

$f = \sum_{k=0}^{\infty} f^{(k)}(0)\frac{X^k}{K!}$, then a segment of this series

$$\sum_{k=0}^{N} f^{(k)}(0) \frac{X^k}{K!}, \tag{58}$$

consisting of first N + 1 terms (where N is set by the user) is transformed to the polynomial of form (55) using the economization procedure.

The algorithms implemented with the aid of the REDUCE system provide computations of approximations in forms (1) or (52), the estimate of the absolute and relative errors and the output of the graphs of the error curves. It is also possible to write a program for the computation of the respective approximation in FORTRAN. In this case the constants of the rational arithmetics are transformed into a standard form with a floating point. If necessary, the computed approximation can be transformed to the form most convenient for the user. Thus, it is possible to compute the numerator and denominator of a fraction in form (1) by the Gorner scheme, and for a fraction of form

(52) by the Clenshaw scheme; it is also possible to transform the rational expression into a continuous fraction or Jacobi fraction.

The ALGOL-like input language of the REDUCE system and convenient methods for solving the problems of linear algebra make the programs rather simple and compact. Thus a program for computing the coefficients of linear Padé-Chebyshev approximations consists of only sixty two lines.

14. EFFECT OF ERROR AUTOCORRECTION FOR NONLINEAR PADÉ-CHEBYSHEV APPROXIMATIONS

Relationships [56] can be considered as a set of equations for the approximation coefficients. Let us assume that the approximations R(x) = P(x)/Q(x) and R(x) = P(x)/Q(x), where P(x) and P(x) are polynomials of degree n, and Q(x) and Q(x) are polynomials of degree m, were obtained by the approximate solution of the above indicated set of equations. Consider an erroneous approximation ΔP(x)/ΔQ(x) where ΔP(x) = P(x) - P(x), ΔQ(x) = Q(x)- Q(x). Let us substitute R(x) and R(x) into [56] and substract the obtained expressions from one another. We see that the following approximate equation is valid

$$\int_{-1}^{1}\left(\frac{P(x)}{Q(x)} - \frac{P(x)}{Q(x)}\right)T_k(x)w(x)\,dx \approx 0,\ k = 0, 1, \ldots, m + n.$$

The above equation and equation (29) yield the following apaproximate equality:

$$\int_{-1}^{1}\left(\frac{\Delta P(x)}{\Delta Q(x)} - \frac{P(x)}{Q(x)}\right)\frac{\Delta Q}{Q}T_k(x)w(x)\,dx \approx 0 \qquad (59)$$

where k = 0, 1, ..., 1, m + n, $w(x) = i/\sqrt{1-x^2}$. If the quantity ΔQ is relatively large (because of ill-conditioning of the set of equations (57)), then equations (59) shows that it is natural to expect that the erroneous approximation would be close to P/Q and, hence, to the approximated function f(x).

The programs described in Sect.13 which use the arithmetics of rational numbers provide the elimination of the rounding-off error and the estimate of the effect of error in the setting of approximate functions on the coefficients of the nonlinear Padé-Chebyshev approximation "in a pure form". In this case, the effect of error autocorrection can be substantiated by a more accurate consideration valid for both nonlinear and linear Padé-Chebyshev approximation and even for linear generalized Padé approximations related to other systems of orthogonal polynomials. This consideration is analogous to the Luke's speculations [5] considered earlier in Sect.[8].

Let us take that the function f(x) is expanded into series (51) and that the rational approximation R(x) = P(x)/Q(x) is sought in form (52).

Let Δb_j be the errors in the coefficients of the denominator Q of the approximation. In the linear case these errors arise in the process of solving the set of equations (53) and in the nonlinear case they arise in solving equations (57). In both cases the coefficients a_i of the numerator of the approximation are defined by equations (54). Whence it follows that

$$\Delta a_i = \frac{1}{2} \sum_{j=0}^{m_1} \Delta b_j (c_{i+j} + c_{|i-j|}), \quad i = 0, 1, \ldots, n, \tag{60}$$

Now it is easy to derive that the erroneous approximation $\Delta P/\Delta Q$ satisfies the relationships

$$\int_{-1}^{1} (f(x)\, \Delta Q(x) - \Delta P(x))\, T_i(x) w(x)\, dx = 0, \; i = 0, 1, \ldots, n. \tag{61}$$

which are analogous to relationships (39) defining the linear Padé-Chebyshev approximations. Indeed, let us use the well known formula for the multiplication of Chebyshev polynomials:

$$T_i(x)\, T_j(x) = \frac{1}{2}\, [T_{i+j}(x) + T_{|i-j|}(x)], \tag{62}$$

where i,j are arbitrary subscripts (see, e.g., [11-13. 33]). With due regard for (62), quantity $f\Delta Q - \Delta P$ can be transformed in the following way:

$$f\Delta Q - \Delta P = \left(\sum_{j=0}^{m}{}' \Delta b_j T_j \right) \left(\sum_{i=0}^{\infty}{}' c_i T_i \right) - \sum_{i=0}^{n}{}' \Delta a_i T_i =$$

$$= \frac{1}{2} \sum_{i=0}^{\infty}{}' \left[\sum_{j=0}^{m}{}' \Delta b_j (c_{i+j} + c_{|i-j|}) \right] T_i - \sum_{i=0}^{n}{}' \Delta a_i T_i .$$

Whence it follows from (60):

$$f\Delta Q -= \Delta P = O(T_{n+1})$$

in other words, the first n + 1 terms of the expansion of the function $f\Delta Q - \Delta P$ into a series in Chebyshev polynomials are absent. This is equivalent to the validity of relationships (61) since Chebyshev polynomials form an orthogonal system. In practical computations, the coefficients c_i are known only approximately, therefore equalities (60) and (61) are satisfied only approximately.

Let us consider the results of the numerical experiments*/ performed by the above described programs within the REDUCE system briefly described earlier in Sect.13. We begin with the example considered earlier in Sect.10 when we constructed the linear Padé-Chebyshev approximation to the function $\cos \frac{\pi}{4} x$ at m = 2, n = 3 in the form (44) on the segment [-1,1] . In order to construct the respective nonlinear Padé-Chebyshev approximation it is necessary to indicate the value of the parameter N determining the number of terms in the partial sum (58) of the Taylor series for the function to be approximated. In the final analysis, the computational error in the given case is determined just by the value of the parameter N.

The coefficients of the approximations of form (44) obtained at N = 15 and N = 20 (nonlinear case) and also the coefficients of the erroneous approximation**/ are:

$a_0 = 0{,}4960471034987563;$

$a_0 = 0{,}4960471027757504;$ $\Delta a_0 = 10^{-8} \cdot 0{,}07230059;$

$a_1 = -0{,}1451091945278387;$

$a_1 = -0{,}1451091928755736;$ $\Delta a_1 = -10^{-8} \cdot 0{,}10522651;$

$a_2 = 10^{-2} \cdot 0{,}5482586543334515;$

$a_2 = 10^{-2}\bullet 0{,}548258121085953;$ $\Delta a_2 = 10^{-9}\bullet 0{,}42224856;$
$a_3 = 10^{-4}\bullet 0{,}5205903601778259;$
$a_3 = 10^{-4}\bullet 0{,}5205902238186334;$ $\Delta a_3 = -10^{-10}\bullet 0{,}13635919;$
$b_0 = 0{,}4960471034987759;$
$b_0 = 0{,}4960471027757698;$ $\Delta b_0 = 10^{-8}\bullet 0{,}\ 07230061;$
$b_1 = 10^{-2}\bullet 0{,}7884201590727615;$
$b_1 = 10^{-2}\bullet 0{,}7884203019999351;$ $\Delta b_1 = -10^{-10}\bullet 0{,}1429272;$
$b_2 = 10^{-4}\bullet 0{,}4977097973870693;$
$b_2 = 10^{-4}\bullet 0{,}4977100977750249;$ $\Delta b_2 = -10^{-10}\bullet 0{,}300388.$

Both approximations have the absolute error $\Delta = 0.4\bullet 10^{-13}$ and the relative error $\delta = 0.6\bullet 10^{-13}$ close to best, the conditioning number of a set of equations (57) in both cases is $0.4\bullet 10^{8}$, the denominator ΔQ of the erroneous approximation goes to zero at $x = x_0 \approx 0.70752...$; the point x_0 is close to the root of the numerator ΔP which at $x = x_0$ takes the value of the order 10^{-8} Such a situation was considered earlier, in Sect.10. If to exclude the small neighborhood of the point x_0, the absolute and relative errors of the erroneous approximation are of the same order as in the "linear case" considered in Sect.10.

Now consider the nonlinear Padé-Chebyshev approximations of form (44) to the function $\tan_4^{-1} x$ at $m = n = 3$ on the segment [-1,1]. The Taylor series in the given case converges quite slowly, and with an increase in N the values of the coefficients of the rational approximations vary quite substantially already in the first decimal places (and in quite a capricious way). The general picture is seen from Table 2 listing the values of the absolute error Δ, absolute error of erroneous approximations***/ Δ_0 (the approximations are compared at N = 15 and N = 20, N = 25 and N = 35, N = 40 and N = 50), and also the values of the conditioning number Cond of the set of linear algebraic equations (57). The relative errors in this case coincide with the absolute ones. The best possible error is $\Delta_{min} = 0.83\bullet 10^{-17}$.

Table 2

N	15	20	25	35	40	50
cond	$0{,}76\cdot 10^{7}$	$0{,}95\cdot 10^{8}$	$0{,}36\cdot 10^{10}$	$0{,}12\cdot 10^{12}$	$0{,}11\cdot 10^{12}$	$0{,}11\cdot 10^{12}$
Δ	$0{,}13\cdot 10^{-4}$	$0{,}81\cdot 10^{-6}$	$0{,}13\cdot 10^{-7}$	$0{,}12\cdot 10^{-10}$	$0{,}75\cdot 10^{-12}$	$0{,}73\cdot 10^{-15}$
Δ_0	$0{,}7\cdot 10^{-4}$		$0{,}7\cdot 10^{-8}$		$0{,}2\cdot 10^{-9}$	

15. SMALL DEFORMATION OF THE FUNCTION TO BE APPROXIMATED AND ACCELERATION OF SERIES CONVERGENCE

Let the function f(x) be expanded in a series in Chebyshev polynomials $f(x) = \sum_{i=0}^{\infty} c_i T_i$; consider the partial sum of this series:

$$\hat{f}_N(x) = \sum_{i=0}^{N} c_i T_i \tag{63}$$

With the aid of formula (62) it is easy to see that the linear Padé-Chebyshev approximation of form (1) or (52) to the function f(x) coincides with the linear Padé-Chebyshev approximation to polynomial (63) at N = n + 2m, i.e., depends on the first n + 2m + 1 terms of the Fourier-Chebyshev series of the function f(x). A similar result is also valid for the approximations of forms (44) and (45) to even and uneven functions, respectively. We note that at N = n + 2m the polynomial $\hat{f}_N$ is the result of applying the algorithm of the linear or nonlinear Padé-Chebyshev approximation to f(x) but with the substitution of m and n by 0 and 2m + n.

An interesting effect considered in [9] consists in that the error of a Padé approximation $\hat{f}_{n+2m}$ depending on n + 2m + 1 parameters can exceed the error of the respective Padé-Chebyshev approximation of form (1) depending only on n + m + 1 parameters. For example, consider the approximation of form (45) to the function $\tan \frac{\pi}{4} x$ on the segment [-1, 1]. For m = n = 3 the Padé-Chebyshev linear approximation to $\tan \frac{\pi}{4} x$ has an error of an order of 10^{-17}, and the respective polynomial approximation (63) has an error of an order of 10^{-11}. This polynomial of the nineteenth degree**** can be considered as a result of the deformation of the function $\tan \frac{\pi}{4} x$ to be approximated. The above deformation does not change the first twenty terms in the expansion of this function in Chebyshev polynomials and hence does not change the coefficients of the respective rational Padé-Chebyshev approximation, but it increases its error by several orders of magnitude. Thus the small deformation of the function to be approached can result in a drastic change of the error order for the rational approximation.

Also, the above indicated effect signifies that the algorithm "extracts" from polynomial (63) some additional information about the next components of the Fourier-Chebyshev series. In other words, in this case the transition from the Fourier-Chebyshev series to the Padé-Chebyshev approximation accelerates the series convergence. A similar effect of the convergence acceleration for the power series is also observed in the transition to the classical Padé approximation [11, 14, 15].

It can readily be seen that a nonlinear Padé-Chebyshev approximation of form (1) to the function f(x) depends only on the first m + n + 1 terms of the Fourier-Chebyshev series for f(x), so that the above considered effect for such approximations is more pronounced.

Since the "tail" of the Fourier-Chebyshev series can be changed in an arbitrary way without changing the rational Padé-Chebyshev approximation, the effect of convergence acceleration can be reached only for series with very regular behavior and for respective "good" functions.

Note also, that the effect of error autocorrection indicates that in deformation of a more general type the variation of the function to be approximated can have only weak effect on the rational approximation of the function (but the approximation coefficients can substantially change). Accordingly, in deformation of functions that can have good rational approximations the approximation error can drastically increase.

There are interesting results which single out the classes of functions that can have effective rational approximation, e.g., a class of functions that are approximated much better by rational fractions than by polynomials (with convergence acceleration) (see, in particular, [10, 50-52]). The above considerations indicate that of special interest are some "individual" properties of the functions that provide their effective rational approximation. There are grounds to believe that such properties are possessed by solutions of some functional and differential equations. Let us note that it was shown [16, 37] that if to take into account the fact that elementary functions satisfy simple differential equations, then it is clear that these functions are approximated better by rational fractions than by polynomials. (We mean here the best approximations. Because of the technical difficulties we consider here only the case where the denominator of the rational approximation is a linear function or, for even and odd functions, a second-degree polynomial).

16. APPLICATIONS TO CALCULATIONS OF FUNCTIONS ON COMPUTERS

The computation of a function value on a computer finally reduces to the performance of a finite set of arithmetic operations with the argument and constants, i.e., to the computation of a value of a certain rational function. Let us indicate some typical applications of methods for constructing rational approximations. A function f(x) is often computed multiply (e.g., in the process of a numerical solution of differential equations) and with a given accuracy. In this case the construction of a rational approximation to such a function with the given accuracy is often performed with the aid of the most "economical" algorithm for computation of f(x). For example, if f(x) is a complicated set of elementary and special functions each of which can be computed with the help of a respective standard program, then the values of the function f(x) can, of course, be computed by these programs. But such an algorithm is often too slow and yields an excessive accuracy.

In turn, the standard programs for computation of special and elementary functions is usually based on rational approximations. Note here that if rational and polynomial approximations to the given function are computed with the same accuracy, the computation of a rational approximation requires a smaller number of operations, i.e., is executed faster (see, e.g. [1, 3, 12, 13, 24, 25, 31].

The coefficients of rational approximations to the main elementary and special functions can be found in numerous reference books of which we indicate here the fundamental work [3] (see also, e.g., [12, 13]). But a computer can possess some specific features so that the development of efficient standard computational programs for the functions can require algorithms and approximations that cannot be found in reference books. In such cases, the programs for constructing rational approximations including the above described PADE program can be of great help (see Sect.9, and also [1,7,9]).

Thus recently great recognition is received by computers (including calculators) with decimal arithmetics. The point is that the use of decimal arithmetics (instead of the binary system typical for computers) allows one to save computational time spent for transformation of numbers from the decimal system into binary and back from binary to decimal. This is especially important if the input-output operations are rather time-consuming, which is characteristic, e.g., for computations in the dialogue mode.

It is easy to see that the algorithms for computing trigonometric functions are independent of the number system used in a computer. To the contrary, in the implementations of the algorithms for computing exponents, logarithms, and related functions (x^y, hyperbolic and inverse hyperbolic functions) the binary system has some substantial advantages over decimal. Thus in the binary system, the computation of logarithms reduces to the search for an approximation on the segment [1/2,1] and not on [1/10, 1]. The fact that 1/10 is closer to zero than 1/2 results in substantial acceleration of the approximation to logarithm in decimal arithmetics.

But the use of decimal arithmetics creates a certain equilibrium between the difficulties of computing logarithms and exponents, on the one hand, and trigonometric and inverse trigonometric functions, on the other hand. This fact is used in the organization of the computation of elementary functions [1]. The idea of the method is rather simple. The computation of various elementary functions by simple reduction algorithms consists in the computation of a rational function of the given form. Roughly speaking, all the major elementary functions are computed by one formula in which, depending on the function to be computed, only the coefficients can be different. As a result, various programs have a common block which can have, e.g., the form of a separate subprogram. This saves the computer memory and programmers' labor and accelerates computation of elementray functions if this single block is implemented in an optimum way. Thus, in Soviet M5000 minicomputers such a block was implemented with the aid of arithmetics with the fixed point (in M5000 computers such srithmetics works much faster than that with a floating point).

The same method also provides efficient computation of elementary functions on computers with octal and hexadecimal number systems [1].

The computation of major elementary functions on the reduction stage consists in the computation of an odd function on a symmetric interval. Then such an odd function is approximated by a Jacobi fraction of the form

$$R(x) = x\left(\alpha_0 + \cfrac{\alpha_1}{x^2 + \alpha_2 + \cfrac{\alpha_3}{x^2 + \alpha_4}}\right) \tag{64}$$

where x is the reduced argument, α_0, α_1, α_2, α_3, and α_4 are constant coefficients depending on the elementary function to be approximated. The Jacobi fraction (64) corrresponds to a rational approximation of the form

$$R(x) = x \cdot \frac{a_0 + a_1x^2 + a_2x^4}{b_0 + b_1x^2 + b_2x^4}, \tag{65}$$

and the coefficients of (64) and (65) can be expressed in terms of one another if to construct the approximations of form (65). Approximation (64) is used in programs computing functions with a conventional accuracy of eight-nine correct signs. For computing functions with a higher accuracy (12-13 correct signs) an expression similar to (54) but dependent on seven coefficients is used (the reduction algorithms being the same).

Thus the function 10^y can be obtained on a computer with the decimal system in the following way. The argument y is represented in the form $y = x + p$, where p is an integer and $-1 < x < 1$. The computation of the function 10^y reduces to the computation of the values of the rational function R(x) in form (64) with the aid of the formula $10^y \approx 10^p \cdot (1+R(x))/(1 - R(x))$, where R(x) is the approximation to the odd functiton tanh(x ln 10/2). In order to construct this approximation, the first version of the PADE program was used. In a similar way logarithm and other functions are computed.

The above considered method for organization of computations of elementary functions turned out to be very efficient for M5000 computers [1]. This method was also implemented on a decimal Mir computer [53] providing higher speed of function computation than the standard "built-in" software. For example, the inverse trigonometric functions are computed with higher accuracy and 15-20 times faster*****/.

17. NONLINEAR MODELS AND RATIONAL APPROXIMATIONS

One of the main problems of mathematical simulation is the construction of analytical formulae (models) which describe approximately the functional dependence between the quantities using the given "experimental" data on the values of these quantities. In particular, let a set of real numbers x_1, x_2, ...,x_ν, which represent the values of "independent" variable, be given. For each x_i-value of this variable the value y_i of the "dependent" variable is determined. Then the problem consists in the construction of such function $y = F(x)$ that its functional dependence is expressed by an analytical formula of a certain form and the following approximate equality is valid $y_i \approx F(x_i)$ for all $i = 1,2,...,\nu$, the $F(x)$ function should take "reasonable values" for points x lying between the given points x_i. In practice, y_i are given, as a rule, with some errors.

As has already been noted, the computation of the functions on a computer reduces, in the final analysis, to the computation of certan rational functions. Therefore, in many instances it is natural to construct an analytical model in the form of rational function (1) where the degrees of the numerator and denominator and the values of coefficients in (1) are being determined in the process of simulation [14]. Of course, in this case we speak about factor models. Multifactor models can be constructed using rational functions of several variables.

The existence of a simple program for construction of rational approximations to continuous functions set on finite segments of the real axis allows the reduction of the model construction to the construction of rational approximations to continuous functions (although in the numerical analysis one usually reduces continuous problems to discrete ones). A model is constructed in two stages. The first stage is the construction of a continuous function $f(x)$ such that $f(x_i) = y_i$. Depending on the user's choice $f(x)$ can be either a linear or cubic spline. A linear spline is a function whose graph coincides with a broken line consisting of linear segments connecting points (x_i,y_i) with coordinates x_i,y_i; the data on cubic spline can be found, e.g., in [41]. The second stage is the construction of a model by the PADE program. Such an approach provides a regular behavior of the model in the whole range of argument variation.

If there are grounds to believe that the initial data fit a sufficiently smooth and regular curve, then it is expedient to apply a cubic spline. If there are grounds to believe that the initial data contain a large number of errors or deviate from theoretically allowed data, then it is expedient to use a linear spline, since in this case the behavior of a cubic spline at intermediate points will be irregular.

The indicated method for constructing models is implemented in the form of a SPLINE-PADE program (written with the participation of I.A. Andreeva). It provides the output of the (pseudo)graphs of splines and rational approximations (together with the initial data) via a printer, which facilitates the further analysis of the models.

Of course, when choosing and analyzing models one should take into account the theoretical requirements to these models associated with the specific features of the problem.

Example. Let points $x_1, x_2, \ldots, x_\nu$ fill uniformly the segment $[\frac{\pi}{4}, \frac{\pi}{4}]$, $x_1 = \frac{\pi}{4}$. $\nu = 32$, $x_\nu = \frac{\pi}{4}$, $y_i = \cos x_i$. A rational approximation of form (45) to a linear spline at $m = n = 2$ yields the approximation to cos x on $[-\frac{\pi}{4}, \frac{\pi}{4}]$ with the absolute error $\Delta = 10^{-3}$. In this case the use of a cubic spline yields an absolute error of $\Delta = 0.35 \cdot 10^{-6}$.

Other approaches to construction of models in the form of rational functions can be found, e.g., in [14].

*/ The number of the points of control can be substantially reduced (e.g. down to 200 points) without any noticeable deterioration of the error estimate.

**/ The computations were performed by A.Ya. Rodionov.

***/ We mean the coefficients of the numerator and denominator expansions of the approximation in powers of x.

****/ As earlier, small neighborhood of the root of the polynomial ΔQ is excluded.

*****/We mean here odd functions, so that $m = n = 3$ in (45) corresponds to $m = 6$, $n = 7$ in (1).

******/It should be noted that the "built-in" functions provide the computations with any decimal places and therefore are executed rather slowly. The subroutines given in [53] are written for computations with a limited number of decimal places.

REFERENCES

1. G.L. Litvinov, R.I. Tolstykh, L.I. Shadurskaya, and E.A. Dmitrieva. Mathematical algorithms and programs for mini computers. Moscow: Finansy i Statistika, 1981 (in Russian).
2. A.N. Tikhonov and V.Ya. Arsenin. Methods for solving ill-posed problems. Moscow: Nauka, 1979 (in Russian).
3. J.F. Hart et al. Computer Approximations. New York: Wiley, 1968.
4. V.V. Voevodin. Computational foundations of linear algebra. Moscow: Nauka, 1978 (in Russian).
5. Y.L. Luke. Computations of coefficients in the polynomials of Padé approximations by solving systems of linear equations. - J. Comp. and Appl. Math., 1980, 6, No.3, 213-218.
6. Y.L. Luke. A note on evaluation of coefficients in the polynomials of Padé approximants by solving systems of linear equations. - J. Comp. and Appl. Math., 1982, 8, No.2, 93-99.
7. I.A. Andreeva, G.L. Litvinov, A.Ya. Rodionov, and V.N. Fridman. PADE-program for camputation of rational approximants. Foundation of algorithms and programs of the NIVTs AN SSSR. Informational catalogue of the program means of the Foundation of Algorithms and Programs of the NIVTs AN SSSR. Moscow, 1985 p.10-11 (in Russian).
8. A.P. Kryukov, A.Ya. Rodionov, and G.L Litvinov. Construction of rational approximations by REDUCE.- In: Proceedings of the ACM-SIGSAM Symposium on Symbolic and Algebraic Computation (SYMSAC'86). Univ. of Waterloo, Canada, 1986, 31-33.
9. G.L. Litvinov and V.N. Fridman. Approximate construction of rational approximations. Proceeding of the Bulgarian Academy of Sciences, 1983, 36, No.1, 49-52. (in Russian).
10. V.M. Tikhomirov. Theory of approximations. In: Modern problems of mathematics. Fundamental directions. vol.14. (Itogi Nauki i Tekhniki VINITI AN SSSR) Mowcow, 1987, pp.103-260 (in Russian).

11. J.Baker, Jr. and P. Graves-Morris, Padé approximants. Moscow: Mir (in Russian).
12. B.A. Popov and G.S. Tesler. Computation of functions on computers. Kiev: Naukova Dumka, 1984 (in Russian).
13. Y.L. Luke. Special mathematical functions and their approximants. Moscow: Mir (in Russian).
14. V.N. Vinogradov, E.V. Gai, and N.S. Rabotnov. Analytical approximation of the data in nuclear and neutron physics. Moscow: Energatomizdat, 1987 (in Russian).
15. J. Zinn-Justin. Strong interactions in dymamics with Padé approximants. - Physics Reports (Section C of Phys.Lett.) 1971, 1, No.3, 55.
16. V.K. Dzyadyk. Approximational methods for solution of differential and integral equations and their application and development. Preprint 86.31, Institute of Mathematics of the Academy of Sciences of the Ukrainian SSR, Kiev, 1986 (in Russian).
17. N.I. Akhiezer. Lectures on the approximation theory. Moscow: Nauka, 1965 (in Russian).
18. E.Ya. Remez. Foundations of numerical methods for the Chebyshev approximation. Kiev: Naukova dumka, 1969 (in Russian).
19. E.Ya. Remez and V.T. Gavrilyuk. Computational development of several approaches to the approximate construction of the solutions of Chebyshev problems with nonlinearly entering parameters. Ukr. Mat. Zh., 1960, 12, No.3, 324-338 (in Russian).
20. V.T. Gavrilyuk. Generalization of the Remez first polynomial algorithm for the problem of constructing fractional-rational Chebyshev approximations. - Ukr. Mat. Zh., 1964, 16, No.5, 575-585 (in Russian).
21. W. Frazer and J.F. Hart. On the computation of rational approximations to continuous functions. -Comm. of the ACM, 1962, 5, pp.401-493, 414.
22. W.J. Cody, W. Frazer, and J.F. Hart. Rational Chebyshev approximation using linear equations. - Numer. Math., 1968, 12, 242-251.
23. A. Ralston. Rational Chebyshev approximation by Remes' algorithms. -Numer. Math., 1965, 7. No.4, 322-330.
24. A. Ralston. Rational Chebyshev approximation. - In: Mathematical Methods for Digital Computers.Vol.2. Ralston A. and Wilf H. (Eds.), New York: Wiley, 1967, 264-284.

25. W.J. Cody. A survey of practical rational and polynomial approximations of functions. - SIAM Review, 1970, 12, No.3, 400-423.
26. A.A. Kalenchuk-Porkhanova. Algorithms and the analysis of errors of the best Chebyshev approximation to a function of single variable. In: Theory of approximations to functions. Proc. Int. Conf. on Theory of approximations to functions.(eds. S.B. Stechkin and S.A. Telyakovskii) Moscow: Nauka, 1977, 213-218 (in Russian).
27. V.V. Ivanov and A.A. Kalenchuk. On the efficiency of the algorithms of polynomial and fractional rational Chebyshev approximations.- In: Proc. Intern. Conf. on constructive theory of functions. Varna, 1-5 July, 1981.Sofia, 1983, 72-77. 28. Ch.B. Dunham. Convergence of the Fraser-Hart algorithm for rational Chebyshev approximation. - Math. Comp., 1975, 29, No.132, 1078-1082.
29. R.B. Barrar and H.J. Loeb. On the Remez algorithm for nonlinear families. - Numer. Math., 1970, 15, 382-391.
30. L. K. Collatz and V. Crabbs. Apptoximation theory. Chebyshev approximations and their applications. Moscow: Nauka, 1988 (in Russian).
31. K. Spielberg. Representation of power series in terms of polynomials, rational approximations, and continuous fractions. - Journal of the ACM, 1961, 8, 613-627.
32. C.K. Clenshaw and K. Lord. Rational approximations from Chebyshev series. - In: Studies in Numerical Analysis./B.K.P. Scaife (ed.)/ London and New York: Acad. Press, 1974, 95-113.
33. S. Pashkovskii. Computational applications of Chebyshev polynomials and series. Moscow: Nauka, 1983.
34. K. Juhász and G. Németh. Padé approximation and its generalizations. - ATOMKI Közlemények, 1980. 22, No.4, 281-300.
35. G. Németh and M. Zimányi. Polynomial-type Padé approximants. - Math. Comp., 1982, 38, No.58, 553-565.
36. V.K. Dzyadyk. The A-method and rational approximation.- Ukr. Mat. Zh., 1985, 37, No.2, 250-252 (in Russian).
37. V.R. Kravchuk. On an efficient approximation to elementary functions by rational polynomials of order (n,1). - Ukr. Mat. Zh., 1985, 37, No.2, 175-180 (in Russian).

38. K. Lantsosh. Practical methods of applied analysis. Moscow: Fizmatgiz, 1961 (in Russian).
39. E.W. Cheney. Introduction to approximation theory. - New York: McGrow-Hill, 1966.
40. D.S. Lubinsky and A. Sidi. Convergence of linear and nonlinear Padé approximants from series of orthogonal polynomials. - Trans. Amer. Math. Soc., 1983, 278, No.1, 333-345.
41. J. Forsight, M. Malcolm, and K. Mowler, Computer methods of mathematical computations. Moscow: Mir, 1980 (in Russian).
42. R. Fletcher. Expected conditioning. - IMA J. Numer. Anal., 1985, 5, No.3, 247-273.
43. T.R. Hopkins. On the sensitivity of the coefficients of Padé approximations with respect to their defining power series coefficients. - J. Comp. Appl. Math., 1982, 8, No.2, 105-109.
44. L. Wuytack. On the conditioning of the Padé approximation problem. - Lect. Notes Math., 1981, 888, 78-89.
45. H.J. Maehly. Rational approximations for transcendental functions. - Proceedings of the International Conference on Information processing, UNESCO, Butterworths. - London, 1960, 57-62.
46. Collection of scientific programs on FORTRAN. 1,2. Moscow: Statistika, 1974 (in Russian).
47. N.S. Bakhvalov. Numerical methods. Moscow: Nauka, 1973 (in Russian).
48. A.C. Hearn. REDUCE User's Manual. - Rand Publ., 1982.
49. V.F. Edneral, A.P. Kryukov, and A.Ya. Rodionov. Language of analytical computations by REDUCE. p.1. Izd.MGU, 1983. p.2. Izd. MGU, 1986 (in Russian).
50. V.A. Popov and P.P. Petrushev. Exact order of best uniform approximation to convex functions by rational functions. - Mat. Sb., 1979, 7, 103, No.2, 285-291 (in Russian).
51. A.A. Gonchar. On velocity of the rational approximation of analytical functions. - Tr. MI AN SSSR, 1984, 166, 52-60 (in Russian).
52. A.A. Gonchar. The velocity of the approximation by rational fractions and function properties. - In: Proc. Int. Congr. Mathem. Moscow, 1966. Moscow: Mir, 1968, 329-356.
53. R.M. Borisyuk and G.L. Litvinov. Program implementation of algorithms for computating elementary functions on Mir computers.

- In: Mathematical algorithms and programs for small computers. - Moscow: Finansy i Statistika, 1981, 103-109 (in Russian).

54. G. Alefel'd and Yu. Herzberger. Introduction into interval computations. Moscow: Mir, 1987 (in Russian).

Mathematics
and Modelling

ISBN 1-56072-104-9

ON PROBABILISTIC PHENOMENA IN PERTURBED SYSTEMS

A.I. Neishtadt

If a small variation of the initial data in a determinate dynamical system results in a dramatic change of the motion, then the behavior of such a system can hardly be distinguished from random. It is the so-called "small causes large consequences" principle. There are numerous examples of such a quasi-random behavior of dynamical systems in classical mechanics and theoretical physics [1-3]. Below we shall consider a quasi-random behavior of some systems that differ from integrable ones by small disturbances. The probabilistic phenomena in such systems arise in the limit where the perturbation value tends to zero.

1. Scattering by a saddle point and the passage through a separatrix.

Below we consider systems where motion may have different modes (e.g., pendulum oscillations and rotations). A small perturbance gives rise to evolution which may result in the transition from one motion mode to another and quasi-random behavior.

Example 1 [4]. Let us consider a one-dimensional system with the potentiaal energy U(q) (Figure 1) and small friction $\varepsilon f(p,q)$, $0 < \varepsilon << 1$. It is clear that in time almost every point will fall into a potential well - A or B. But into which of them?

The phase portrait of unperturbed ($\varepsilon = 0$) system is shown in Figure 2. Separatrices l_1 and l_2 divide it into regions G_1, G_2, and G_3. On the phase portrait of the perturbed system (Figure 3) the initial conditions that make a point to fall into one or another well alternate (Figure 3). At $\varepsilon \to 0$ the determinate approach to the problem becomes senseless since a small change in the initial conditions may cause the point to fall into another well whereas the initial conditions are always known only within a certain accuracy. In [5], this phenomenon was called scattering at a saddle point singularity.

It is reasonable to consider the fall of a point into one or another well as a random event. The probability $P_1(x)$ of a fall of a point x of the phase plane into a well A is defined as a fraction of points from the small neighbourhood of x that falls into A at $\varepsilon \mid 0$:

$$P_1(x) = \lim_{\delta \to 0} \lim_{\varepsilon \to 0} \frac{\text{mes } \Omega_1(x, \delta, \varepsilon)}{\text{mes } \Omega(x, \delta)}, \tag{1.1}$$

where $\Omega(x,\delta)$ is the δ-neighborhood of the point x, $\Omega_1(x, \delta. \varepsilon)$ is the set of points of this neighborhood that fall into A. $P_2(x)$ is defined in a similar way.

It turns out that if the initial energy $h > 0$, the probabilities of the fall of the point into A and B exist and can be calculated by the formulae

$$P_1 = \Theta_1 / (\Theta_1 + \Theta_2), \quad P_2 = 1 - P_1$$

$$\Theta_\nu = - \oint_{l_\nu} Pf(p,q)dt, \quad \nu = 1,2.$$

The respective integrals are taken along the sepatrices l_ν of the unperturbed system in unperturbed motion. The respective probabilities do not depend on the initial point x and are determined by the values of f(p,q) at the critical value of energy, $h = 0$.

In the problem under consideration, of ineterst is the variation of energy h with time. Far from separatrices of the unperturbed system such energy variation can be approximately described with the aid of averaging: the equations of perturbed motion in the "action-angle" variables of the unperturbed system should be averaged over angles; instead of the variable "action" one can use the energy h [6]. But on separatrices the equations of perturbed motion have singularities. Then how one can describe the transition through separatrices?

Figure 4 illustrates three solutions of an averaged system. The solution $h_3(\varepsilon t)$ in the region $G_3 = \{h>0\}$ starts at $t = 0$, where it has the value equal to the energy of the initial point x, and then, at a certain $t = t_*$, goes to the separatrix ($h = 0$). The $h_1(\varepsilon t)$ and $h_2(\varepsilon t)$ solutions are glued to the above considered solution in the regions G_1 and G_2, respectively, they start at $t = t_*$ on the separatrix: $h_{1,2}(\varepsilon t_*) = 0$. It turns out that the motion of most points that fall into the well A (or B, respectively) for times $1/\varepsilon$ is approximately described by gluing of the solution h_3 and h_1 (or h_3 and h_2, respectively). The exception is made by the points that go too close to the saddle C and remain there for quite a long time; the measure of their set tends to zero at $\varepsilon \to 0$. It can be said that in this case the motion is described with probability P_1 by gluing of h_3 and h_1 and with probability P_2 by gluing of h_3 and h_2. A similar loss of determinacy occurs if to substitute friction by a smooth variation of the potential resulting in falling of the point into one or another well (e.g., in case where a point is moving along the curve with two minima (Figure 1) in a smoothly increasing gravitational field).

Let us consider a general situation in which such phenomena may arise [7]. There is a set of equations

$$\dot{p} = -\frac{\partial H}{\partial q} + \varepsilon f_1, \quad \dot{q} = \frac{\partial H}{\partial p} + \varepsilon f_2, \quad \lambda = \varepsilon f_3, \tag{1.2}$$

where $0 < \varepsilon < 1$, $(p,q) \in R^2$; $\lambda \in R^m$; $H = H(p,q,\lambda)$, $f_i = f_i(p,q,\lambda,\varepsilon)$. The functions f_i are assumed to be infinitely differentiable. An unperturbed ($\varepsilon = 0$) system for p and q is Hamiltonian with the Hamiltonian dependent on the parameter λ. For definiteness we shall take that the phase portrait of the unperturbed system for all λ is of the same form as in Example 1 in Figure 2 (but it is also possible to consider other potraits with saddle nondegenerate singular points connected by separatrices).

The separatrices l_1 and l_2 divide an unperturbed phase plane into three regions: G_1, G_2, and G_3 (Figure 2). Under the effect of perturbation, the points from the region G_3 intersect the separatrix and are captured either by G_1 or G_2. The capture of points into one or another region had to be considered as a random event. The probabilities of such events for the initial point $x = (p, q, \lambda)$ are given by relationship (1,1).

The averaged system in each of the regions has the form

$$\dot{h} = \frac{\varepsilon}{T} \oint \left[\frac{H}{p} f_1^0 + \frac{H}{q} f_2^0 + \frac{H}{\lambda} f_3^0 \right] dt \tag{1.3}$$

$$\dot{\lambda} = \frac{\varepsilon}{T} \oint f_3^0 dt,$$

where $T = \oint dt$ is the period of unperturbed motion, $f_i^\circ = f_i(p, q, \lambda, 0)$, $i = 1, 2, 3$, the integrals are taken along the solution of the unperturbed system where $H(p , q, \lambda) = h$.

We assume that the Hamiltonian H is normalized in such a way that at the saddle point (and therefore on sepratrices) $H = 0$. Let us define the quantities

$$\Theta_\nu(\lambda) = - \oint_{l_\nu} \left[\frac{\partial H}{\partial p} f_1^0 + \frac{\partial H}{\partial q} f_2^0 + \frac{\partial H}{\partial \lambda} f_3^0 \right] dt, \quad \nu = 1,2, \tag{1.4}$$

where integrals are taken along separatrices in unperturbed motion. These are improper integrals since the motion along the separatrix takes infinite time, but, as can readily be verified, for the chosen normalization of the Hamiltonian they converge. Below we take the quantities Θ_ν to be positive. The value $(-\Theta_\nu)$ is close to the energy variation on a segment of the perturbed trajectory close to the unperturbed separatrix. Therefore for most initial conditions the positiveness of Θ_ν guarantees sufficiently fast approach to it in the region G_3 and sufficiently rapid removal from the separatrix in the regions G_1 and G_2. The following statements about the motion of a point at $0 \le t \le 1/\delta$ are valid for the initial condition (p_o, q_o, λ_o) where $(p_o, q_o) \in G_3$ at $\lambda = \lambda_o$.

1°. Let us denote by $h_3(\varepsilon t)$, $\lambda_3(\varepsilon t)$ the solution for an averaged system in the region G_3 with the initial condition $(H(p_o, q_o, \lambda_o), \lambda_o)$. Let this solution at certain $t = t_* < 1/\varepsilon$ go into the separatrix: $h_3(\varepsilon t_*) = 0$. Then at $0 \le \varepsilon t \le \varepsilon t_*$ the change in H, λ along the real motion is described by the solution h_3, λ_3 with the accuracy $0(\varepsilon)$.

2°. Let λ_* be a value of the parameter λ at the moment when the average solution goes into the separatrix: $\lambda_* = \lambda_3(\varepsilon t_*)$. Then the probabilities of the capture of the point $x = (p_o, q_o, \lambda_o)$ by regions G_1 and G_2 are calculated by the formulae

$$P_1(x) = \frac{\Theta_1(\lambda_*)}{\Theta_1(\lambda_*) + \Theta_2(\lambda_*)}, \quad P_2(x) = 1 - P_1(x). \tag{1.5}$$

3°. Let $h_1(\varepsilon t)$, $\lambda_1(\varepsilon t)$ and $h_2(\varepsilon t)$, $\lambda_2(\varepsilon t)$ be the solutions of the averaged system in the regions G_1 and G_2 with the initial condition lying on the "separatrix" that are glued to h_3: $h_{1,2}(\varepsilon t_*) = 0$, $\lambda_{1,2}(\varepsilon t_*) = \lambda_*$ (see Figure 4). Then for most initial points x captured by the region G_v, the change of H, λ along the motion at $\varepsilon t_* \le \varepsilon t \le 1$ is described by the solution h_v, λ_v with the accuracy $O(\varepsilon\,|\ln\,\varepsilon|\,)$. The measure of the exceptional set of the initial points whose motion cannot be described in such a way tends to zero with a decrease in ε faster than any power of ε (it is said that there exists $O(\varepsilon^\infty)$).

Thus in order to describe the motion one should use the averaged system up to the attainment of the separatrix, caclulate the probability for a point to be captured by one or another region on the separatrix, and then use the averaged system again starting with the separatrix in the region in which the captured point is located. Such a scheme was first used in the analysis in [5].
Similar proofs of statements 1°-3° can be found in [7]*.

Some comments to the derivation of probability formulae (1.5). In the region G_3, the phase point describes almost closed curves, returning repeatedly on a ray $C\xi$ which is the bisectrix of the angle formed by the separatrices at the saddle point (Figure 5). We denote by h' the energy value of the point at the moment of its last outgoing into this ray; in this case it is obvious that $\lambda \;|\approx \lambda_*$ (for most initial conditions an exception here is made by the points remaining for a long time in the vicinity of the saddle). Let us introduce the notation: $\Theta_\nu^* = \Theta_\nu(\lambda_*)$, $\nu = 1, 2$; $\Theta_3^* = \Theta_1^* + \Theta_2^*$. The further motion of the point is determined by the value of h'.

<u>Proposition</u>. If $k\varepsilon^{3/2} < h' < \varepsilon\Theta_2^* - k\varepsilon$, then the point is captured by $G_2^{\cdot}$. It is necessarily that $h' < \varepsilon\Theta + k\varepsilon$. Here $k = \text{const} > 0$.

□ Figure 5 shows a segment of the perturbed trajectory l_2' close to l_2 and leading to the capture of the point by the region G_2. The energy variation along this segment is equal approximately to $(-\Theta_2^*)$. A similar energy variation is also observed along the segment of the trajectory l_3' close to $l_1 \cup l_2$ and leads to the capture by G_1; it is equal approximately to $(-\varepsilon\Theta_3^*)$. It can be shown that the error of these approximations is of the order of $\varepsilon^{3/2}$. Hence, within the small quantities of higher orders, we have the following picture. The value of h' belongs to the segment $d_3 = (0, \varepsilon\theta_3^*)$. If h' turned out to be at the segment $d_1 = (0, \varepsilon\Theta_1^*) \subset d$, then the point is captured by G_1, and if it is on $d_2 = d_3 \backslash d_1$, then by G_2.

The flow of the phase volume through the segments d_1 and d_2 is approximately equal to their lengths**. Therefore it is natural that the volumes of the sets of phase points from the neighborhood of the initial point x that flow through these segments are proportional to the segment lengths. In accordance with probability definition (1.1) we have

$$P_1 = \frac{\text{length of } d_1}{\text{length of } d_3} = \frac{\Theta_1^*}{\Theta_1^* + \Theta_2^*}, \quad P_2 = 1 - P_1,$$

which had to be proved. □

The above consideration allows one to calculate the probabilities in all similar problems (for other types of phase portraits and other combinations of the signs of quantities Θ_ν).

The probability of a capture depends on the initial point through the parameter value λ only at the moment of its outgoing to the separatrix. It is natural to consider the quantities

$$Q_1(\lambda) = \frac{\Theta_1(\lambda)}{\Theta_1(\lambda) + \Theta_2(\lambda)}, \quad Q_2(\lambda) = 1 - Q_1(\lambda)$$

as the probabilities of capture by G_1 and G_2 at the moment of the outgoing to the separatrix with the given value of λ. These values can be calculated independently of the initial point. Then the probabilities for the given intial point to be captured are calculated by the formula

$$P_\nu(x) = Q_\nu(\lambda_*), \; \nu = 1,2.$$

Let us mention some problems where the above indicated phenomena can occur:

- the formation of the 3:2 resonance between the axial and orbital rotation of Mercury [8];
- the tide theory of the formation of pair-orbital resonances in the system of Saturn satellites [9, 12, 34];
- the formation of the Kirkwood gaps at the 3:1 resonance in the main zone of asteroids [13. 14].
- scattering of quasiparticles (which are conduction carriers in metal and semiconductors) in electromagnetic fields [5];
- propagation of short radiowaves in wave-guide channels of the ionosphere [15];
- the motion of charged particles in a field of evolutioning wave [16, 17];
- electronic cyclotron resonance heating of plasma in magnetic traps [18, 19];
- the somersault of a solid with an immobile point under the action of small perturbations [20];
- the evolution in the Lorentz system with large Rayleigh numbers [21].

Let us consider the next example in more detail.

Example 2 [8, 16]. The equation of the pendulum motion under the action of small perturbations (a constant external moment, dissipative moment, and a smooth frequency variation) has the form

$$\ddot{x} + \omega^2(\lambda) \sin x = -\varepsilon[L + K\dot{x}], \ \dot{\lambda} = \varepsilon; \ L, K = \text{const}.$$

The phase cylinder of an unperturbed problem is shown in Figure 6. The separtrices l_1 and l_2 divide the phase portrait into the regions of rotation, G_1, and counterrotation, G_2, and oscillations G_3. Let $L \geq 0$, $K \geq 0$, $\omega' = d\omega/\delta\lambda \geq 0$, $L^2 + K^2 \neq 0$. Then the points from the region of direct rotation under the effect of perturbation approach the separatrix and either are captured by oscillations, or start experiencing counter-rotation. The above introduced integrals along the separtrices can readily be calculated:

$$\Theta_{1,2} = 8K\omega + 8\omega' \ \ 2\pi L.$$

If $\Theta_2 < 0$, then the probability of the transition from the rotation to oscillations is given by the formula $Q = (\Theta_1 + \Theta_2)/\Theta_1$.

If $\Theta_2 > 0$, this probability equals unity (since the transition to counterrotations is impossible). Finally we have

$$Q = \begin{cases} \dfrac{8(\omega' + K\omega)}{4(\omega' + K\omega) + \pi L}, & L > 4(\omega' + K\omega)/\pi \\ 1, & L \leq 4(\omega' + K\omega)/\pi. \end{cases} \tag{1.6}$$

The motion in each region up to the respective separatrices is described by the averaged equation.

Using formulae (1.6) with $\omega' = 0$ [8], it is possible to explain the origin of the above mentioned resonance in the Mercury rotation.

Note 1. In the above considered problems, of interest are also multiple passages through the separatrix but, unfortunately, they have not been studied as yet. In particular, let the disturbance in Example 1 be dependent on slow time $\lambda = \varepsilon t$ and change its sign periodically. Then random walks arise: the point returns back into the region G_3 and then is captured again either by the region G_1 or the region G_2. What is the probability of being captured two times in a row by region G_1? Is not it equal to $Q_1(\lambda_*)Q_1(\lambda_{**})$ (where λ_* and λ_{**} are sequential moments of slow time of the outgoing of the trajectory point to the separatrix calculated for the average system) so that the sequential captures can be considered as independent random events? It would be very important to know the answers to these questions.

Note 2. A particular case of the systems of form (1.2) is Hamiltonian systems smoothly dependent on time ($H = H(p,q,\lambda)$, $\lambda = \varepsilon t$) and Hamiltonian systems with fast and slow motion ($H = H(p, q, y, x)$; p and y are pulses, and q and $\varepsilon^{-1}x$ are coordinates). The variable called "action" of the unperturbed system (λ = const or y, x = const) is an integral of the averaged system and, therefore, also an adiabatic invariant of the initial system. The asymptotic behavior of the variation of the adiabatic invariant in the transition through the separatrix was calculated in [22-26]***. Summation of these quasi-random variations in the multiple transitions results in diffusion of the adiabatic invariant. The results of the computer experiment for such diffusion can be found elsewhere [27].

2. CAPTURES BY RESONANCES

The motion of a multyfrequency integrable system under the action of small perturbations can often be described by the differential equations of the form [1]:

$$\dot{I} = \varepsilon f(I,\varphi,\varepsilon), \quad I \in R^n$$

$$\dot{\varphi} = \omega(I) + \varepsilon g(I,\varphi,\varepsilon), \quad \varphi \in T^m.$$

Here a small positive parameter ε characterizes the perturbation magnitude. Under sufficiently general assumptions, the approximate description of the evolution of slow variables I at times of the order of $1/\varepsilon$ for most initial conditions can be made with the use of the following averaged system

$$\dot{J} = \varepsilon F(J),$$

where F(J) is the average of the function $f(I, \varphi, 0)$ over the torus T^m. But such a description can be used for all the initial data which is associated, in particular, with the captures by resonances. In the course of evolution, frequencies ω slowly vary and at some moments of time an integer commensurability of frequencies arises - a resonance. For some relationships between the phases, a capture can occur - the point continues moving in such a way that the above commensurability is preserved. Such a motion cannot be described by averaging the motion over the whole torus T^m: the solution of the averaged system and the I-component of the solution of the initial systems diverge for time $1/\varepsilon$ by a value of the order of unity. At $\varepsilon \to 0$, the initial conditions for the trajectories with the capture by the resonance tend to locate densely everywhere in the phase space (see Figure 34 in [28]). Therefore we have to consider a capture as a quasi-random phenomenon.

The measure of the set of the initial data for the phase points captured within the time interval $0 \le t \le 1/\varepsilon$ in the typical cases turns out to be a quantity of the order of $\sqrt{\varepsilon}$ [28, 29]. But the measure of the set of the initial data for the phase points experiencing capture within the time interval $0 \le t \le 1/\varepsilon^{3/2}$ can turn out to be a value of the order of unity. Indeed, if a beam of trajectories multiply intersects a resonance surface and if in each intersecting the fraction of the order of $\sqrt{\varepsilon}$ of the phase points is captured by the resonance (probably only for some time), then after $1/\sqrt{\varepsilon}$ intersections a substantial fraction of the beam (of the order of unity) will experience capture. A respective example is considered in the next section.

2.1. An Example of Captures in Hamiltonian System

We shall construct a perturbed Hamiltonian system possessing the following property. The averaged system has an integral whose value in the perturbed motion changes by a value of the order of unity for a set of the initial data of the measure of the order of unity within the time of the order of $1/\varepsilon^{3/2}$ because of multiple passages through a resonance and captures of points by this resonance. In the region where there is no resonances such an integral of the averaged system is an adiabatic invariant: its value in the perturbed motion experiences only oscillations $O(\varepsilon)$ within times of the order of $1/\varepsilon^2$ (and even much longer times).

As usual in such examples, the problem of the resonance in a two-frequency system with the nonzero frequencies is reduced to an equivalent problem about the resonance (inversion of the zero frequency) in a one-particle system.

Let us consider in the region $1 < R < 4$ a set of equations

$$\dot{R} = \varepsilon b(\lambda) \sin \varphi, \quad \dot{\lambda} = \varepsilon, \quad \dot{\varphi} = R - a(\lambda). \tag{2.1}$$

Here R and λ mod 2π are polar coordinates on the plane of slow variables and a and b are triply continuously differentiable functions of λ having the period 2π, $2 < a < 3$. This Hamiltonian system with one degree of freedom (with respect to R and φ) dependent on the smoothly varying parameter λ. Such systems arise in the study of adiabatic invariants. The averaged system is of the form

$$\dot{R} = 0, \quad \dot{\lambda} = \varepsilon.$$

The motion in such a system occurs over the circumferences R = const with the angular velocity ε (Figure 7).

Such a system has a resonance - the frequency of the variation of the phase φ goes to zero at $R = a(\lambda)$. The trajectories of the averaged system intersect the resonance surface (in our case, a curve).

Let us study the capture of points by the resonance and their outgoing from it. Let us rewrite the system in the form of the pendulum equation with a torque and smoothly varying parameters

$$\frac{d^2\varphi}{d\tau^2} = b(\lambda) \sin \varphi - a(\lambda), \quad \frac{d\lambda}{d\tau} = \sqrt{\varepsilon}, \ \tau = \sqrt{\varepsilon}\ t.$$

Here a prime denotes a derivative with respect to λ. Such a reduction of the problem has already been used in a series of studies [30-33].

The phase portrait of unperturbed (λ = const) pendulum at $|a'(\lambda)| > |b(\lambda)|$ is shown in Figure 8a, and for $|a'(\lambda)| < |b(\lambda)|$ in Figure 8b (for the case $a' > 0$). The phase portrait shown in Figure 8b has a saddle singular point and a separatrix loop passing through it and limiting the oscillation region; the area of this region is denoted as $S(\lambda)$.

The capture by the oscillation region (i.e., the capture by resonance) occurs at those points of the resonance curve where $S'(\lambda) > 0$, and the outgoing from the oscillation region (i.e. the outgoing from the resonance) occurs at points where $S'(\lambda) < 0$ [28].

Since (2.2) is a Hamiltonian system with smoothly varying parameters, the "action" variable, I, in the oscillating region of unperturbed pendulum (Figure 8b) is an adiabatic invariant of perturbed motion [1]. According to Sect. 1, the adiabatic invariance of I can be used for the description of the motion in the oscillation region till the separatrix. Let us also take into account that the "action" variable for the point from the oscillation region is an area limited by the unperturbed trajectory (passing through this point) divided into 2π. Therefore in the adiabatic approximation the moments of going to the separatrix and outgoing from it, $I = I_o = \text{const}$, are determined from the relationship $S(\lambda) = 2\pi I_o$. Hence, if the capture of the point by the resonance occures at $\lambda = \lambda_*$ and $S'(\lambda_*) > 0$, then the exit from the resonance will occur at $\lambda = \lambda_{**} > \lambda_*$,

$$S(\lambda_{**}) = S(\lambda_*),\ S'(\lambda_{**}) < 0 \text{ with}$$
$$S(\lambda) > S(\lambda_*) \text{ at } \lambda \in (\lambda_*, \lambda_{**}).$$

The solutions of the initial perturbed system with the portions where points are captured by the resonance are described by gluing of the segments of the circumferences $R = \text{const}$ and the segments of the resonance curve $R = a(\lambda)$; the motion along these segments occur by the law $\lambda| = \varepsilon$. The moment when the point leaves the resonance curve is determined by the moment of reaching it with the aid of the adiabatic approximation and the analysis of the function $S(\lambda)$, as was described earlier. Let us consider the accuracy of this approximation. Let U be a circle on the plane R, λ. Let the singular point of the pendulum (if at all) for the above described approximate solutions with the initial conditions from $U \times S^1$ at $0 \le \varepsilon t \le 1$ be nondegenerate at the moment of the attainment of the resonance curve and at the moment of leaving it $|b(\lambda)| > c_1^{-1}$, $|a'(\lambda) - |b(\lambda)|| > c_1^{-1}$; here C_1 and C_i (later on) are positive constants. Then for most initial conditions $(R_o, \lambda_o, \varphi_o) \in U \times S^1$ at $0 \le t\ 1/\varepsilon$ along the solution of the initial system (with the accuracy

$0(\sqrt{\varepsilon}|\ln\ \varepsilon|))$ we have that either $R = \text{const}$ or the variation in R is described by one of the above described glued solutions. The exceptional set of the initial conditions for which such estimates are invalid has the measure $O(\varepsilon)$. These estimates are a result of the union of estimates made in [12] and [33] (see also [7] and Sect. 1).

In the regions $1 < R < 2$ and $3 < R$ 4 (where there is no resonance) R experiences along the motion at $0 \le t \le 1/\varepsilon$ only oscillations $O(\varepsilon)$.

We shall take that the functions a and b and numbers $\lambda_1, \lambda_2 \in (0, \pi)$ and $\lambda_3 \in (\pi, 2\pi)$ possess the following properties:

a) a increases monotonically at the segment $(0, \pi)$ and decreases monotonically at the segment $(\pi, 2\pi)$,

b) $\lambda_1 < \lambda_2$, $S(\lambda_1) = S(\lambda_2)$; $S'(\lambda_1) > 0$. $S'(\lambda_2) < 0$, $S(\lambda) > S(\lambda_1)$ at $\lambda \in (\lambda_1, \lambda_2)$,

c) $a(\lambda_3) = a(\lambda_2)$, $a'(\lambda_{1,2}) > 0$, $a'(\lambda_3) < 0$,

d) $a'(\lambda_{1,2}) < |b(\lambda_{1,2})|$, $|a'(\lambda_3)| > b(\lambda_3)$.

An example of functions a and b is shown in Figure 9. The function a in Figure 9 is linear in the vicinity of the segment $[\lambda_1, \lambda_2]$; therefore in this neighbourhood the area of the oscillation region S can be represented as monotonically increasing function of the value $|b(\lambda)|$.

From the properties (a) and (b) it follows that on the resonance curve the capture by the resonance occurs at λ close to λ_1, and at λ close to λ_2 the point goes out from the resonance, the points captured at $\lambda \approx \lambda_1$ outgo from the resonance at $\lambda \approx \lambda_2$, $R = a(\lambda_2)$. The bold curve in Figure 7 indicate the trajectory portion where the capture occurs. The circumference $R = a(\lambda_2)$ intersects the resonance curve also at $\lambda \approx \lambda_3$. In virtue of property (d), at λ close to λ_3 no capture by resonance is possible.

Let us introduce the notation $R_i = a(\lambda_i)$, $i = 1, 2, 3$. Let us show that for all the points $(R_o, \lambda_o, \varphi_o)$ such that $|R_o - R_2| < c_2^{-1}$ in shift occurring in virtue of (2.1) for time π/ε for the motion within the region $|R - R_2| < c_2^{-1}$ the following condition is fulfilled

$$R(t) \ge R_0 - c_3 \sqrt{\varepsilon}\ , \tag{2.3}$$

where $R(t)$ is the value of R along the trajectory at the moment of time t.

The procedure of the error averaging from [39] yields the estimate $R(t) = R_o + O(\sqrt{\varepsilon}) + \alpha$, where α is the change in R on the connected time interval within which the point moves in the region

$|R - a(\lambda)| = O(\sqrt{\varepsilon})$, $|R - R_2| < c_2^{-1}$, $0 \leq \lambda \leq \pi$.

At this time interval $a(\lambda)$ increases whence follows relationship (2.3).

The above given description of the capture by a resonance shows that in the region $|R - R_1| < c_4^{-1}$ there is a set W of the points of a mesure larger than $c_5^{-1}\sqrt{\varepsilon}$ which in the shift for time π/ε, in virtue of system (2.1), turn out to be in the region $|R - R_2| < c_4^{-1}$, $c_4 > 2c_2$ and $c_4 > 4(R_2 - R_1)$. From (2.3) it follows that in the further motion these points remain in the range $R > R_1 + c_4^{-1}$ for time c_5^{-1}/ε.

Let us denote by W_j the set obtained from W_o due to the shift in virtue of system (2.1) for time $(-j\pi/\varepsilon)$, $j = 0, 1, \ldots, N$, $N = [c_5^{-1}\pi^{-1}/\sqrt{\varepsilon}] - 1$. The previous estimates show that the sets W_j do not intersect one another. Therefore for $W = \bigcup_{j=0}^{N} W_j$ we have mes $W = \sum_{j=0}^{N}$ mes $W_j > c_6^{-1}$.

For the initial data from W the solution on the time interval $0 \leq t \leq c_5^{-1}/\varepsilon^{3/2}$ has at least one portion where R increases by more than c_4^{-1} for time $O(1/\varepsilon)$. Thus, we have constructed for system (2,1) a set of the initial data of the measure of order of unity for which the integral R of the averaged system changes by a value of the order of unity due to the capture by the resonance for time of the order $1/\varepsilon^{3/2}$.

In the regions $1 < R < 2$ and $3 < R < 4$ (where there is no resonances) the value of R experiences along the motion only oscillations of the order of ε at the times of the order of $1/\varepsilon^2$, and, if the functions a and b are infinitely differentiable, then also at times of the order of $1/\varepsilon^l$ at any $l > 2$.

<u>Note.</u> Let us consider along with (2.1) the following Hamiltonian system

$$\dot{R} = \varepsilon h(R) b(\lambda) \sin \varphi, \quad \dot{\lambda} = \varepsilon, \quad \dot{\varphi} = R - a(\lambda) + \varepsilon \frac{dh}{dR} b(\lambda) \cos \varphi \qquad (2.4),$$

where h(R) is doubly continuously differentiable function equal identically to unity at $2 \leq R \leq 3$ and to zero at $1 \leq R \leq 1.5$ and $3.5 \leq R \leq 4$. The region $1.5 \leq R \leq 3.5$ for system (2.4) is invariant. By the Poincaré theorem of return, almost all the points with the initial conditions from the above constructed set **W** have to return, sooner or later, to the initial value of R. Therefore the trajectories of all these points should have after the capture portion a portion of length not shorter than c/ε where no capture by the resonance occurs but R changes by a value of the order of unity due to its multiple passages through the resonance.

2.2. An Example of Captures in a System with Dissipation

Multiple passages through the resonance in the systems with dissipation can also result in quite unusual phenomena. In particular, an averaged system on a plane has two asymptotically stable limit cycles around the origin. The points captured by the resonance pass along two branches of the resonance curve from the larger cycle into the attraction region of the smaller one and vice versa (Figure 10, the respective equations are written below). Then in the perturbed system there occur quasi-random walks between the limit cycles of the averaged system. An example of an appropriate location of the cycles and the branches of the resonance cúrve is given by the system of equations:

$$\dot{R} = \varepsilon(R - R_0)\{-(R - R_1)(R - R_2)(R - R_3) + b(R)\cos\lambda\sin\varphi\}$$

$$\dot{\lambda} = \varepsilon\{1/2 - 1/2\sin\varphi\,[a_1(R)(1 + \cos\lambda) + a_3(R)(1 - \cos\lambda)]\}$$

$$\dot{\varphi} = \sin\lambda$$

$$a_1(R) = \frac{(R - R_2)(R - R_3)}{(R_1 - R_2)(R_1 - R_3)} + 2\frac{(R - R_1)(R - R_3)}{(R_2 - R_1)(R_2 - R_3)}$$

$$a_3(R) = 2\frac{(R - R_1)(R - R_3)}{(R_2 - R_1)(R_2 - R_3)} + \frac{(R - R_1)(R - R_2)}{(R_3 - R_1)(R_3 - R_2)} \qquad (2.5)$$

$$b(R) = 2(a_1(R) + a_3(R) + 1)(R_3 - R_1)^3$$

$$R_i = \text{const} > 0,\ R_0 < R_1 < R_2 < R_3.$$

Here R and λ mod 2π are polar coordinates on the plane of slow variables, the motion is considered in the invariant region $R \geq R_o$. The parabolas $y = a_1(R)$ and $y = a_3(R)$ are shown in Figure 11.
The averaged system has the form

$$\dot{R} = -\varepsilon(R - R_0)\{-(R - R_1)(R - R_2)(R - R_3), \; \dot{\lambda} = \varepsilon.$$

For this system the circumferences $R = R_1$ and $R = R_3$ are asymptotically stable limit cycles, the circumferences $R = R_o$ and $R = R_2$ are unstable cycles.

The resonance curve (a straight line in our case) has two branches: $\lambda = 0$ and $\lambda = \pi$. In the region $|\lambda| = O(\sqrt{\varepsilon})$ the motion is described by a set of equations

$$\varphi'' = 1/2 - a_1(R) \sin \varphi + O(\varepsilon)$$

$$R' = \sqrt{\varepsilon}\, (R - R_0)\{-(R - R_1)(R - R_2)(R - R_3) + b(R) \sin \varphi + O(\varepsilon)\}$$

where a prime denotes the derivative with respect to $\tau = \sqrt{\varepsilon}\, t$.

Unperturbed ($\varepsilon = 0$, R = const) pendulum equation for φ has a oscillation region at $|a_1, (R)| > 1/2$. The variation in R in a perturbed motion for points captured by this region is described within the accuracy $O(\sqrt{\varepsilon}|\ln \varepsilon|)$ at times τ $1/\sqrt{\varepsilon}$, by averaging over oscillations of the unperturbed pendulum:

$$R' = \sqrt{\varepsilon}\, (R - R_0)\{-(R - R_1)(R - R_2)(R - R_3) + 1/2\; b(R) / a_1(R)\}.$$

Here the account is made of the fact that the average of sin φ over oscillations is $(1/2)\, a_1^{-1}(R)$. From the form of b(R) in (2.5) it follows that for the captured points $R' > 0$ at $R_1 \leq R < R_3$.

The motion of points in the oscillation region (i.e. in the resonance), the capture by it and the liberation from it are described with the aid of the adiabatic approximation, as in Sect.2.1. The points captured by the resonance at $R \approx R_1$, $\lambda \approx 0$ are moving along the ray $\lambda = 0$ towards increasing of R and leave the resonance at $R \in (R_2, R_3)$.

In a similar way the branch of the resonance line $\lambda = \pi$ is considered. The points entered the resonance at $R \approx R_3$, $\lambda \approx \pi$ are moving along the ray $\lambda = \pi$ towards smaller R values and leave the resonance at $R \in (R_1, R_2)$.

A typical phase point of the constructed system is for most of its time in the vicinity of asymptotically stable cycles of the averaged system. It experiences some hoppings between them in the time intervals that seems to be quasi-random. The average length of these intervals is presumably of the order of $1/\varepsilon^{3/2}$, the transition between the cycles takes the time of the order of $1/\varepsilon$.

2.3 On Adiabatic Invariants of Multifrequency Hamiltonian Systems

Let us consider a Hamiltonian system dependent on the parameter λ which, at each fixed parameter value, is quite integrable and admits the variables "action-angle" , I, φ. If a parameter varies smoothly with time ($\lambda = \varepsilon t$), then the variation of I, φ is described by a Hamiltonian system with the Hamiltonian of the form $H = H_0(I,\lambda) + \varepsilon H_1(I,\varphi,\lambda)$ [1]. The respective averaged system has the form $\dot{J} = 0$, i.e., in the approximation of the averaging method there is no evolution of the "action" variables.

If a system has one degree of freedom and $\partial H_0/\partial I \neq 0$, then in the perturbed motion I varies only by $O(\varepsilon)$ for the time $1/\varepsilon$; such values are called adiabatic invariants [1]. Further on, if the dependence of the Hamiltonian on λ is periodic, then, according to the Arnol'd theorem of perpetual adiabatic invariance [4], in a nonlinear system the quantity I remains perpetually in the $O(\varepsilon)$, the neighborhood of its initial value (the nonlinearity condition: $\partial^2 H_0/\partial I^2 \neq 0$, where H is the average of H_0 over λ).

For two and more degrees of freedom new phenomena arise. For I $\approx$ const, $\partial H_0(I,\lambda)/\partial I$ slowly varies, because of the variation in λ, and the system goes to resonances. Due to the capture by resonance the "action" I for some initial conditions can deviate for time $1/\varepsilon$ from its initial value by about unity. In the general case, the measure of the set of such initial conditions is small (of the order of $\sqrt{\varepsilon}$) so that the "action" is said to be an almost adiabatic invariant [26]. But because of entering the resonance and numerous passages through it, the value of the "action" can change for time $1/\varepsilon^{3/2}$ by a value of unity for a substantial (of the measure of the order of unity) set of initial conditions. The aim of the present consideration is the construction of a respective example.

Let the initial Hamiltonian have the form

$$H = 1/2(p_1^2 + p_2^2) + 1/2\ \omega^2(\lambda)Q_1^2 + V(Q_2), \text{ where}$$
$$Q_1 = q_1 \cos \alpha(\lambda) + q_2 \sin \alpha(\lambda),$$
$$Q_2 = -q_1 \sin \alpha(\lambda) + q_2 \cos \alpha(\lambda),$$
$$V(Q_2) = 1/2\ Q_2^2 + 1/4\ Q_2^4.$$

Here p_1, p_2, q_1, and q_2 are the conjugated canonical variables, α and ω are still undefined sufficiently smooth 2π periodic funtions of λ. Introducing the coordinates Q_1 and Q_2 and the conjugated pulses P_1 and P_2, at λ = const, we can rewrite the Hamiltonian in the form

$$H = H_1 + H_2,$$
$$H_1 = 1/2\ P_1^2 + 1/2\ \omega^2 Q_1^2,\ H_2 = 1/2\ P_2^2 + V(Q_2).$$

The variables are separated and the system can be integrated. Let I_j, φ be the action-angle variables for the Hamiltonian H_j, $j = 1,2$. For definiteness, we take that the angle φ_j is measured from the point where $Q_j = 0$ and $P_j > 0$. We have

$$H = \omega I_1 + E(I_2),\ I_1 > 0,\ I_2 > 0,$$

$$P_1 = \sqrt{2I_1\omega}\ \cos \varphi_1,\ Q_1 = \sqrt{2I_1/\omega}\ \sin \varphi_1,$$
$$P_2 = P_2(I_2, \varphi_2),\ Q_2 = Q_2(I_2, \varphi_2),$$

$$Q_2(I_2, \varphi_2) = \sum_{k=0} f_k\ (I_2) \sin((2k + 1)\varphi_2),$$
$$P_2(I_2, \varphi_2) = \Omega(I_2)\partial Q_2/\partial\varphi_2,$$

where E is the Hamiltonian H_2 expressed in terms of I_2, and $\Omega = \delta E/\partial I_2$ is the frequency of the unperturbed motion for H_2. The functions E, Q_2, and P_2 are expressed in terms of the elliptic integrals and elliptic functions. Their explicit form is not important; what is important is that with a growth in I_2 from 0 to $+\infty$ the frequency $\Omega(I_2)$ monotonically decreases from 1 to 0 and $\delta\Omega/\delta I_2 < 0$, and that $f_o \neq 0$.

Introducing variables I_j and φ_j in the initial system with $\lambda = \varepsilon$, we obtain that their variation is described by the Hamiltonian

$$H = \omega I_1 + E(I_2) + \varepsilon\alpha' [P_1Q_2 - P_2Q_1] + \varepsilon\omega' \frac{I_1}{2\omega} \sin 2\varphi_1$$

where a prime denotes the derivative with respect to λ.

The expansion of H in a Fourier series in φ_1 and φ_2 contains only the harmonics with arguments $\varphi_1 \pm (2k + 1)\varphi_2$ and φ_1. Therefore the resonance $\omega(\lambda) - \Omega(I_2) = 0$ is isolated from other resonances of the first approximation. We make use of this fact to reduce the problem to the case of a single resonance considered in Sect.2.1.

Let us introduce the notation $D_o = (0.3, 0.9)$ and $D_1 = (0.4, 0.8)$. Let $\omega(\lambda) \in D_1$ for all λ. At $\Omega(I_2) \in D_o \backslash D_1$ all the harmonics of the Hamiltonian H are nonresonance. In this region the first step of the Lindschtedt method [26] allows one to make the symplectic (different from identical on $O(\varepsilon)$) change of the variables $(I_1, I_2, \varphi_1, \varphi_2) \to (\hat{I}_1, \hat{I}_2, \hat{\varphi}_1, \hat{\varphi}_2)$ such that in a new Hamiltonian the dependence on phases $\hat{\varphi}_j$ enters only the terms $O(\varepsilon^2)$. And thus $\dot{\hat{I}}_j = O(\varepsilon^2)$ and for time $1/\varepsilon^{3/2}$ the change in $\hat{I}_j$ will be $O(\sqrt{\varepsilon})$. Therefore, if at the initial moment we have for a phase point $\Omega(I_2) \in D_1$, then after time $1/\varepsilon^{3/2}$ we obtain $\Omega(I_2) \in D_o$.

If $\Omega(I_2) \in D_o$, then $(2k + 1)\Omega(I_2) \notin D_o$ at $k = 1,2,\ldots$ Therefore at $\Omega(I_2) \in D_o$ the first step of the Zeipel method [26] allows one to make a symplectic (different from identical on $O(\varepsilon)$) change of variables $(I_1, I_2, \varphi_1, \varphi_2) \to (\bar{I}_1, \bar{I}_2, \bar{\varphi}_1, \bar{\varphi}_2)$ such that in a new Hamiltonian the terms of the order of ε depend on phases $\bar{\varphi}_j$ only via the combination $\bar{\varphi}_1 - \bar{\varphi}_2$.

The change of the new variables (the bar over them is omitted) is described by the Hamiltonian (in the old notation)

$$H = \omega I_1 + E(I_2) + \varepsilon\alpha'| \sqrt{\frac{I_1}{2\omega}} (\omega + \Omega) f_0 \sin(\varphi_2 - \varphi_1) + O(\varepsilon^2)$$

Now, let us introduce new canonic variables $\psi = \varphi_2 - \varphi_1$, $\chi = \varphi_1$, $R = I_2$, $J = I_1 + I_2$. The Hamiltonian takes the form

$$H = \omega J + [E(R) - \omega R] + \varepsilon\alpha'| \sqrt{\frac{J - R}{2\omega}} (\omega + \Omega) f_0 \sin\psi + O(\varepsilon^2)$$

Whence it is seen, in particular, that for time $1/\varepsilon^{3/2}$ the value changes only by $O(\sqrt{\varepsilon})$.

On the plane with polar coordinates R and λ the resonance condition $\omega(\lambda) = \Omega(R)$ determines a closed resonance curve $R = a(\lambda)$, $0 \leq 1 < 2\pi$ (Figure 7). To describe the motion in the vicinity of the resonance we introduce $\rho = (R - a(\lambda))/\sqrt{\varepsilon}$ and a new time $\tau = \sqrt{\varepsilon}\, t$. The motion equations at $\rho = O(1)$ take the form

$$\frac{d\psi}{d\tau} = \frac{\partial F}{\partial \rho}, \quad \frac{d\rho}{d\tau} = -\frac{\partial F}{\partial \psi} - a'(\lambda)$$

$$\frac{d\lambda}{d\tau} = \sqrt{\varepsilon}, \quad \frac{dJ}{d\tau} = O(\varepsilon^{3/2}), \quad \frac{d\chi}{d\tau} = O(1/\sqrt{\varepsilon})$$

$$F = 1/2\, g(\lambda)\rho^2 + b(\lambda,J)\sin\psi + \sqrt{\varepsilon} F_1(\rho,\psi,\lambda,J) + O(\varepsilon)$$

$$g(\lambda) = \left(\frac{\partial^2 E}{\partial R^2}\right)_{R = a(\lambda)}, \quad b(\lambda,J) = \alpha \sqrt{2(J - a)\omega}\; f_0(a), \quad F_1 = O(1)$$

Rejecting two last terms and fixing λ and J in the Hamiltonian F, we arrive at the following system (a "pendulum") for ψ:

$$\frac{d}{d\tau}\left(\frac{1}{g(\lambda)}\frac{d\psi}{d\tau}\right) = -b(\lambda,J)\cos\psi - a'(\lambda), \qquad (2.6)$$

$$\lambda = \text{const},\ J = \text{const}$$

The further consideration is analogous to that of Sect.2.1 and therefore we describe it in brief. The phase portrait of "pendulum" (2.6) is as is shown in Figure 8 (for $a' > 0$). Let $S(\lambda, J)$ on this portrait be an area of the oscillation region in variables ψ, $\rho = \frac{1}{g}\frac{d\psi}{d\tau}$. If $S' > 0$, the capture to the resonance occurs, and if $S' < 0$ the rejection from the resonance occurs.. The motion of the captured particles proceeds in such a way that within the accuracy $O(\sqrt{\varepsilon}\ \ln \varepsilon)$, a new adiabatic invariant is preserved - the area limited by the unperturbed trajectory. This property allows one to determine the moment when the point leaves the resonance (within the accuracy $O(\sqrt{\varepsilon}\ \ln \varepsilon)$ with respect to parameter λ from the moment of entering the resonance. Choosing in an appropriate way the functions α, ω and the constant J_o, it is possible to make the funtions $a(\lambda)$ and $b(\lambda, J)$ and numbers λ_1, λ_2, λ_3 to satisfy the conditions (a)-(d) of Sect.2.1. Then the points entered the resonance at $|\lambda - \lambda_1| < c_1^{-1}$, $|J - J_o| < c_1^{-1}$, $|R - a(\lambda_1)| < c_1^{-1}$ leave it at $|\lambda - \lambda_2| < c_2^{-1}$, $|R - \alpha(\lambda_2)| < c_2^{-1}$ (Figure 7) and then, within time $c_3^{-1}/\varepsilon^{3/2}$, cannot enter the resonance again. For these points of the "actions" of the initial system $I_1 = R$ and $I_2 = J - R$ experience in the course of the capture a change exceeding c_4^{-1}. For time $1/\varepsilon$ a set of points is captured by the resonance whose phase volume exceeds $c_5^{-1}\sqrt{\varepsilon}$. Therefore for time $c_3^{-1}/\varepsilon^{3/2}$ a set of the points of the phase volume exceeding c_5^{-1} is captured, which actually was required. Thus, in multifrequency Hamiltonian systems with smoothly varying parameters the adiabatic invariance of the variable "action" can be destroyed for time of the order of $1/\varepsilon$ owing to captures of points by resonance.

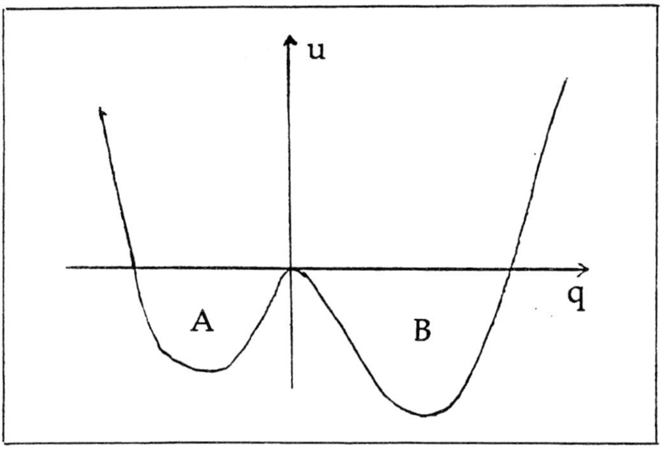

Figure 1.

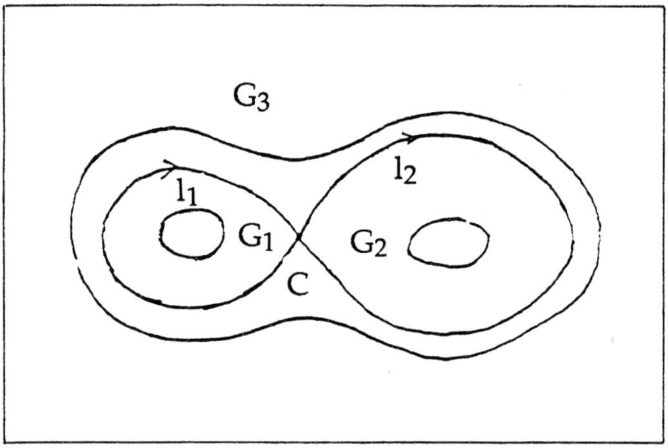

Figure 2.

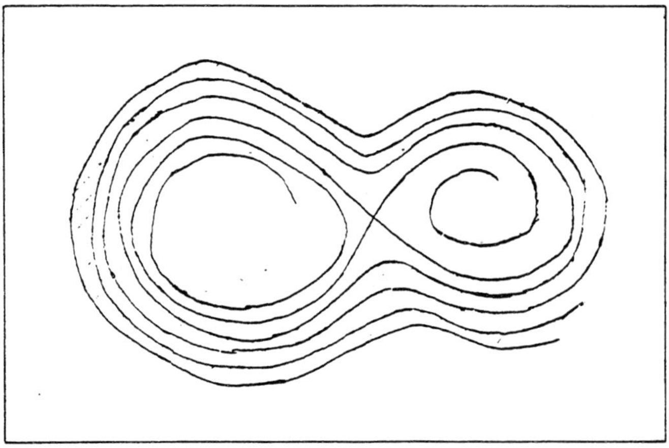

Figure 3.

Figure 4.

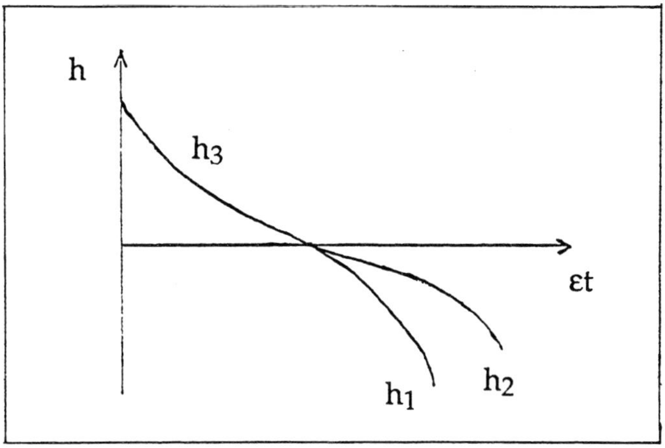

Figure 5.

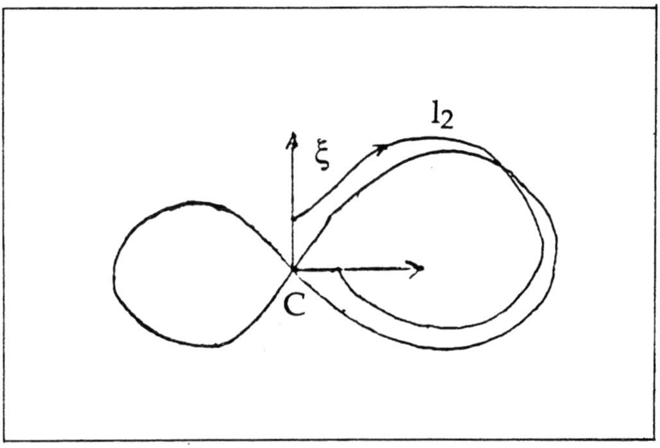

Figure 6.

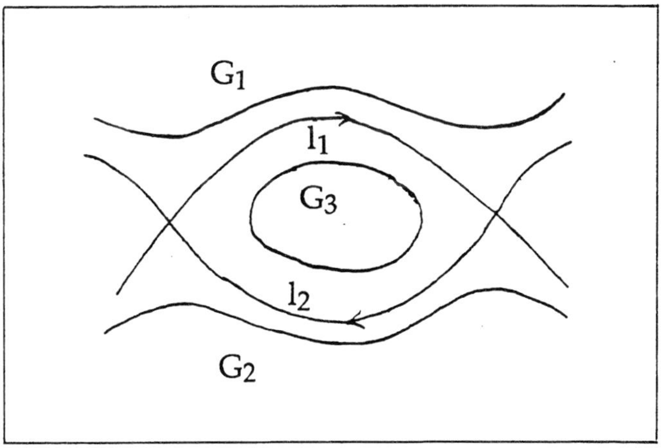

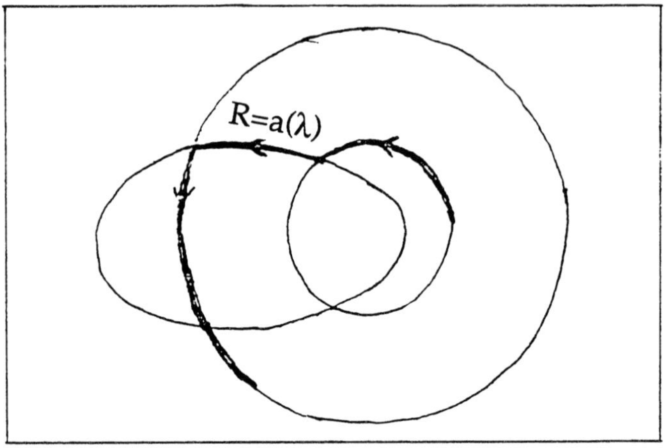

Figure 7.

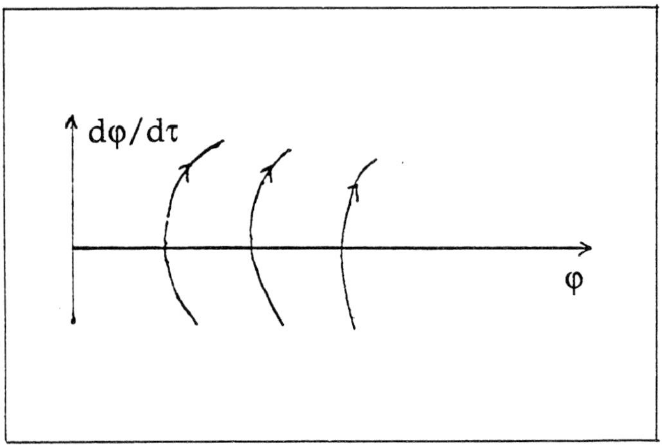

Figure 8a.

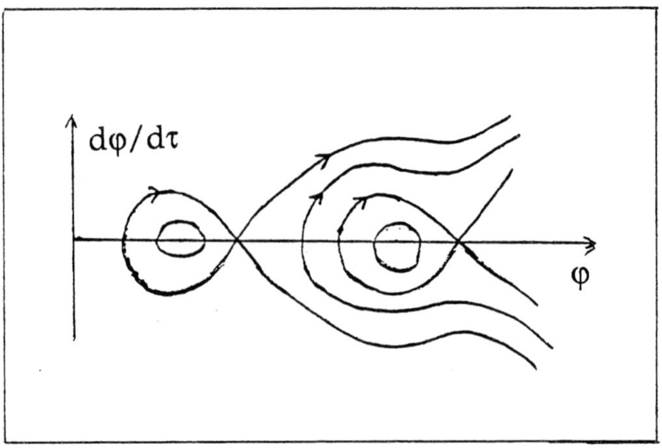

Figure 8b.

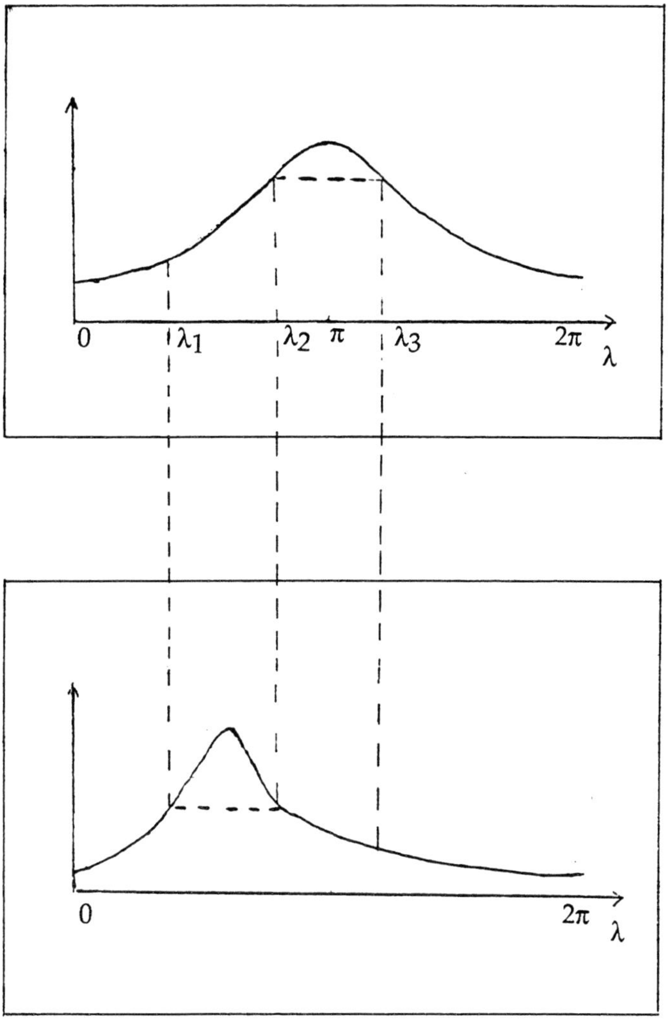

Figure 9a.

Figure 9b.

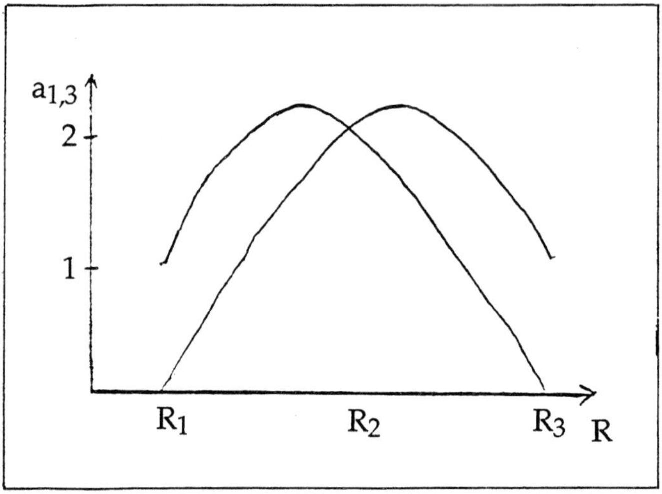

Figure 10

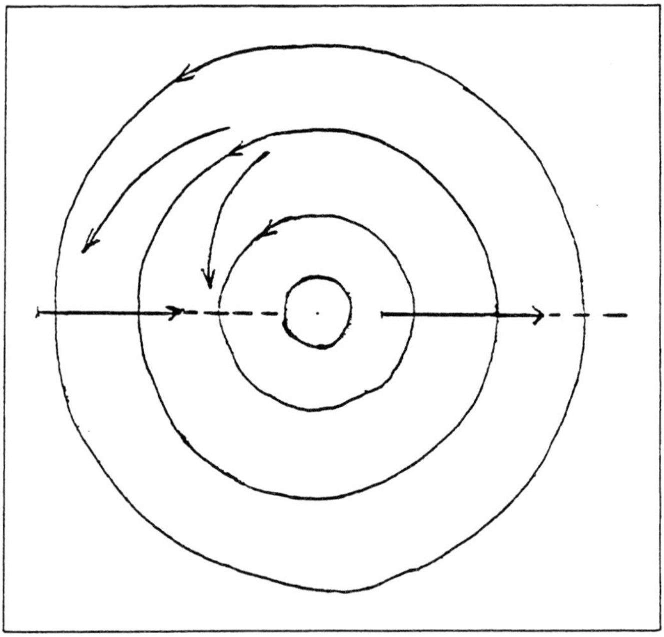

Figure 11.

REFERENCES

1. V.I. Arnol'd. Mathematical methods of classical mechanics. Moscow: Nauka, 1977 (in Russian).
2. G.M. Zaslavskii|. Stochasticity of dynamical systems. Moscow: Nauka, 1984 (in Russian).
3. A. Likhtenberg and M. Liberman. Regular and stochastic dynamics. Moscow: Mir, 1984 (in Russian).
4. V.I. Arnol'd. Usp. Mat. Nauk, 1063, 18, No.6.
5. Lifshits, A.A. Slutskii|, and V.M. Nabutovskii|. Dokl. AN SSSR, 1961, 137, No.3.
6. N.N. Bogolyubov and Yu.A. Mitropol'skii|. Asymptotic methods in the theory of nonlinear oscillations. Moscow: Nauka, 1978 (in Russian).
7. A.I. Nei|shtadt. Thesis for Cand. degree., Moscow, MGU, 1976 (in Russian).
8. P. Goldrech and S. Peale. Astron. J., 1966, 71, No.6.
9. A.T. Sinclair. Monthly Notices Roy. Astron. Soc., 1973, 78, No.4; 1974, 66, No.1.
10. R. Allan. Astron. J., 1969, 74, No.3.
11. R. Greenberg. Astron. J., 1973, 78, No.4.
12. A.I. Nei|shtadt. PMM, 1975, 39, No.4.
13. J. Wisdom. Icarus, 1985, 63, No.2.
14. A.I. Nei|shtadt. Dokl. AN SSSR, 1987, 295, No.1.
15. A.V. Gurevich and E.E. Tsedilina. Superdistant propagation of short radiowaves. Moscow: Nauka, 1979 (in Russian).
16. A.S. Bakai|. Atomnaya Energiya, 1966, 21, No.6.
17. A.S. Bakai| and Yu.P. Stepanovskii|. Adiabatic invariants. Kiev: Naukova dumka, 1981 (in Russian).
18. V.V. Soloviev and D.R. Sklyar. ZhETF, 1986, 90, No.2.
19. A.I. Nei|shtadt and A.V. Timofeev. ZhETF, 1987, 93, No.11.
20. A.I. Nei|shtadt. Izv. AN SSSR, MTT, 1980, No.6.
21. L.A. Pokrovskii|. TMF, 1985, 67, No.2.
22. A.V. Timofeev. ZhETF, 1978, 75, No.10.
23. A.I. Nei|shtadt. Fizika Plazmy, 1986, 12, 8.
24. J.R. Cary, D.F. Escande, and J.L. Tennyson. Phys. Rev. A, 1988, 34, No.5.
25. A.I. Vasil'ev and M.A. Guzev. TNF, 1986, 68, No.3.

26. A.I. Nei|shtadt. PMM, 1987, 51, No.11.
27. C.R. Menyuk. Phys. Rev. A., 1985, 31, No.3.
28. V.I. Arnol'd, V.V. Kozlov, and A.I. Nei|shtadt. INT VINITI, Sovr. Probl. Matem. Fundam. Napravl. v.3. Izdatel'stvo VINITI, 1985 (in Russian).
29. V.I. Bakhtin. Funkt.Analiz i Ego Pril., 1986, 20, 2.
30. B.V. Chirikov. Dokl. AN SSSR, 1959, 125, No.5.
31. A.M. Moltchanov. Icarus, 1968, 8, No.2.
32. A.D. Morozov. ZhVM i MF, 1973, 13, No.5.
33. A.I. Nei|shtadt. Dokl. AN SSSR, 221, No.2, 1975.
34. J. Henrard. Celest. Mech., 1982, 27, No.1.

Footnotes

*In [7], the author obtained for the measure of the exceptional set of initial points the estimate $O(\varepsilon)$ instead of the above indicated $O(\varepsilon^{\infty})$.

**In a Hamiltonian system the flow of the phase volume through a curve is equal to the difference of the Hamiltonian values at its ends.

***In [22] the author considered the passages through the separatrix for a pendulum with smoothly varying frequency, in [25] the systems of a pendulum type were considered, in [23, 24] the authors studied arbitrary Hamiltonian systems smootly dependent on time, and, finally, in [26] the author considered systems with fast and slow motion with two degrees of freedom.

Mathematics
and Modelling

ISBN 1-56072-104-9

MODERN PROBLEMS OF SPECTRAL THEORY OF DISORDERED SYSTEMS

V.A. Chulaevskii

INTRODUCTION

Mathematical models of systems having no translational symmetry in the exact meaning of the word but being homogeneous "on the average' attract ever growing interest in many fields of the mathematical physics. In such cases it is usually implied that one studies not an individual system but a whole ensemble, i.e., a multiparametric family with a defined measure, topology, etc. In the present article we consider ensembles with a defined normalized measure. Thus, it becomes possible to consider "typical" systems and study the properties which are possessed by almost any (with respect to the fixed measure) system of the ensemble. According to the accepted terminology, we consider here a disordered system although the term is not quite adequate.

An example of a disordered system is an imperfect crystal with "randomly" distributed various defects such as interstitials, dislocations, vacancies, etc. or even a perfect crystal in an "occasional" nonuniform external field. More exactly, one has to consider a quantum particle, say, an electron, in such a crystal. We are interested here mainly in the properties of the particle spectrum and its wave functions. In mathematical terms, we are interested in the spectral properties of the Schrödinger operator describing the particle behavior in a disordered medium in the single-particle approximation. Although the electronic properties of the medium should be described within the framework of a multiparticle problem, this is beyond the framework of the present article.

It should be noted that systems possessing no translational symmetry have been studied for quite a long time, but at present inhomogeneity of such systems is payed much more attention in system modeling. If earlier the consideration was based on the assumption of a small inhomogeneity (local or rapidly attenuating perturbation, perturbation of a small norm, etc), nowadays it is just the existence of inhomogeneity itself that makes the basis of the description of important qualitative effects.

1. ANDERSON LOCALIZATION IN ONE-DIMENSIONAL RANDOM MEDIA

In a series of studies by S.A. Molchanov, I.Ya. Gol'dsheid, and L.A. Pastur [1],[2], a fruitful approach has been developed to studying spectral properties of one-dimensional Schrödinger operators

$$H(\omega) = - \alpha^2/ \alpha x^2 + V(x,\omega)$$

and their difference analogues

$$(H(\omega)f)\,(n) = f(n - 1) + V(n,\omega)f(n) + f(n + 1)$$

in the situation where $V(\bullet, \omega)$ is a typical implementation of a homogeneous Marcovian process. In particular, in case of a difference operator, one can take that the values of V at different points are independent and uniformly distributed random quantities. Such problems arise in physics of one-dimensional and quasionedimensional materials. The well known physicists Anderson and Mott stated in their early works that the spectrum of operators with a "stochastic potential" is of pure point nature and that all the eigenfunctions decrease exponentially (in terms of physics, all the states of a particle are exponentially localized). The following statement clarifies the sense of the term "stochastic potential".

Theorem 1. (Cf [3]) Let K be a compact Riemann manifold, X_t be a standard diffusion process on K related to the Riemann metrics (Brownian movement on K), F be a smooth function for which at least one of differentials D^nF, $n \geq 1$ at each point $x \in K$ is nondegenerate. Then the spectrum of the Schrödinger operator

$$H = - \alpha^2 / \alpha t^2 + F(X_t)$$

is of pure point nature and dense everywhere on the semiaxis (inf Spec H, $+\infty$) with probability 1, whereas all the eigenfunctions decrease exponentially.

A similar statement is true [4] also for a difference random Schrödinger operator whose values at different points are independent uniformly distributed random quantities. Their distribution has a sufficiently regular density with respect to the Lebesgue measure. Note also that the continuity condition for the distribution of the potential V at a certain point plays a very important role. Moreover, if to introduce a singleparametric family of potentials gV, then for any infinitesimal $g \neq 0$ the spectrum remains to be of pure point nature. In other words, if the potential V has extremely strong mixing properties, the localization occurs independently of the "stochastic component" amplitude of the potential.

2. FRACTAL SPECTRUM OF A PARTICLE IN A LIMIT-PERIODIC MEDIUM

Now let us consider an example of a disordered system which, in a certain sense, makes the antipodes with a system having strong mixing. It is well known that the Schrödinger operator with a periodic potential is characterized by an absolutely continuous spectrum and all the solutions of the HB = EF equation have the form

$$\Phi(x,E) = \exp\,(ip(E)x)\, f(x,E),\ f(x + T,E) = f(x,E),$$

where T is the potential period. Now a natural question arises - is there any class of potentials that although are not periodic but satisfy a certain weaker condition of almost-periodicity for which the spectrum and eigenfunctions of the respective Schrödinger operator have similar properties? The answer is positive. The respective potentials are usually called limit-periodic potentials and admit the following representation

$$V(x) = \sum_{n=1}^{\infty} a_n U_n \ (x/T_n), \ \ U_n(x + 1) = U_n(x),$$

where $T_n/T_{n-1} = m_n > 1$ are integers, and the coefficients a_n decrease rather quickly.

Theorem 2. ([4]). Let us fix a sequence of periodic functions U_n of the period T limited in the set and the sequence of periods $T(N) = Tm_1$, $\ldots m_n$, $m_k \in \mathbb{N}$. Let the sequence a(n) satisfy the conditions

$$e^{BN} \sum_{M=N+1}^{\infty} |a(n)| \xrightarrow[N \to \infty]{} 0.$$

at any $B > 0$. Then: (a) the spectral measure of the operator H_v is absolutely continious with respect to the Lebesgue measure and doubled; (b) the eigenfunctions have the form

$$\Phi_{\pm}(x,E) = \exp\ (\pm ip(E)x)\ f_{\pm}(x,E),$$

where $f_{\pm} \in \Omega(V)$ and p(E) is continued to an analytical function conformly mapping the upper half-plane into itself.

Such a high rate of approximating is required not accidentally: condition (2) indicates only the asymptotic nature of the a(n) coefficients whereas the initial segment of the series can be arbitrarily large. But if to impose a constraint on the values of all the coefficients a(n), then the superexponential rate of approximating can be considered as one of possible results.

Theorem 3. Let the Fourier coefficients a(n) of potential (1) satisfy the condition

$$(\Pi m_1 \ldots m_{N=1})^B \sum_{M=N+1}^{\infty} |a(n)| \leq 1/4, \ B > 0. \tag{3}$$

If there exists $q \in (1, 2 - 2^{-9})$ such that

$$2q^3/(1-q)^2 \leq B,$$

then: (a) The PSs of opertors $H_0 + V_N$ converge in the space $L_{loc}^{q'}(\mathbf{R})$ for all $q' \in [1,q)$. The limit function p(E) is a real part of the analytical function, holomorphous in the upper half-plane. (b) The spectral measure of the operator $H_0 + V$ is absolutely continuous with respect to the Lebesgue measure.

An interesting property of Schrödinger operators with limit--periodic potentials is the presence of an everywhere enumerable dense lacunary set in the spectrum of such operators. In other words, a carrier of the spectrum measure is a nowhere dense perfect Cantorian set. This property is inherent in "typical" limit-periodic potentials, whereas for degenerate potentials the spectrum can completely fill a certain segment. At first the statements on typicalness of a Cantor spectrum had the form of existence theorems, but later on it became possible to describe the set of operators with the Cantor spectrum in a more explicit form, using the method of the inverse problem of the scattering theory [6],[7]. Let us remind that in a series of studies performed by V.A. Marchenko and I.S. Ostrovskii| [8] it has been established that the class of Schrödinger operators with periodic potentials allows the inner description in terms of spectral data of a specific form that received the name of Marchenko-Ostrovskii| spectral data. L.A. Pastur and V.A. Tkachenko [8] have proved recently that the class of limit-periodic potentials with absolutely continuous spectrum is the natural closure of the class of periodic potentials in the metrics of the Besicovich function space. Not giving here the complete formulation of this result, we should like to note that it can be used for constructing the limit-periodic Schrödinger operators with an anywhere dense Cantor spectrum of the given structure.

Note that the absolute continuity of the spectrum of the Schrödinger operator with the limit-periodic potential gV(x) of the above described form is preserved for any large g.

3. ANDERSON LOCALIZATION IN QUASIPERIODIC SYSTEMS

Above, we considered examples of potentials V which, in one case, possessed very weak stochastic properties (strong mixing), and in the other extremely weak properties (small perturbations of periodic functions). In both cases the spectral type of a Schrödinger operator with potential gV(x) did not change for any $0 < g < \infty$. Much more interesting examples of disordered systems are given by quasiperiodic potentials, e.g.,

$$V(x,\alpha) = A\cos(\alpha + x) + B\cos(\alpha + x\sqrt{2})$$

or

$$V(n,\alpha) = \cos(\alpha + 2\pi n\sqrt{2}).$$

The uniqueness of this class of systems consists in that the spectral properties of an operator with the potential gV for small g are close to the properties of operators with periodic potential; for sufficiently large g such properties are close to the properties of operators with a strong stochastic potential. Moreover, of great importance is also the fact whether the operator is differential (unbounded) or finite-difference (bounded).

Theorem 4. (see [9]). Let V be a doubly continuously differentiable function of period I with exactly two nondegenerate critical points per period (the local maximum and local minimum) and ω be an irrational number with the coefficients of the expansion into a continued fraction

$$\omega = [k_1; k_2; \ldots] = \cfrac{1}{k_1 + \cfrac{1}{k_2 + \ldots}}$$

that satisfy the estimate of growth, $k_s < cs^2$. In this case there exist a limit

$$\lim_{s\to\infty} 1/s \ln k_s < \infty$$

Then at sufficiently small positive ε the spectrum of the difference operator

$$(Hf)(n) = \varepsilon(f(n-1) + f(n+1) + V(\alpha + n\omega)$$

is of pure point nature, the eigenfunctions decrease exponentially, and the carrier of the spectral measure is a Cantorian set dense nowhere with the complement with the Lebesgue measure $m(\varepsilon) \to 0$ at $\varepsilon \to 0$.

4. UNSOLVED PROBLEMS

It seems that the fact that the qualitative spectral analysis of difference (and moreover differential) Schrödinger operators with the simplest quasiperiodic potential, e.g., $V(n) = g \cos(n\sqrt{2})$ for all the values of the "coupling constant" g has not been completed as yet and can be considered as a challenge to mathematicians. This example is also of a separate interest, but what is more important is that it itself contains most difficulties and qualitative effects observed in the case of quasiperiodic operators of a more general form. Moreover, the case of a multidimensional (even two-dimensional) quasiperiodic Schrödinger operator remains to be almost unstudied since all the known approaches use, in one way or another, the specific features of the one-dimensional problem. Of special interest is the case where the potential is an almost periodic function with the finite number of solutions. Such examples arise in the description of electronic properties of quasicrystals (The Penrose model of a quasicrystal, etc.). Whereas the one-dimensional models of quasicrystals allow almost a trivial complete analysis [10], their multidimensional analogs are almost unstudied.

CONCLUSION

The spectral theory of disordered systems is an extremely rapidly developing field of mathematical physics. The success here is as great as fragmentary. The main result obtained in the recent years reduces to the establishment of the fact that both the probabilistic approach and the approach with the use of the compact topological groups and almost periodic functions are somewhat one-sided and incomplete. It seems that new advances in this field would be associated with more complete use of the language of dynamical systems and renorm-group methods.

REFERENCES

1. I.Ya. Gol'dsheid, S.A. Molchanov, and L.A. Pastur. "A pure point spectrum of a random one-dimensional Schrödinger operator." Funkts. Analiz i Ego Prilozheniya, 1977, v.II, No.1, 1-10.
2. S.A. Molchanov. "Structure of eigenfunctions of disordered systems." Izv. AN SSSR, Ser. Matem., 1978, 12, No.1, 69-101.
3. H. Kunz and B. Souillard. "Sur le spectre des operateurs aux differences finies aleatoires." Communications in Mathematical Physics, 1980, 78, 201-246.
4. V.A. Chulaevskii|. "On perturbations of a Schrödinger operator with a periodic potential." UMN, 1981, 36, 5, 143-144.
5. V.A. Chulaevskii|. "Spectra of disordered systems. Limit- periodic systems." Preprint of NIVTs AN SSSR, PUshchino, 1986 (in Russian).
6. V.A. Chulaevskii|. "Inverse spectral problem for limit-periodic Schrödinger operators." Funkts. Analiz i Ego Prilozheniya, 1984, 18, No.3, 63-66.
7. L.A. Pastur and V.A. Tkachenko. "To the spectral theory of one-dimensional Schrödinger operator with the limit-periodical potential." DAN SSSR, 1984, 279, 1050-1054.
8. V.A. Marchenko. Sturm-Liouville operators and their applications. Kiev: Naukova Dumka, 1977.
9. Ya.G. Sinai. "Anderson localization for the one-dimensional Schrödinger operators with quasi-periodic potentials." Proceedings

of Int. Congress on Mathematical Physics, Marseille, France, 1986; World Scientific Publiching Co., Singapore, 1987, 370-432.

10. F. Delyon and D. Petritis. "Absence of localization in a class of Schrödinger operators with quasi-periodic potentials." Communications in Mathematical Physics, 1986, 103, 441-444.

Mathematics
and Modelling

ISBN 1-56072-104-9

METHODS OF BIFURCATION ANALYSIS OF DYNAMICAL SYSTEMS. APPROACH BASED ON CONTINUATION BY PARAMETER

A.I. Khibnik

INTRODUCTION

One of the key problems of the qualitative analysis of mathematical models in the form of ordinary differential equations containing parameters is the study of invariant sets determining the structure of a phase portrait such as equilibrium, limit cycles, separatrix manifolds, etc. It is reasonable to study various ivariant sets separately only if their parameters have fixed values. If the parameters vary smoothly and continuously, the invariant sets vary also smoothly and at the bifurcation values of the parameters transform one into another. Thus the connected families of invariant sets arise. Such families should be considered in the direct product of the phase space and the space of the parameters of the model.

The aim of the present investigation is the construction of the scheme for numerical study of connected families of equilibria and limit cycles on the basis of bifurcation theory methods and the classical idea of the continuation by a parameter. The article consists of two chapters: the first one is devoted to the discussion of the suggested method and the second one to problems of its numerical implementation. The main ideas of this study were briefly stated earlier [10, 11, 18].

The scheme of a continuous parametric analysis described below resulted from numerous discussions with R.M. Borisyuk of problems concerned the "technology" of the numerical investigation of models. The program implementation of this scheme was started together with R.M. Borisyuk and Yu.A. Kuznetsov. The work was also inspired by an active interest shown by A.D. Bazykin. Numerous useful remarks were made by Yu.S. Il'yashenko. To all them the author expresses his deep gratitude for their kind attention and support.

A deep influence on the formation of author's scientific interests and on his studies of the algorithmic problems of the bifurcation theory was exerted by A.M. Molchanov and E.E. Shnol'to whom the author dedicate this article with most sincere gratitude.

1. SCHEME OF CONTINUOUS PARAMETRIC ANALYSIS

1.1. Preliminary Notes

Let us consider an automonous system of ordinary differential equations

$$\dot{x} = F(x, \alpha), \tag{1.1}$$

where $x \in \mathbf{R}^n$ is the vector of phase variables, $\alpha \in \mathbf{R}^m$ is the vector of parameters, $F(x,\alpha)$ is the vector-function having a sufficient number of derivatives with respect to x and α. We call a phase-parametric space P of the system (1.1) the direct product of its phase space and the space of parameters.

The qualitative study of system (1.1) usually implies:

(1) the description of the topological structure of the phase portrait at different parameter values;

(2) the partition of the parameter space into sets with topologically different types of phase portraits;

(3) the description of the set of bifurcation parameter values;

(4) the description of the phase portrait bifurcations due to the parameter variation in the vicinity of the bifurcation values.

In the qualitative study, the leading part is played by the bifurcation theory methods since they provide the description of the structure either of the whole phase portrait or of some its parts at the parameter values close to bifurcation ones. The phase portrait obtained in the bifurcation study can then be continued by continuity to other parameter values far from bifurcation ones.

To be able to apply the bifurcation methods to concrete systems one should be able to solve (with the aid of numerical algorithms) the following problems:

1^0. To determine the bifurcation values of parameters.

2^0. To determine the structure of the partition of the neighborhood of the given bifurcation point in the space of parameters into sets corresponding to different types of phase portraits and to construct these portraits themselves in the vicinity of the bifurcation point.

3^0. To continue by continuity the determined phase portraits or their main elements.

Bifurcations of the phase portraits of system (1.1) occur in the phase neighbourhood of some of its invariant sets, the so-called "bifurcation carriers" [2]. Among such invariant sets are equilibria, limit cycles, invariant manifolds of equilibria and limit cycles, etc. The search for bifurcation values of the parameters in problem 1^0 can be reduced to the analysis of various invariant sets of the system, their dependence on parameters, and the detection of situations where the bifurcation conditions are fulfilled for these invariant sets. Thus, we will find the bifurcation values of the parameters and, at the same time, single out the elements of the phase portrait experiencing bifurcation.

The essense of problem 2^0 consists in the description of all the invariant sets arising from the bifurcation carrier at slight parameter variation. Finally, problem 3^0 is the continuation by continuity of various invariant sets determined in the bifurcation analysis.

Thus, problems 1^0-3^0 can be considered together as a problem of constructing and studying of a continuous family of invariant sets in the space P of the phase variables and parameters of the system. Of practical interest is also a study of an individual connectivity components in the indicated family which can be separated, e.g., by a certain given equilibrium position or a limit cycle. Often, the problem is formulated even more specifically which is associated with the limitation of the class of the invariant sets under consideration, e.g., the study of a connected family of equilibria, connected family of limit cycles, the joint analysis of the equilibria, the limit cycles, and the separtix contours*, etc. The indicated classes of invariant sets, i.e., of equilibria, cycles and separtix contours, will be the objects of our further consideration.

Note. Everywhere in this article the connectivity of the family of invariant sets implies a linear connectivity, i.e., the existence of a path in the family (of the image of a continuous mapping of an interval on a numerical line) connecting any its two elements. The concept of continuity is associated here with the concept of the topological limit. In other words, two invariant sets are close** if each of them intersect in P with a sufficiently small neighbourhood of each point of another set.

Let us formulate the problem in more detail and describe the suggested scheme for the study by way of example of the family of equilibria.

1.2. Family of Equilibria. Problem Formulation

The family of equilibria of system (1.1) is a set of the solutions of the system of equations

$$F(x,\alpha) = 0. \quad (1.2)$$

It is a smooth manifold in P (possibly with a boundary) of dimensionality m.

Let a certain equilibrium x^0 be given at $\alpha = \alpha^0$. It is necessary to study the connectivity component in the set of the solutions of system (1.2) containing the point (x^0,α^0). Let us denote this component by **M** and called it a connected family. We shall consider the connected families of equilibria possessing the so-called simple bifurcation structure. This implies the following.

1. Family M is stratified.

A family **M** is called stratified if it is a union of smooth mutually disjoint manifolds - strata - possessing the following property: the closure of each stratum in M consists of the stratum itself and a finite union of strata of lower dimensionality. The codimensionality of a stratum in M is called its rank. An arbitraty element from M has a rank equal, by definition, to the rank of the stratum containing this element. An element of rank k is also called a k-element or a k-point. The set of elements of M having rank not lower than k is called a subfamily of rank k.

2. Local properties of elements of a stratum are weakly equivalent.

We call a local family at point $(x,\alpha) \in P$ a confinement of system (1.1) to the direct product of some neighborhoods of the points x and α in the phase and parametric spaces, respectively. The local families for two elements of M, (x^1, α^1) and $(x,^2 \alpha^2)$, are weakly equivalent [2] if there exist a homeomorphism of sufficiently small neighborhoods of the points α^1 and α^2 in the space of parameters such that the phase portraits of system (1) for the respective values of the parameters in some small fixed neighborhoods of points x^1 and x^2 are orbitally topologically equivalent. The equivalence of the elements of the stratum (in the above indicated sense) allows us to consider its type, i.e., the topological structure of the bifurcation diagram and (local) phase portraits of the stratum elements.

3. Strata in M Are Singled Out by the Degeneracy and Nondegeneracy (Tipicalness) Conditions

Degeneracy conditions are the conditions of the equality type and nondegeneracy (typicalness) conditions are those of inequality type imposed on the system parameters. It is implied that these conditions arise in the classification of local families of relatively weak equivalence in general (or some special) families of vector fields.

Each set of degeneracy conditions singles out a subset in M; its closure (in M) is called a class. The codimensionality of a class in M, i.e., the number of determining it (independent) degeneracy conditions is called a class rank. The nondegeneracy conditions single out in the class an open subset whose connectivity components are strata.

Each set of degeneracy conditions has a respective set of nondegeneracy conditions. The violation of the nondegeneracy conditions, i.e., the transformation of some inequalities into equalities, yields new degeneracy conditions and new classes. The family M itself is a class of rank O.

The elements of a class satisfying the nondegeneracy conditions are called *regular*; they form a regular part of the class. The remaining elements (called *singular*) form a singular part of the class.

4. Partition of Family M into Strata is Correct (by Adjacency)

A stratum or a class A is adjacent to a stratum B of a lower rank (which is denoted as $A \to B$) if $A \subset B$ (closure in M). We say that the class A is adjacent to the class B if $A \subset B$. The adjacency $A \to B$ is called correct if the rank of A exceeds by unity the rank of B. The partition of M into strata (stratification of M) is called correct (by adjacency) if from adjacency of strata $A \to B$ it follows that either this adjacency is correct or there exists a stratum C such that $A \to C \to B$. The equivalent definition of a correct partition consists in that the boundary of any stratum of rank k consists of strata of rank $k + 1$ and adjacent strata of higher ranks. If partition is correct, then each stratum of a nonzero rank is adjacent to a certain stratum of rank lower by unity and to (some) strata of all the lower ranks.

5. Number of Strata in Partition of M Is Finite

We call a *bifurcation structure* of M the partition of M into strata and classes that satisfies conditions 1-3. The bifurcation structure is called *simple* if conditions 4 and 5 are also fulfilled, i.e., the partition is correct and the number of strata is finite. The description of the bifurcation structure includes the description of all the classes and strata (degeneracy conditions for classes and types of strata) and also the relationships between them - inclusions and adjacencies.

Let us formulate the problem of studying the bifurcation structure of a family M in the case where this structure is simple. Hereinafter it will be assumed that system (1.1) is of the general position unless otherwise stated.

Example. Let $m = 1$ and family M have a form shown in Figure 1.

This family has four strata of rank O (A_1, A_2, A_3, and A_4) and three strata of rank 1 (B_1, B_2, and B_3). The strata of rank O correspond to hyperbolic equilibria. The strata B_1 and B_2 correspond to equilibria with $\lambda_1 = 0$ (λ is an eigenvalue) and the stratum B_3 to that with $\lambda_{1,2} = \pm i\omega$. The strata B_1 and B_2 form one class and the stratum B_3 another class of rank 1; together they form a subfamily of rank 1. The family M itself is a class and a subfamily of rank O.

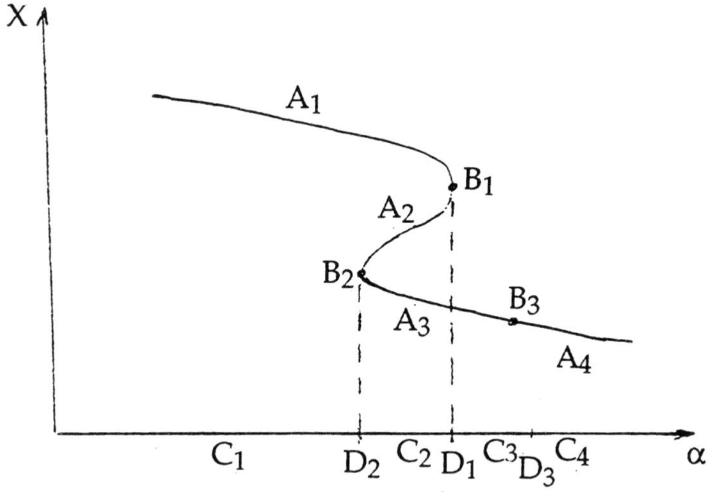

Figure 1. Illustrating the partition of a family of equilibria and the axis of parameter values into strata.

The adjacency diagram of strata has the form

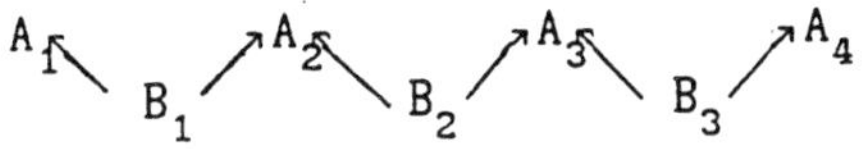

In the systems of general positions with an arbitrary number of parameters the strata of the zeroth rank correspond, similar to the above considered example, to hyperbolic equilibria and those of a nonzeroth ranks to nonhyperbolic equilibria. The sub- family of rank 1 can contain two classes of rank 1: $M_1^{(0)}$ with the degeneracy conditions $\lambda_1 = 0$ and $M_1^{(1)}$ with the degeneracy condition $\lambda_{1,2} = \pm i\omega$. A subfamily of rank 2 can have five classes of rank 2: (1) $M_2^{(0,0)}$ with the condition $\lambda_1 = 0$ and the simplest degeneracy of nonlinear terms, (2) $M_2^{(1,0)}$ with $\lambda_{1,2} = \pm i\omega$ and the simplest degeneracy in nonlinear terms, (3) $M_2^{(0,0)}$ with $\lambda_{1,2} = 0$, (4) $M_2^{(0,1)}$ with $\lambda_1 = 0$ and $\lambda_{2,3} = \pm i\omega$, and (5) $M^{(1,1)}$ with $\lambda_{1,2} = \pm i\omega$, $\lambda_{3,4} = \pm i\omega$.

The adjacency digram for the above indicated classes has the form:

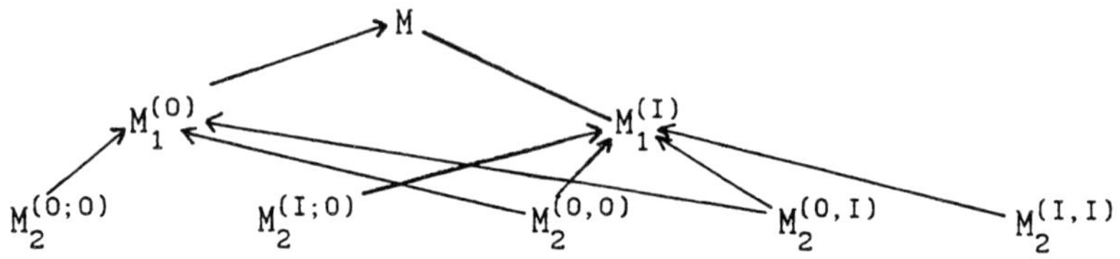

The study of the bifurcation structure of a family of equilibria in concrete systems should be based on the theoretical results known for bifurcation equilibria in the case of general position (1-3). Such results include the formulation of the degeneracy conditions defining the classes and the nondegeneracy conditions that single out the strata, and also the topology of the bifurcation diagrams and phase portraits for each stratum. At present there is the complete classification of degeneracies of an arbitrary codimensionality for the cases $\lambda_1 = 0$ and $\lambda_{1,2} = \pm i\omega$. The case $\lambda_{1,2} = 0$ has been studied in detal for m = 2 and "almost" in detail for m = 3 (the word "almost" implies that there are some unstudied subcases). The degeneracies of codimensionality 2 that correspond to three (o, ±iω) and four (±iω, ±iω) critical eigen values*** have also been studied almost completely.

Thus the degeneracies of different codimensionalities have been studied with a different degree of completeness, so that in some concrete systems with m > 1 one can encounter theoretically uninvestigated cases although it is not known a priori. In the algorithm considered below the value of m is not limited a priori by assuming that an "unstudied" case is one of natural failures of the algorithm.

Prior to proceeding to the description of the algorithm for the analysis of the bifurcation structure of a connected equilibrium family we should like to emphasize the relation between the formulated problem and the initial problem of the qualitative study of system (1.1). The partition of a family **M** into strata generates in projecting **M** into the space of parameters the partition of this space into the so-called parameteric strata (mutually disjoint sets). In Figure 1 such strata are the intervals C_i (i = 1, 2, 3, 4) and points D_j (j = 1,2,3). It is seen that the correspondence between the strata in **M** and the parametric strata is not necessarily one-to-one correspondence. The final result of the qualitative study includes the determined parametric strata and phase portraits in each of them. Such phase portraits are composed by local phase portraits corresponding to the strata of family **M**. Of course, the study of one connected family does not exhaust the problem of the qualitative investigation. The study of other families, if any, complements and refines the obtained results.

1.3. The Idea of Algorithm. Paths and Cuts

The suggested approach to the study of a (simple) bifurcation structure of a family of equilibria M can be compared with a travel over M with the aim to reveal and describe all the classes and strata. The initial point of such a travel is the given family element, as a rule, corresponding to a certain rough equilibrium. The rout of the motion is built by the so-called paths - one-dimensional curves in M such that each curve lies completely within a certain class. Let us single out in a class some parameters by fixing their values (the number of remaining free parameters should exceed by unity the class rank). The continuity of the motion over M is attained because each next path passes through an already passed point. The correct adjacency of strata in M allows the search for classes and strata of rank k + 1 by constructing paths in the classes of rank k and determining the points of the intersection of these paths with the classes of the next rank. Whence the possibility of continuous motion from degeneracies of low codimensionality to degeneracies of high codimensionality. In a similar way, it is possible to return continuously from degeneracies of high codimensionality to degeneracies of low codimensionality using in this process (and it is very important) the information about the adjacent classes and strata contained in degeneracies of high codimensionality. The suggested scheme of studying the bifurcation structure of the family M is based just on such "shuttle" motion associated with the construction of paths in classes of different ranks.

In this Section we shall introduce the concepts of a path, cut, net, and route. In the following Section we shall formulate the basic schemes for the construction of routes. These schemes do not determine the route uniquely, thus leaving a certain choice for the investigator. It is assumed that the construction of a route (i.e., the choice of new paths is carried out in the interactive mode with due regard for the information obtained at the previous stages. On the whole, the problem of studying the bifurcation structure is considered here just as a search without any guarantees of its final solution at m > 1. Our main goal is

to suggest the appropriate methods for the study of such problems and indicate the ways of their use.
Let us give necesary definitions.

A path in M is called any smooth one-dimensional curve lying completely in a certain class. The path has a rank k and is called a k-path if it lies in the class of rank k and contains only a finite number of singular points of this class. The notion of a regular and singular points of the class are extended to a path lying in this class: a point of a k-path is regular if its rank equals k and singular if it exceeds k.

A path of rank k is called nondegenerate if it intersect transversally all the classes of ranks higher than k and is called degenerate otherwise. Nondegeneracy of a path of rank k signifies that all its singular points have rank k + 1 and are the points of transversal intersection of this path with classes of rank k + 1. A degenerate k-path is called weakly degenerate if the points where the transversality condition is violated are either points of rank k + 2 or points of quadratic tangency of the path with the stratum of rank k + 1.

A route is a sequence of paths such that each path, beginning with the second one, has common points with one of the previous paths. A linear route is a route in which each pair of sequential paths has a nonvoid intersection.

The paths implicitly set in M as an intersections of classes with certain manifolds in P are called cuts. A parametric k-cut is any smooth manifold of dimensionality k in the space of parameters. A phase-parametric k-cut is a manifold in P of dimensionality n + k which is a direct product of the phase space and the parametric k-cut. Phase-parametric k-cuts are also called k-cuts or cuts of rank k. The cut of a family M is the intersection of M with a k-parametric cut.

We say that a cut is transversal to a family M if it transversally intersect all the classes and strata in M. A cut can be transversal to a certain subset of the family M. A nontransversal cut S_{k+1} of rank k + 1 is called almost transversal if the transversality condition is violated at the finite number of points of the following three types:

1) points of rank k = 2;

2) points of rank k + 1 which are the points of the quadratic tangency of the cut S_{k+1} with a certain class of rank k + 1;

3) points of rank k which are the points of quadratic tangency of S_{k+1} with a certain class of rank k.

Below, if not otherwise stated, the cuts are assumed to be transversal or almost transversal.

Using cuts, we shall define a k-path as a smooth branch in the transection of a class of rank k with a cut of rank k + 1. For the unique setting of the parth of rank k, one should indicate in addition to the generators of its class and cut also one k-point on it. A path formed by a cut transversal to the family M is nondegenerate. An almost transversal cut can be both nondegenerate and weakly degenerate.

Consider a set of points in the transection of a cut of rank k + 1 with all the classes of rank k. We shall call this set a k-net (a net of rank k) or a homogeneous net. The connectivity component in a homogeneous net is a connected homogeneous net . For a transversal cut the K-net is a union of all the nondegenerate k-paths lying in this cut. In case of almost transversal cut of rank k + 1 the respective k-net is a union of nondegenerate and weakly degenerate paths and also of isolated tangential points of the cut tangency with strata of rank k. The singular points of paths forming a homogeneous net are called net vertices and the segments of paths between the vertices are called arcs. Each arc of the k-net lies completely in a certain stratum of rank k. Isolated points of rank K in the k-net are considered as degenerate arcs.

Let us introduce the concept of a standard k-cut as an (n + k)-dimensional plane in P parallel to a certain k-dimensional coordinate subspace in the space of parameters. The set of parameters corresponding to this subspace is called a set of active parameters. If a set of active parameters α_{i_1} , ..., α_{i_k} is given, then the corresponding standard k-cut is denoted as (α_{i_1} ,...α_{i_k})-cut or S_k (α_{i_1} , ..., α_{i_k}). In order to set a standard k-cut completely it is necessary to indicate also one point in it.

Hereinafter we consider only standard cuts and therefore the word "standard" is omitted. A cut S_{k+1} of rank k + 1 can be represented as a single-parametric family of parallel cuts of rank k parametrized by one of the active parameters of the initial cut S_{k+1}. Such a family is called nondegenerate if the following conditions are fulfilled:

1) the cut S_{k+1} is transversal to M,

2) all cuts of the family are transversal or almost transversal to M,

3) to almost transversal cuts there corresponds a finite number of the so-called critical values of the family parameter.

Example. Let us consider a system of differential equations on a plane (the principal family for degeneracy of codimensionality 2 "two zero eigenvalues" [1,2]):

$$\dot{x}_1 = x_2$$
$$\dot{x}_2 = \alpha_1 + \alpha_2 x_1 - x_1^2 + a x_1 x_2, \qquad a = \pm 1$$

In this system the family of α_2-cuts dependent on α_1 is degenerate. At the same time, the family of α_1-cuts dependent on α_2 is nondegenerate.

Indeed, the family of equilibria

$$M = \{(x_1, x_2, \alpha_1, \alpha_2) : x_2 = 0,\ \alpha_1 + \alpha_2 x_1 - x_1^2 = 0\}$$

contains two classes of rank I: $M_1^{(o)}$ and $M_1^{(I)}$, and one class of rank 2: $M_2^{(0,0)}$. The class $M_1^{(0)} = \{(x_1,x_2,\alpha_1,\alpha_2): x_2 = 0,\ \alpha_2 = 2x_1,\ \alpha = -2x_1^2\}$ contains equilibria with $\lambda_1 = 0$, the class $M_1^{(I)} = \{(x_1,x_2,\alpha_1,\alpha_2) : x_1 = x_2 = \alpha_1 = 0, \alpha_2 \leq 0\}$ contains equilibria with $\lambda_{1,2} = \pm i\omega$, the class $M_2^{(0,0)} = \{x_1 = x_2 = \alpha_1 = \alpha_2 = 0\}$ contains equilibria with $\lambda_1 = \lambda_2 = 0$. The straightforward checking indicates that all the α_1-cuts and also all the α_2-cuts (except of α_2-cuts corresponding to $\alpha_1 = 0$) are transversal to the above indicated classes. The cut $\alpha_1 = 0$ is tangential to class $M_1^{(0)}$ at the origin, i.e., is not transversal to the family M, and moreover, it contains the whole class $M_1^{(I)}$, i.e., neither is it almost transversal. Whence follows the degeneracy of the family of α_2-cuts.

Below we consider only nondegenerate single-parametric families of parallel cuts.

Let us establish the correspondence between each cut of the family M and the set of classes and strata intersecting this family. This set, together with the relationships of adjacency and inclusion, is called a bifurcation structure of the cut of M. In a similar way, it is also possible to define the bifurcation structure of a cut of any subfamily in M. In particular, the bifurcation structure of the k-net, the (k + 1)-cut of a subfamily of rank k, is the partition of this set into paths, vertices, and arcs together with the relationships of their inclusion and adjacency.

Bifurcation structures for two cuts of **M** (or subfamilies in **M**) are called the same or equivalent, if there is a one-to-one correspondence between the classes and strata of both structures preserving the types of the classes and strata and the relationships of adjacency and inclusion.

The cut of a family or a subfamily called stable if its bifurcation structure is preserved, i.e. remains to be equivalent to the initial one for all sufficiently close parallel cuts. (Parallel cuts have the same sets of active parameters and their closeness signifies that the respective values of inactive parameters are close). The transversal cuts are stable.

Let us consider (nondegenerate) single-parametric family of parallel cuts. Let us denote the family parameter by γ. By definition of a nondegenrate family, it consists of transversal and almost transversal cuts, the latter correspond to the critical values of the parameter γ and the former corresponds to the intervals between these values. Since the transversal cuts are rather rough, the bifurcation structure of the cut in the above indicated intervals remains unchanged. Almost transversal cuts are not rough, and the bifurcation structure is not changed in the transition through critical values of the parameter γ. This is reflected in the changes of the bifurcation structure of the homogeneous net, namely, in the appearance and disappearance of vertices and arcs.

To three types of almost transversal cuts (see p.) there correspond three variants of the change in the bifurcation structure of the homogeneous k-net depending on the variation of the family parameter: i/ the decomposition of a vertex of rank $k + 2$ into two (usually of different types, i.e., belonging to different strata) vertices of rank $k + 1$ and connecting their arcs; ii/ the decomposition of a multiple vertex of rank $k + 1$ into two (conventional) vertices of the same ranks and types and a connecting arc, and iii/ the formation of a closed arc from an isolated point of rank k. In the reverse change of the parameter the above indicated vertices and arcs merge together and disappear.

Let us take a certain finite set of transversal cuts of the family **M**. Consider the union of homogeneous nets for all cuts of the given set; this union is called an inhomogeneous net. We say that an inhomogeneous net covers a connected family **M** if it intersects all the strata of this family and is connected. The study of the bifurcation structure of the family M can be reduced to the construction of a covering inhomogeneous net (of course, if such a net exists).

In order to construct a covering net a following recurrence scheme can be used. First, a homogeneous net of rank m - 1 is constructed in the whole family M. By definition, such a net should contain all the strata of rank m. Then M is represented as a nondegenerate single-parametric family of parallel cuts of rank m-1. The set of the transversal cuts of this family is divided into the connectivity components corresponding to the intervals of the family parameter values between the critical values. In each of the connectivity components we choose an arbitrary, so to say, "typical" cut. Then for each such cut a similar procedure is performed, i.e., a net of rank m - 2 is represented in the form of a family of parallel cuts, and then typical cuts are chosen etc., until the cuts of the minimum nonzero rank are obtained. As a result of such a construction, we arrive at a partially ordered set of transversal cuts with respect to inclusion that yields a covering net.

An important role in the implementation of the above described scheme is played by the search for critical values of the parameter for a family of parallel cuts. The solution of this problem is associated with the search for points M at which the transversality condition is violated. Let us assume that we study a cut S_{k+2} of rank k + 2 reprersented as a family of parallel cuts $S_{k+1}(\gamma)$ of rank k + 1 dependent on the parameter γ (γ is one of the active parameters of cut S_{k+2}). Two types of points of the nontransversal intersection, I and II, are defined in terms of the (k + 1) net of the cut S_{k+2}. A point of type I is a vertex of the given set, and points of type II are points on arcs where the parameter γ reaches its extremal values. The points of type II or, more exactly, points of the tangency of the cuts and the strata of the same rank, are called <u>extremal or degeneracy points with respect to the parameters</u> corresponding to the given cut. Points of type III at which the cuts $S_{k+1}(\gamma)$ are tangential to the strata of rank k are absent in the (k + 1) net. In order to find them it is useful to consider the so-called <u>extremal paths</u>, i.e., the paths consisting of extremal points alone. An extremal path of rank k lies at the transaction of a class of rank k with a cut of rank k + 2 and can be singled out by an additional nontransversality condition - degeneracy in the given set of k parameters. The points of family III searched for in the family of cuts $S_{k+1}(\gamma)$ are the points on the extremal k-paths in the cut S_{k+2} at which the parameter γ reaches its extremal values. The initial points for the construction of the extremal k-paths can be extremal points of the path of rank k in the cuts of the family $s_{k+2}(\gamma)$.

An open problem in the above described scheme is whether it is possible to represent any transversal cut in the form of nondegenerate single-parametric family of parallel cuts of a rank lower by unity.

1.4. Motion Over Family of Equilibria

Let us assume that there is a procedure which ensures for each element $p \in M$ the following:

(a) a class of the maximum rank containing an element p and thus its rank k;

(b) the type of the given element (topology of the bifurcation diagram and the phase portraits for the respective local family);

(c) classes of rank k - 1 containing an element p.

Knowing these data, one can formulate possible schemes of motion over the family M with the aim to study its bifurcation structure. As elementary acts we shall consider a constuction of a path in the indicated cut and class, the separation there of singular and extremum points, and for all the singlular points also the description of all paths passing through it in the given cut of paths.

The consistent construction of paths such that the initial point of each following path is a point on one of the already constructed paths sets the route. The union of all the paths making the route is a connected subset in a certain homogeneous or inhomogeneous net. The construction of a route approximately corresponds to the intuitive idea about the "continuous motion over M" with the addition that the graphically represented point can "instantaneously" return to one of the already passed points. Let us note that in our approach the freedom of each individual "motion" is constrained by standard cuts.

The route is usually constructed by the recurrence method: the choice of a following path and an initial point on it depends on the already constructed paths and their properties. The rules determining this choice are associated with the problem under consideration and, generally speaking, can be rather diverse. We describe here several types of routes which, in a sense, are considered as basic. These routes are intended for the study of a fixed cut, the search for degeneracies of high codimensionalities, the "descend" from the degeneracies of high codimensionalities to rough equilibria and their simplest bifurcations.

We should like to indicate the triviality (from the standpoint of the problem to be solved) of the case $m = 1$. In the system of the general position the connected family of equilibria consists of the unique path of rank 0. Therefore we consider the case where $m > 1$.

Route of Type F (of Word "Fixed")

The route of type F is used for the construction of a connected homogeneous net in the given cut. Let us assume that a certain transversal cut of rank $k + 1$ and a point p^0 of rank k in the transection of this cut with M are given. It is necessary to construct a homogeneous net containing the point p^0.

The scheme of constructing a route of type F follows from the definitions of a route and a homogeneous net. At the first step, a path passing through the point p^0 is constructed; such a path is unique. On it, all the singular points are separated and all the paths passing through such points are determined (locally). At the second and the following steps, we choose one of the singular points and a path containing this point but not constructed as yet; then we determine this path completely. The construction of a route of type F is terminated when a list of the revealed but not passed as yet paths is exhausted. The number of paths in this route does not exceed a number of strata in M. Pay attention to a characteristic trend: at the beginning of a route of type F each passed path "opens" new singular points and new paths passing through them. On the contrary, closer to the route end, the paths pass through already known singular points and "close" the paths revealed in their analysis but not constructed as yet.

The scheme of constructing a route of type F can be applied only if in a cut of rank $k + 1$ not one but several points of rank $k + 1$ or k were set from the very beginning. In this case we arrive, generally speaking, at several connectivity components of a homogeneous net; the number of such components does not exceed the number of the initially given points.

Route of Type U (of Word "Up")

A route of type U is intended for the determination of points of the maximum rank (i.e., of degeneracies of maximum codimensionality) in the family M. This route consists of the routes of gradually increasing ranks and thus is linear.

Let a point $p^0 \in M$ of rank 0 and a certain sequence of the addition or "inclusion" of active parameters be given; for simplicity we assume that it is a sequence $\alpha_1, \ldots, \alpha_m$ that defines the sequence of embedded cuts

$$S_0(S_1(\alpha_1)(S_2(\alpha_1,\alpha_2)(\ldots(S_m(\alpha_1, \ldots ,\alpha_m) = P$$

containing the point p^0. Each of these cuts has a corresponding homogeneous net; their union forms an inhomogeneous net denoted by G.

The route of type U consists in the upward motion over G. The first path is of rank O and passes through the point p^0 in the cut $S_1(\alpha_1)$. As usual, we single out all the singular points on this path. The second path is of rank 1 and lies in the cut $S_2(\alpha_1,\alpha_2)$; its initial point is one of the singular points of the first path. Note also that for the second path this point is regular. In a similar way, the kth path of the U-type route is of the rank k - 1, lies in the cut $S_k(\alpha_1, \ldots, \alpha_k)$, and passes through one of the points of rank k-1 determined on the previous path. The route is ended upon the construction of m paths or when the last constructed path does not contain any singular points.

When describing the route of type U, we assumed that the sequence of adding active parameters was given a priori. But in fact, the choice of these parameters can be chosen during the construction of the route, i.e., at each its step.

The U-type route can be modified both towards narrowing and extending the set in G that is covered by this route. Let us consider two variants.

In a short variant, called the route of type $\breve{U}$, we construct at each step not the whole route but only its segment from the initial point to the first singular point (in the chosen direction). As soon as such a point is found, the construction of the given path is terminated, a new parameter is "switched on", and a construction of a path with a rank higher by unity is started. As a result of the motion along the route of type $\breve{U}$ we obtain a piecewise smooth curve whose smooth portions lie in the strata (and classes) of consistently increasing rank.

An example of the $\breve{U}$-type route used for the search for degeneracy of codimensionality three is a "quadruple equilibrium" (the so-called

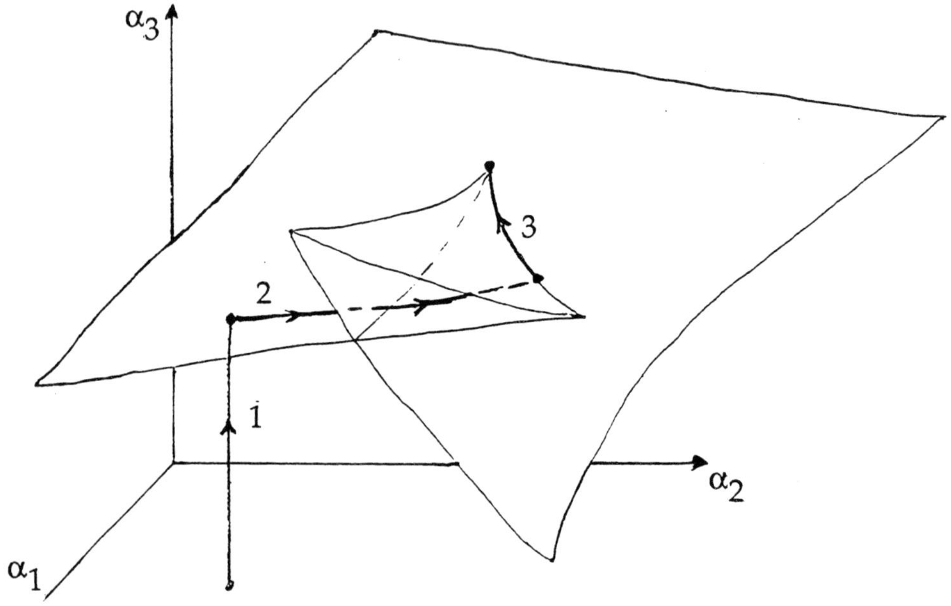

Figure 2. Illustrating the search for degeneracy of codimensionality three ("a swallow tail") with the use of a Ʊ-type route. 1, 2,and 3 indicate the subsequent paths in the 1-, 2-, and 3-cuts.

"swallow tail") shown in Figure 2 that depicts the projection of the family of equilibria and of the route there in the space of parameters.

The idea of extending the U-type route consists in the following. Studying consistently the cuts $S_1, S_2, \ldots, S_m$, one has to construct in each of them instead of one path a certain net of paths. Let us define the route of type UF in such a way. Let a certain set of points of rank k be determined in a cut S_k (at the beginning it is the only point of rank 0). Let us take these points as initial and implement the F-type routes in the cut $S_{k+1}, S_{k+1} > S_k$. We arrive at one or several connectivity components of a homogeneous k-net. In this set we separate all the singular points (points of rank k + 1). In a similar way we study also the cut S_{k+2}, $S_{k+2} > S_{k+1}$, etc.

A net of paths obtained upon the passage of the UF-type route is connected. Any point of this net can be obtained from the initial point p^0 with the aid of a sequence of paths each of which is of a rank equal to or higher by unity than the previous one. At the same time the given net, generally speaking, does not coincide with the inhomogeneous net G corresponding to the set of cuts S_0, S_1, ..., S_m.

Let us make one more remark of a general character. The experience indicates a certain similarity in the routes of types F and UF: with the progress in the construction of the route the paths start intersecting more often. In other words, the number of singular points revealed in each following cut is larger at the beginning of the route and then start decreasing (ideally, down to unity). This reflects an interesting fact that in many instances the whole bifurcation structure of the family **M** is "generated" by one or several degeneracies of the maximum codimensionality. This fact was first noted by A.M. Molchanov [7,8] who simulated critical modes of biological systems. A large number of degeneracies of the maximum codimensionality seems to indicate that the set of the varying parameters of the system does not correspond to its "complexity" (i.e. to "complexity" of the dynamical modes and their bifurcations) and that it therefore should be extended.

Route of Type D (of Word Down).

This route is reverse to the U-type route. Here a set of cuts $S_m = P$, S_{m-1}, ..., S_1, $S_0$is assumed to be given. This corresponds to graduate "switching-off" of the active parameters, i.e. $S_k \supset S_{k-1}$. The cut S_o corresponds to the given fixed values of all the parameters: $\alpha = \alpha^* = (\alpha_1^*$, ..., $\alpha_m^*)$. Moreover, it is assumed that the initial point $p^0 \in$ **M** of the route is given which, in this case, is of rank m.

A route of type D consists in the downward motion over an inhomogeneous net G corresponding to the indicated cuts: the rank of each following path is lower by unity than the rank of the previous one. We take as the first path any path of rank m - 1 in the cut S_m passing through point p^0. Then we separate on this path all the points of its intersection with the cut S_{m-1}; it is assumed that they are of rank m-1. Then the cut S_{m-1} is considered and a point of rank m - 1 is chosen from the determined points; through this point a path of rank m - 2 is constructed, etc. The route is terminated when m paths have been constructed or when a running path of rank k does not contain any points of the intersection with the cut S_k.

The reverse route with respect to the route of type $\breve{U}$ is the route of type $\breve{D}$, a shortened variant of the route of the D-type. In this new route we construct at each step not the whole route but only a segment between the initial point and the first point of the intersection with the next cut. The scheme for constructing a route of type DF, the extension of the D-type route, is similar to that of type UF: each paths is replaced by the F-type paths in the same cut, and the points of the intersection of the paths of these routes with the next cut are taken as the initial points for construction of next F-type routes.

One more modification of a route of the D-type or, more exactly, of the DF-type, is associated with the construction of a covering net. In this case we cannot restrict ourselves, as a rule, to the ordered set of cuts with respect to embedding and have to consider a partially ordered set. Such a set arises in accordance with the scheme stated at the end of Sect.1.3. Namely, in a running cut of rank k we construct a net as in the DF-type route. Then this cut is represented in the form of a single-parametric family of cuts of rank k - 1; among them we choose "typical" cuts and fix the points of the intersection of this net with the cuts. A similar procedure is performed for each of thus separated cuts. Obviously that the given scheme generates a new route called a DFF-type route.

Let us emphasize once again that constructing a covering net, it is necessary to single out at each path not only singular points but also extremum points and construct extremum paths. This relates also to the motion over M both upwards and downwards. Let us also emphasize that for the construction of a covering net it is necessary to make several "up-and-down" passages over M using the UF- and DFF-type routes.

1.5. Connected Family of Limit Cycles

The above introduced concepts and methods can be used for studying not only equilibria but also limit cycles, at least if m = 1,2. We shall discuss here some problems associated with such studies.

Let us consider a union of limit cycles of system (1.1) in the phase-parametric space P. It is a manifold of dimensionality m + 1. Let a certain limit cycle Γ^0 be given at the parameter value α^0. It is necessary to study the connectivity component in the union of limit cycles containing an element (Γ^0, α^0). Let us denote this component by C and call it a connected family of limit cycles.

Let us assume that the family C has a bifurcation structure similar to the bifurcation structure of the family M. The dimensionality of classes and strata of rank k in C is equal to m - k + 1. The path in C is a smooth connected single-parametric family of limit cycles lying completely in one of the classes and singled out by the transection of this class with its cut having a rank higher by unity. The path in C, being a geometrical set in P, is a two-dimensional manifold.

If the bifurcation structure in the family C is simple, then all the above described schemes of motion over M can be applied also to this family without any changes. These are the study of a connected set of k-paths in the (k + 1) cut, a consistent increase and decrease of the cut dimensionality and, therefore, of codimensionality of the degeneracies studied there.

Despite the above indicated similarity, the families of equilibria and limit cycles have also some substantial differences. First, the family C is not defined explicitly in terms of the right-hand side of system (1.1) like it was for equations (1.2) for equilibria. Instead, it is given in an implicit form in terms of the condition for the existence of a periodic solution of system (1.1) or the condition for the existence of an immobile point of the Poincaré mapping. This is also reflected in the difficulties met in the numerical study of families of limit cycles (see Ch.2).

Second, degeneracies determining the bifurcation structure are different for the families of equilibria and limit cycles. In the case of a general position there are three types of degeneracies of limit cycles of codimensionality I (i.e., three classes of rank I). They are determined by conditions (1) $\mu_1 = I$, (2) $\mu_1 = -I$, and (3) $\mu_{1,2} = e^{\pm i\varphi}$, respectively ($\mu$ is the cycle multiplicator). The classes of rank two associated with degeneracies of codimensionality two for cycles can be divided into three groups: i/ three classes corresponding to the simplest additional degeneracies in nonlinear lengths for each of the indicated linear degeneracies, ii/ six classes corresponding to double linear degeneracies

$$\mu_{1,2} = 1;\ \mu_{1,2} = -1;\ \mu_1 = 1 \wedge \mu_2 = -1;\ \mu_1 = 1 \wedge \mu_{2,3} = e^{\pm i\varphi};$$
$$\mu_1 = -1 \wedge \mu_{2,3} = e^{\pm i\varphi};\ \mu_{1,2} = e^{\pm i\varphi}1 \wedge \mu_{3,4} = e^{\pm i\varphi}2\ ;$$

iii/ an enumerable number of classes corresponding to a pair of resonance multiplicators $\mu_{1,2} = e^{\pm 2\pi p/g}$ (p and q are integers, q > 2). The

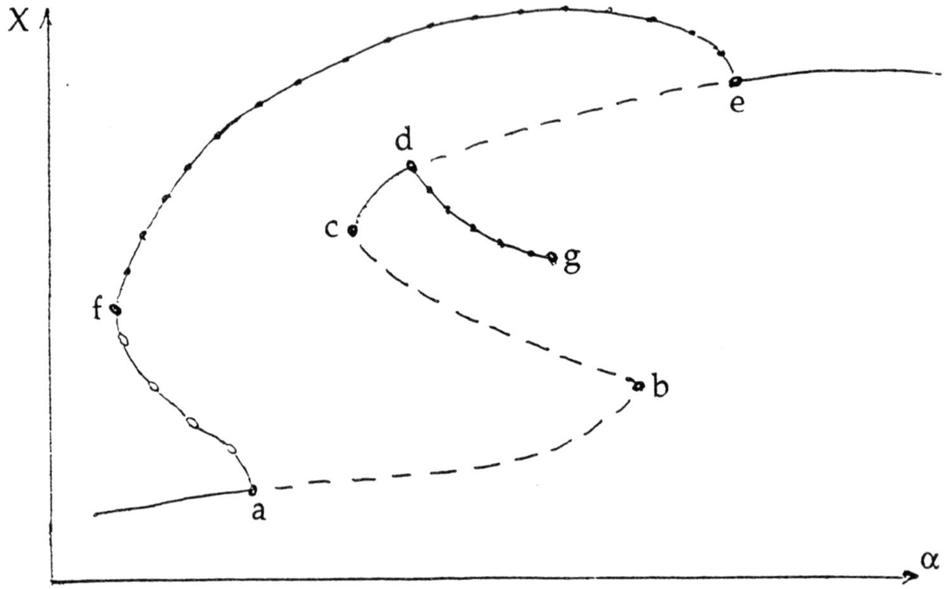

Figure 3. The connected family of equilibria, limit cycles, and separatrix contours (schematic depiction). Notation: ——— denotes stable equilibria, -------- denotes unstable equilibria, -•-•-•-•-denotes stable limit cycles, and -o-o-o- denotes unstable limit cycles. To point g there corresponds a separatrix loop. Letters a, b, c, d, e, f, and g indicate degeneracies of codimensionality 1.

character of degeneracies in the family of limit cycles is such that even in the case of a general position the connected set of O-paths in the I-cut can contain more than one path (compare with a similar situation for equilibria). The possibility of intersecting different O-paths is associated with the bifurcation of the formation of a double-period cycle in the passage by the cycle multiplicator of the value -1.

Finally, not all the degeneracies that influence the bifurcation structure of the family C lie in this family. The point is that with the parameter variation, the limit cycle can tend (in the sense of a topological limit) not only to a limit cycle but also to a nonrough equilibrium or to a separatrix contour, whereas the equilibria can tend only to equilibria. In other words, a family of equilibria is closed in the phase-parametric space whereas the family of limit cycles can be nonclosed. The consequence of this is the possible existence in P of a large number of "small" connected families of limit cycles each of which should be found and studied separately.

The above considerations make the joint study of equilibria, limit cycles and separatrix countours quite natural. An example of a connected family of equilibria, cycles, and contours for m = 1 is shown in Figure 3.

To provide the closure of the family of equilibria, cycles, and separatrix contours in the phase-parametric space it is possible to include into it only such cycles and contours whose length does not exceed a certain given value. We do not know whether the condition of the boundedness of the cycle and contour lengths is really sufficient for the closure. Yet it seems that it is really so. In any case, this condition excludes from the consideration the infinite series of bifurcations caused by doubling of the cycle period (which agrees with the requirement of a simple bifurcation structure (i.e., a finite number of strata), bifurcations of long-period resonance cycles on a torus, and the "blue sky" catastrophe.

Footnotes

*A separatrix contour is a closed invariant curve composed by a finite number of equilibria (in particular, by one) and their one-dimensional separatrices.

**In the Hausdorff metrics.

***In these cases the difficulties arising in bridging the gaps in the theory are of the fundamental character.

CHAPTER 2. PROBLEMS OF NUMERICAL IMPLEMENTATION

2.1. Main Problems

The implementation of the algorithm for studying the connected families considered in the first chapter reduces to the solution of two main problems - a global one consisting in the construction of paths and search on them for special elements and a local one consisting in studying the properties of individual elements of the family. (The expressions "local" and "global" relate to the parametric space).

The construction of a k-path can be reduced to the construction of a smooth curve in a certain N-dimensional space (or on an M-dimensional manifold). In the simplest cases such a space is the (k + 1)st cut of the phase-parametric space. The choice of the space and the system of equations describing the representing curve depends, to a large extent, on the properties of the class containing the given path. The construction of the curve itself given by a nondegenerate* set of nonlinear equations can be considered as a standard numerical problem [14, 19, 20]. Thus, the main difficulty here is the choice of the equations determining the curve described by the equations and the calculation of the functions setting these equations.

A class of rank k in the family partition is determined with the aid of K degeneracy or bifurcation conditions (of equality type). Moreover, a class is characterized by the nondegeneracy or typicalness conditions of the inequality type that single out regular elements in it. The relation between the bifurcation and typicalness conditions consists in that the transformation of inequalities into equalities under the typicalness conditions provides the formation of new bifurcation conditions that single out classes of higher ranks (i.e., corresponding to degeneracies of a higher codimensionality).

The transformation of one of the inequalities in a singular path element under the typicalness conditions into an equality can be used for the search of special elements. Indeed, if the typicallness condition is formulated explicitly as nonvanishing of a certain set of functions on the curve, then the search for singular elements reduces to the search for zeroes of these functions in the motion over the curve. The given procedure offers no essential difficulties if the functions are smooth and their zeroes are simple. Thus also in the problem of finding singular path elements the main point is an appropriate choice of the functions determining the typicalness conditions and their calculation. Since these functions, as has been indicated earlier, enter also the formulations of the bifurcation conditions, they are called the bifurcation functions.

The above suggested scheme of the search for singular elements relates only to internal singular path elements. The boundary elements should be considered separately. If functions that set the curve and the typicalness conditions on it can be additionally smoothly defined in the vicinity of the boundary point, then the problem reduces to the previous one. Otherwise in order to find the boundary elements some special procedures have to be used. Let us note that an inverse problem, e.g., a construction of a path with the origin at the boundary element, can also be nontrivial; some difficulties here can be encountered on the initial portion of the path.

Thus the solution of the problem of a global parametric analysis is built by two parts - "individual" and "standard". The first one consists in setting a curve representing the given path in the connected family, indication of the typicalness conditions and possible character of boundaries. The second one consists in the construction of a curve and the determination of the zeroes of some functions on it.

The solution of the problem of the local parametric analysis includes some procedures:

1) the determination of the class and stratum in which the given family element lies;

2) the description of the bifurcation diagram for the family element and the construction of the respective local phase portraits;

3) the description of the cut paths adjacent to the given singular element; the construction, if necessary, of an asymptotic representation of these paths in the vicinity of a singular element.

The performance of all these procedures is associated, as a rule, with the reduction of the system with the aid of some transformations to a specific form, the so-called "model system"** [3], for which all the necessary data are considered to be known (the reduction is performed locally in the vicinity of the family element under consideration with a required accuracy). To each class there corresponds its own model system.

Indeed, the use of the model systems of differential equations provides the investigation of equilibria alone. In the analysis of limit cycles instead of the model differential equations the model mappings are used: to a specific form then the Poincaré mapping is reduced. In the analysis of the separatrix contours it is possible to use simultaneously model differential equations and model mappings - the former for the study of the equilibria entering the contour and the latter for the analysis of sequence and correspondence mappings along the contour and separatrices entering it.

In the solution of the first problem - the determination of the class and stratum containing the given element - of great importance are the coefficients obtained for model systems. Very often they can be calculated straightforward without using the reduction procedure. In order to solve the second and third problems it is also necessary to know a concrete form of the reducing transformations.

The above indicated global and local problems are not absolutely independent. The bifurcation and typicalness conditions are usually formulated in terms of the model systems coefficients. Therefore the implementation of the global problem includes as an important constituent part the procedures that were used in the implementation of topic (1) of the global problem.

The local problem also consists of the general and individual parts. For equilibria, the general part in the reduction to the model system is the separation of the central manifold, the reduction of the system onto this manifold, and then its reduction to the required form with the help of normalizing transformations. The individual part (for the class) is a form of the model system, the order to which the system should be normalized, the type of the normalizing transformations, and the formulae for checking the typicalness conditions and the determination of the stratum containing the given element. All the above said relates also to the limit cycles within the accuracy of the replacement of the differential equations by mappings. As to the separatrix contours, it is very difficult to separate the general and individual parts of the local problem.

Now let us consider the following problem: which classes can be contained by the family under consideration in the given concrete system, what degeneracies and singular elements should be sought in it?

Generally speaking, different types of systems are characterized by different types of the sets of degeneracies. Moreover, formally equal degeneracies are of different codimensionalities in different classes of systems. Thus, a triple equilibrium is a degeneracy of codimensionaly two for the systems of the general position and of codimensionality one for the systems with the reflection symmetry.

The answer to the above formulated problem should indicate the system class to which the system under consideration belongs. The answer to this question determines the part of the bifurcation theory that should be invoked for its study, and, therefore, also the assumptions that should be made in the developing of respective algorithms. The latter note about algorithms signifies that the above indicated individual parts of the local and global problems are also strongly dependent on the system type. Thus, each type of the systems (those of the general position, those with certain symmetry groups, those with the first integrals, etc.) dictates the development of its own algorithms. The use of inappropriate algorithms, i.e., errors in the determination of the type of the system under consideration can give rise to most diverse failures. Thus the system of the nonlinear equations setting the representing curve for a certain nondegenerate path can turn out to be degenerate.

Up to now we mentioned only one aspect of the determination of the system type that can be conditionally called the phase aspect. It concerns the problem of the codimensinality of degeneracies and the rank of the respective classes in the bifurcation structure of the family. The correct indication of the system type signifies that the ranks of all the family classes coincide with the standard values for the systems of the given type.

There is one more aspect of the problem - parametric - associated with the appropriateness of the parameter set of the system. The point is that in virtue of the specific parameter choice for the given system its characteristic feature can turn out to be not an intersection but a tangency of the respective cuts and classes determining degenerate paths. ...... yields the one-dimensional equation ($x \in \mathbf{R}$)

$$\dot{x} = xf(x, \alpha) \tag{2.1}$$

containing the indicated degeneracy on the manifold $x = 0$, $f(0,\alpha) = 0$. In such cases, the obtained bifurcation diagrams are also somewhat different from the standard ones corresponding to the principal families. The model systems in such cases are obtained from the standard ones with the aid of irreversible (at some points) change of the parameters. Thus, in addition to the system type, it is necessary to indicate the character of its dependence on the parameters. This should also be taken into consideration in the development or the choice of the respective numerical algorithms.

Indeed, in the real systems one often encounters a mixed situation where some bifurcations occur like in the systems of a general position, others like in the systems with symmetry, still others like in the case of the above indicated specific entering of parameters. Fortunately, the situation is simplified by the fact that various "deviations" from the general case occur not randomly but on certain separated invariant manifolds which can be indicated in an explicit form. Thus, in the system with a symmetry such a manifold is a symmetry plane, and in equation (2.1) it is the point $x = 0$. All this signifies that for the investigation of one system we may need several algorithms oriented at different types of systems and parameter entering. The explicit indication of the invariant manifolds and the singularities arising on them provides the use of adequate algorithms in each situation. In principle, the general algorithm for studying a system would envisage the possibility of setting and use of such a priori information about the system.

We have discussed only the main problems associated with the development of algorithms of the continuous parametric analysis. The further details essentially depend on the type of a system, the invariant manifolds studied in it, and also on some other properties of the system, e.g., on the presence in it of small parameters. Below we consider a variant of the numerical implementation oriented to the global parametric analysis of equilibria and limit cycles in the case of the general position.

2.2. Numerical Analysis of Equilibria

In this Section we describe the general structure of numerical algorithms of the global parametric study of equilibria in the systems of the general position. These algorithms are implemented within the LINLBF [101] program complex for the construction of individual k-paths in the family of equilibria for k = 0, 1, 2, 3.

When studying equilibria, the notions of a path and a curve representing it are taken to be equivalent. The curve corresponding to the k-path is constructed in the (k+1) cut of the phase-parametric space, i.e., in the (n + k + 1)-dimensional space. It is given with the aid of n + k equations. Of them, n equations are of type (1.2) and define equilibria; such equations are called basic. Additional k equations define the bifurcation conditions. The 0-path, i.e., the curve of rough equilibria, is given by the basic equations alone.

In order to construct a curve corresponding to a certain k-path it is necessary: i/to indicate a class of rank k containing this path; ii/ to separate a set of active parameters α_{i_1} , ..., $\alpha_{i_{k-1}}$; iii/ to set an initial point of the path. Knowing the class, we can determine a set of k bifurcation functions whose vanishing yields additional equations for setting the curve and a set of (other) bifurcation functions for checking the typicalness conditions and the search for singular elements. The active parameters and the phase variables form a set of the spatial coordinates in which the curve is constructed. We note that all these coordinates are considered as equivalent. The initial point of the path determines the initial point of the curve and also the fixed values of all inactive parameters.

The construction of a curve implies a consistent search for the points of the curve that correspond to the "motion over it", i.e. to a monotonous and smooth variation of the natural parameter - the length of the curve arc. The found points should lie on the curve we search for with the required accuracy and density. A relative point density should be higher there where the curve has its maximum curvature. The latter requirement allows us to hope that in constructing the curve, not any of its portions will be omitted. . In addition to the above indicated curve points, it is also necessary to find some additional ones, namely, those of self-intersection, those where the functions set on the curve are going to zero, and those where the above variables attain their extremum values.

The equations defining the curve should be nondegenerate and smoothly defined in its vicinity, which signifies the existence of a vector tangential to the curve considered as a solution of the linearized set of equations (such a vector would exist everywhere except of some isolated points of the curve). The functions that set additional singular points on the curve should be smooth, and their zeroes should be simple.

Each new (regular) point of the curve is found in two stages: at the first stage a new point is predicted and on the second this point is refined. The prediction is based on the consideration of the tangential vector to the curve at the previous point. The refinement is performed by the Newtonian procedure in the plane corresponding to the fixed value of one of the variables. The function zeroes and the extrema on the curve are determined by two combined methods - that of chords (for prediction) and the Newton method (for refinement).

Note. The above described scheme of the motion along the curve is implemented in the BEETLE program written by E.V. Nikolaev and the author in 1987, and also in the PITCON program /20/. It is also suggested to write a new version of the LINLBF complex on the basis of the BEETLE program. The initial version was besed on the CURVE program [4] implementing a close but somewhat different scheme. The prediction of a new point in the CURVE program is made along the chord passing through two previously determined points, and its refinement is performed by the Newtonian procedure modified for systems having a one-dimensional manifold of solutions. At each step of this procedure, the system, is linearized, as usual, and then its normal solution is computed, i.e., from all possible Newtonian shifts we choose that which is minimal in norm. With the aid of this algorithm we determine one of the curve points not separated a priori in any special way. As a rule, thus determined point is close to the predicted one, but in some situations the refinement procedure results in a noticeable drift along the curve.

Thus, the final step in finding the paths in the family of equilibria is the determination of necessary bifurcation functions. As was indicated above, these functions should be smooth functions of their arguments defined in the whole neighborhood of the curve. Also, they should be computed rapidly and with a small computational error, since it is very important for the speed of the program execution and the reliability of the obtained results. The above considerations dictate the refusal, when possible, of setting the bifurcation functions in terms of the eigenvalues of the linearization matrix of system (1.1) and the use of the matrix itself.

Let $A = \partial F/\partial x$ be a linearization matrix of system (1.1) in an equilibrium, $P(\lambda)$ be a characteristic polynomial of the matrix A, and λ_j ($j = 1, ..., n$) be its roots, i.e. the eigenvalues of the matrix A. The resultant of two polynomials $u(\lambda)$ and $V(\lambda)$ is denoted by $Res(U(\lambda), V(\lambda))$.

Now let us consider "linear" degeneracies of the equilibrium related to the spectrum degeneracy of the matrix A. There are two simple (of codimensionality 1) degeneracies: $\lambda_1 = 0$ and $\lambda_{1,2} = \pm i\omega$. The first degeneracy corresponds to the condition

$$\det A = 0, \tag{2.2}$$

the second to the condition

$$\mathrm{Res}\ (P(\lambda), P(-\lambda)) = 0. \tag{2.3}$$

Indeed, the determinant of the matrix A is equal to the product of eigenvalues and is zero then and only then when among the eigenvalues there are some equal to zero. Vanishing of the polynomials $P(\lambda)$ and $P(-\lambda)$ signifies that they have common roots, i.e., that there exist two eigenvalues such that their sum is zero. Such eigenvalues are either complex conjugated (then they are pure imaginary), or real (then they have different signs). Formally, it is only in the first case that (2.3) is equivalent to the condition $\lambda_{1,2} = \pm i\omega$. But the second case is also of interest because of the bifurcations of the saddle separatrices.

Two of three linear degeneracies of codimensionality two, $\lambda_1 = \lambda_2 = 0$ and $\lambda_1 = 0$, $\lambda_{2,3} = \pm i\omega$, are set by a pair of equations (2.2)and (2.3). The latter linear degeneracy of codimensionality two, $\lambda_{1,2} = \pm i\omega$, $\lambda_{3,4} = i\omega_2$, is defined as a condition of selfintersection of the manifold corresponding to a pair of pure imaginary eigenvalues and given by equations (1.2) and (2.3). But, since at the point of selfintersection these equations, or, more precisely, their linear parts, are linearly dependent, the addition to them of any equation that expresses the condition of selfintersection results in the formation of a degenerate system of equations. Thus, it is impossible to describe two pairs of pure imaginary eigenvalues using the combination of equation (2.3) and any other bifurcation condition, as was made earlier. We do not know whether there exists an alternative pair of conditions defining in a nondegenerate way of the given curve in the same space.

Note 1. The above indicated degeneracy can be lifted by increasing the dimensionality of the space in which the curve is constructed. Namely, the number of independent variables is complemented with ω_1 and ω_2, and instead of equation (2.3) and the condition of self-intersection the use is made of two complex equations:

$$\begin{aligned} P(iw_1) &= 0 \\ P(iw_2) &= 0 \end{aligned} \tag{2.4}$$

which are equivalent to four real ones.

Note 2. The condition of multiple eigenvalues can naturally be related to linear degeneracies. Formally, it is not bifurcation condition, but it sets a "node-focus" boundary important for many applications. The condition of multiple eigenvalues is set by the equation

$$\text{Res}\ (P(\lambda),\ P'(\lambda)) = 0 \tag{2.5}$$

Note that this equation, similar to equation (2.3), enters the bifurcation conditions for some bifurcations of the saddle separatrices of codimensionality two.

Two so-called nonlinear degeneracies of codimensionality two are defined by combining one of conditions (2.2) or (2.3) with a respective additional condition expressed in terms of the coefficients of the nonlinear terms of the Taylor expansion. For the zero eigenvalue such an additional condition can be set in the form

$$\frac{d^2}{d\xi^2} \left. \frac{(e^*, F(x^0 + \xi e))}{(e^*, e)} \right|_{\xi=0} = 0, \tag{2.6}$$

where X^o is an equilibrium, e and e^* are eigenvectors of the matrix A and the conjugated matrix A^* corresponding to the zero eigenvalue. The pair of conditions (2.2) and (2.6) defines a triple equilibrium.

If there is a pair of pure imaginary numbers, the additional bifurcation condition has the form

$$l_1 = 0, \tag{2.7}$$

where l_1 is the coefficient of the cubic term in the corresponding complex model equation [3]:

$$\dot{Z} = (\beta_1 + iw)Z + l_1 Z^2 \bar{Z}, \tag{2.8}$$

which is called the first Liapunov quantity. It is calculated with the aid of rather a cumbersome procedure carried out in three stages:
i/ the reduction of the linear part of the system to a canonical form;
ii/ the computation of the quadratic approximation of the two-dimensional central manifold and the reduction of the initial system onto this manifold within the accuracy of cubic terms;
iii/ the normalization of the obtained (complex) equation on the central manifold up to cubic terms and the computation of a necessary coefficient.

In practice, all these procedures reduce to the computation of necessary derivatives of the right-hand sides of the system (or of some intermediate functions) up to the third order and the solution of a series of linear systems of algebraic equations (for more detail see [12]). Unfortunately we do not know any simpler (in the computational sense) bifurcation condition equivalent to (2.7).

Note. The algorithm for the computation of the first Liapunov quantity is an essential part of the ASIMPC [10] and BIFOR2 [12] programs for the analysis of the bifurcation of cycle generation (the

Andronov-Hopf bifurcation). In the LINLBF complex, the computation of the first Liapunov quantity is carried out by a respective subroutine from ASIMPC.

Thus, we have introduced a set of bifurcation functions into (2.2)-(2.7). They provide the determination of the curves which correspond to the 1-, 2- and some 3-paths in the family of equilibria. In order to determine the 1-paths we use one of conditions (2.2) or (2.3): for the determination of the 2-paths the pairs of conditions (2.2) and (2.3), (2.2) and (2.6), and (2.3) and (2.7) are used in combination with condition (2.4); for the determination of the 3-paths the triads of conditions (2.2), (2.3) and (2.6) or (2.7) are used. Note here, that all the introduced bifurcation functions can be additionally defined in the vicinity of the curve under study, although it is not always very simple. They are also used for checking the typicalness conditions and the search for singular path elements.

Note. In order to determine other 3-paths different from the above indicated ones it is necessary to compute and equate to zero the coefficients of the model systems corresponding to the degeneracies of codimensionality two. In a similar way it is also possible to determine the 4-paths for all those degeneracies of dimensionality three for which the model systems are known. Note here that both coefficients of the model systems and some of their functions*** can play the role of bifurcation functions. The procedures proper for computing such coefficients are similar to those used for the computation of the first Liapunov quantity.

The formation of the equations defining the extremum paths in the family is illustrated by the following example. Suppose we have to construct an extremal I-path in a class corresponding to a pair of pure imaginary numbers such that it corresponds to degeneracy with respect to parameter α_{i_1}. The extremal path is constructed in a certain $(\alpha_{i_1}, \alpha_{i_2}, \alpha_{i_3})$ cut (i.e. in a (n + 3)-dimensional space) and is given by a system of equations (1.2), (2.3) and

$$\det B = 0, \tag{2.9}$$

where B is the matrix of partial derivatives with respect to $x_1, \ldots, x_n$.

α_{i_1} of equations (1.2) and (2.3). In the general case, the extremal path in the class can be determined with the use of an additional equation similar to (2.9). In other words, we equate to zero the determinant of the matrix of partial derivatives of the complete system of equations (basic and additional) defining the given class with respect to the phase variables and "degenerate" parameters. It should be noted that equation (2.2) defining the I-path is constructed just in such a way; it can also be interpreted as the extremal O-path. Upon the determination on the curves of the points where the parameters take their extremal values they can be used as initial ones for the construction of extremal paths.

In the LINLBF complex all the necessary bifurcation conditions and the functions for checking the typicalness conditions and the search for specific elements are formed automatically in accordance with the indicated class in which the path is sought. Thus the user should specify the system of equations itself, the class under study (i.e., the type of degeneracy), the set of active parameters, and the initial point of the path.

Concluding this section we should like to note that one of the key problems in the numerical construction and study of the paths is the computation of necessary derivatives. The point is that we have to differentiate not only the right-hand sides, given by the formulae, but also various functions of them, given by various computational procedures, which sometimes are so complicated that they themselves may include the computation of derivatives inside themselves. In many instances one has to compute not only the first but also higher-order derivatives. For differentiation we use the standard difference formulae. A step of numerical differentiation is chosen with the use of the standard values associated with the rounding-off error. Or else the step is chosen with the aid of a special algorithm [5]. And nevertheless differentiation remains to be the least developed computational procedure. But the problem becomes even more difficult if limit cycles are studied.

2.3. Numerical Analysis of Limit Cycles

In this section we describe the structure of numerical algorithms used in the global parametric analysis of limit cycles in systems of general positions implemented in the LINBAS complex of programs [10]. The complex is designed for the construction of separate k-paths in the family of limit cycles for k = 0, 1, 2.

The study of a family of limit cycles reduces (with the aid of the sequence mapping g defined on a certain secant plane L) to the study of a family of immobile points of this mapping. Of great importance is the fact that the degeneracies of limit cycles are also defined in terms of degeneracies of immobile points of the sequence mapping****.

We shall take that the procedure of numerical construction of a sequence mapping already exists. It includes the computation of the system trajectory (1.1.) between the initial point of the secant plane and the point of the next intersection with this secant (in the same direction) and the refinement of intersection point [9]. Then one can concentrate on the general problem of the parametric study of immobile points of smooth mappings.

Note that using an immobile point of the sequence mapping, it is possible to reconstruct uniquely the whole cycle by constructing the trajectory outgoing from this point. In the search for an immobile point of the sequence mapping we automatically determine also the cycle period.

Note 1. Sometimes an unknown cycle period is added to the system parameters, and one considers instead of the sequence mapping the mapping of the phase space into itself for one period [15]. The initial point of the trajectories for constructing the given mapping is then chosen at a certain fixed hyperplane.

Note 2. We assume here that the whole connected family of limit cycles can be studied with the aid of one secant plane. If it is not the case, then the path in the family should be divided into separate portions. Then for each portion its own secant plane is used. In principle, it is also possible that the secant plane is changing "dynamically" in the process of motion over the family, e.g., a continuous evolution of the plane that it is orthogonal to the field vector at each determined immobile point [9].

Note 3. The study of limit cycles of a complicated geometric form, e.g., of multiturn cycles with close turns reduces to the study of periodic points of the sequence mapping g. If the periodic trajectory returns to the intial point on the secant plane at its lth intersection with this secant, then this point is a periodic point of period l of the mapping g or an immobile point of the mapping g^l, the lth degree of mapping g. The mapping g^l will also be called the sequence mapping; it is characterized by an additional integer parameter l.

Note 4. From the computational standpoint the search for limit cycles with the aid of sequence mappings is possible for systems in which the rates of variable change are of close orders. For systems with "fast" and "slow" variables such a procedure yields a substantial computational error, and therefore the use is made of some alternative methods, e.g., the reduction to a boundary problem [13, 21].

Now consider the problem of studying a connected family of immobile points of a smooth mapping

$$y \longmapsto G(y,\alpha), \tag{2.10}$$

where $y \in \mathbf{R}^{n'}$, $\alpha \in \mathbf{R}^m$. In case where G is the sequence mapping of system (1.1), the phase space (2.10) is a set of points of a secant plane, i.e., $n' = n - 1$.

The numerical construction of paths in a family of immobile points of smooth mappings is carried out in the same way as the construction of paths in a family of equilibria of differential equations. Namely, immobile points are determined by the set of equations

$$G(y,\alpha) = y, \tag{2.11}$$

similar to system (1.2). The system (2.11) sets in the I-cut of the phase-parametric space of the mapping the G curve of immobile points which corresponds to the O-path. Complementing system (2.11) with bifurcation conditions and increasing, at the same time, the number of active parameters, it is possible to determine paths in various classes.

The construction of paths in the family of immobile points of mappings of type (2.10) in the case of general position is performed by the LINBFP complex [10] similar to the LINLBF program complex. The LINBFP complex made the basis for the LINBAS complex for studying limit cycles.

Let us formulate the bifurcation conditions for immobile points of mapping (2.10). Let $A = \partial G/\partial y$ be a linearization matrix for mapping (2.10) at an immobile point (called for the sequence mapping a monodromy matrix), $P(\mu)$ be a characteristic polynomial of the matrix A, μ_j ($j = 1, \ldots, n'$) be its roots called multiplicators.

Then there are three simplest degeneracies (of codimensionality one) of immobile points:

the first one, $\mu_1 = 1$, is given by the condition

$$\det(A - E) = 0; \tag{2.12}$$

the second, $\mu_1 = 1$, by condition

$$\det(A + E) = 0; \tag{2.13}$$

and the third one, $\mu_{1,2} = e^{\pm i\varphi}$, by the condition

$$\mathrm{Res}(p(\mu), p(1/\mu)) = 0. \tag{2.14}$$

where E is the unit matrix.

Combining the pairs of conditions (2.12)-(2.14), it is possible to set five degeneracies of codimensionality two. Equations (2.12) and (2.13) define two degeneracies, $\mu_{1,2} = 1$ and $\mu_1 = 1$, $\mu_{2,3} = e^{\pm i\varphi}$, equations (2.13) and (2.) also two degeneracies, $\mu_{1,2} = -1$ and $\mu_1 = -1$, $\mu_{2,3} = e^{\pm i\varphi}$, and, finally, equations (2.12) and (2.13) define one degeneracy, $\mu_1 = 1$, $\mu_2 = -1$.

In setting one more linear degeneracy of codimensionality two, $\mu_{1,2} = e^{\pm i\varphi}$, $\mu_{3,4} = e^{\pm i\varphi}$, the same problem arises as in the case of two pairs of pure imaginary eigenvalues for an equilibrium (see Sect.2.2). Similar to the case of Sect.2.2, adding quantities φ_1 and φ_2 to independent variables, one can also set the given degeneracy by a pair of complex equations

$$P(e^{i\varphi_1}) = 0$$

$$P(e^{i\varphi_2}) = 0. \tag{2.15}$$

Finally, there is a series of linear degeneracies of codimensionality two related to the "resonance" points on a unit circumference $\mu_{1,2} = e^{\pm 2\pi i p/q}$ (p and q are integers)*****. The condition for the resonance p:q is set by one complex equation

$$P(e^{2\pi i p/q}) = 0, \tag{2.16}$$

equivalent to two real ones. One more possible definition of the resonance p:q is the combination of (2.14) with one of equations (2.16).

There are also three degeneracies of codimensionality two associated with the nonlinear degeneracies additional to (2.12)-(2.14). The nonlinear degeneracy for the case $\mu = 1$ is defined similar to (2.6)

$$\frac{d^2}{d\xi^2} \left. \frac{(e^*, G(y^0 + \xi e, \alpha))}{(e^*, e)} \right|_{\xi=0} = 0 \tag{2.17}$$

where y^0 is an immobile point, e and e^* are the eigenvectors of the matrices A and A^* corresponding to $\mu = 1$. The nonlinear degeneracies for two other cases are also completely analogous to condition (2.7) for an equilibrium, i.e., the coefficients before the cubic terms of the respective model mappings are equal to zero.

$$U \longmapsto (B - 1)U + \nu U^3, \quad U \in \mathbb{R}, \tag{2.18}$$

$$Z \longmapsto (B + e^{i\varphi})Z + \kappa Z^2\bar{Z}, \tag{2.19}$$

The above coefficients are computed by the same scheme as the first Liapunov quantity for an equilibrium, only in the first case the central manifold is one-dimensional.

Thus we listed all the degeneracies of the immobile points of codimensionalities one and two and indicated (for the majority of them) the bifurcation conditions. Note here that these conditions are written in a sufficiently invariant form. Combining the above given conditions in "triads", it is also possible to define degeneracies of codimensionality three. In order to describe other degeneracies of codimensionality three it is necessary to compute the coefficients of the respective model systems. All the above introduced bifurcation conditions can be additionally smoothly defined in the neighborhoods of the respective curves.

For immobile points of the mappings, as well as for equilibria, of interest are the curves on which degeneracy with respect to the parameters occurs for each bifurcation (extremal paths). The specification of such curves is described in Sect.2.2.

Concluding this section, we return to the limit cycles and note that the differentiation of the sequence mappings necessary for the setting of bifurcation conditions and the construction of curves is a nontrivial problem since the sequence mapping itself is determined via the numerical construction of trajectories and contains a noticeable computational error. In order to avoid the error caused by a "nonsynchronous" change integration step in determining close trajectories for the computations of derivatives by the difference formula, it is useful to integrate the initial trajetory and the trajectories with the perturbed initial conditions (or system parameters) simultaneously [9] (see also [14]). Along with the numerical differentiation using the difference formulae, the first derivatives of the sequence mapping can also be completed by so-called "analytical" differentiation [9] consisting in the solution of the system of equations in variations. These equations with the set of various initial conditions should be integrated simulataneously with the initial system. Although such integration is carried out numerically, the computational error of the derivative substantially reduces (and the time of computation, as a rule, increases). Note also, that it is just because of the problems associated with the accuracy of computation of derivatives (especially of the second and higher orders) that the set of degeneracies that can be studied by the LINBAS programs is rather limited.

CONCLUSION

The scheme of continuous parametric analysis suggested in the present study combines the approaches developed in the bifurcation theory of dymanical systems with the concept of continuation by parameter (or, as is often said, continuation by continuity). The bifurcation theory in this case plays a double role: that of the language describing bifurcation manifolds in the space of phase variables and the system parameters and that of the method for studying its qualitative properties at the parameter values close to bifurcation ones. The continuation by parameter is the means of numerical study of the system in the wide range of the parameter variation, and also the means of the search for and construction of various bifurcation manifolds. The combination of the bifurcation theory with the continuation by parameter manifests itself in both the choice of the object for the study (a connected family of invariant sets) and in the methods of its study (with the aid of constructing a connected path nets passing through various classes and strata of the family).

Despite the fact that initially we formulated the problem of studying connected families with simple bifurcation structure, the above developed approaches can be useful also for a wider class of problems associated with the study of manifold partition into classes and strata. An important role in their applicability is the connectivity of manifolds and the correct adjacency of the strata.

We left aside many problems arising in the qualitative study of mathematical models in natural sciences and technology. We mention here those of them which are close to the problem solved here: how to detect various connected families of invariant sets, how to construct the complete phase and parametric portraits from the results of the investigation of these connected families, how to perform parametric studies of other invariant sets different from equilibria, limit cycles and separatrix contours, e.g., of invariant tores, what is the relation between the parametric study of invariant sets and the established modes of the system.

The parametric analysis of the system can be complemented with the so-called phase analysis. It implies the study of the the phase-portrait structure of a system at fixed parameter values. Namely, the determination of the mutual arrangement of already investigated invariant sets, the construction of invariant manifolds (or separate trajectories) "connecting" them, the search for and investigation of the established modes, etc. The methods of the phase analysis provide the data for answering some of the above formulated problems.

The construction of the scheme of a continuous parametric analysis has numerous gaps and unsolved problems of both mathematical and computational nature. As has been noted earlier, a series of bifurcations has not been studied well enough, the results of such investigation are not sufficient for their application in the numerical analysis.

As to the numerical implementation, we are closer to the beginning than to the end of our way. Some standard programs****** for the solution of individual problems of the local and global parametric study of equilibria, limit cycles and saddle separatrices were developed in the NIVTs AN SSSR (Scientific-research Computational Center of the USSR Academy of Sciences) [5, 6, 9, 10]. Standard programs for computation of close problems are described elsewhere [12, 13, 15-17,19]. But a large number of the problems of the parametric analysis are still waiting for their numerical implementation. A separate group among them are formed by problems associated with the study of systems containing, explicitly or implicitly, small parameters.

With the accumulation of programs for the solution of partial problems, the problem of their combination arises, or, more exactly, of the development on their basis of a dialogue system of programs within which one can "travel" more or less freely over the family of invariant sets studying their bifurcation structure and including the obtained results into the general picture of the qualitative properties of the model.

REFERENCES

1. V.I. Arnol'd. Additional chapters of the theory of ordinary differential equations. - Moscow: Nauka, 1978 (in Russian).
2. V.I. Arnol'd, V.S. Afrai|movich, Yu.S. Il'yashenko, and L.P. Shil'nikov. "Bifurcation theory." In: Modern problems of mathematics. Fundamental directions. v.5 (Advances of science and technology VINITI AN SSSR) Moscow 1986, pp.5-218 (in (in Russian).
3. A.D. Bazykin, Yu.A. Kuznetsov, and A.I. Khibnik. "Bifurcation diagrams of the dynamical systems on plane. Information material." Pushchino: ONTI NTsBI AN SSSR, 1985 (in Russian).
4. N.K. Balabaev and L.V. Lunevskaya. "Motion along a curve in the n-dimensional space (Software. Algorithms and programs in FORTRAN, issue 1)." Pushchino: ONTI NTsBI AN SSSR, 1978 (in Russian).
5. R.M. Borisyuk. "Stationary solutions of a system of parameter--dependent ordinary differential equations (Software. Algorithms and programs in FORTRAN, issue 6)." Pushchino: ONTI NTsBI AN SSSR, 1981 (in Russian).
6. Yu.A. Kuznetsov. "One-dimensional separatrices of a system of parameter-dependent ordinary differential equations. Software. Algorithms and programs in FORTRAN, issue 8)." Pushchino: ONTI NTsBI AN SSSR, 1983 (in Russian).
7. A.M. Molchanov. "Mathematical models in ecology. Role of critical modes." In: Mathematical simulation in biology. Moscow: Nauka, 1975, pp.133-141 (in Russian).
8. A.M. Molchanov. "Critical points of biological systems (mathematical models)." Ibid, pp.142-153.
9. A.I. Khibnik. "Periodic solutions of a system of n differnetial equations. (Software. Algorithms and programs in FORTRAN, issue 5)." Pushchino: ONTI NTsBI AN SSSR, 1979 (in Russian).
10. A.I. Khibnik and E.E. Shnol'. "Programs for qualitative studies of differential equations. Information material." Pushchino: ONTI NTsBI AN SSSR, 1982 (in Russian).
11. A.I. Khibnik. "On computational algorithms for studying nonlinear oscillations." In: Qualitative methods of the theory of nonlinear

oscillations (Proceeding of 9th Int. Conf. on Nonlinear Oscillations, v.2). Kiev: Naukova Dumka, 1984 (in Russian).
12. B. Hassard, N. Kazarinov and I. Van. Theory and applications of bifurcation of cycle formation. Moscow: Mir, 1985 (in Russian).
13. E.J. Doedel and J.P. Kernevez. AUTO: Software for continuation and bifurcation problems in ordinary differential equations (including the AUTO 86 User Manual), Report, Applied Mathematics, California Institute of Technology, 1986.
14. E. Hairer, S.P. Norsett, and G. Wanner. Solving ordinary differential equations. I. Nonstiff problems. - Springer-Verlag, 1987 (Springer Series on Computational Mathematics, v.8).
15. M. Holodniok and M. Kubiček. "DERPER - an algorithm for the continuation of periodic solutions in ordinary differential equations." J. Comp.Phys., v.55, 1984, 254-267.
16. W. Jansen and U. Kriegel. "DEAN - a differential and difference equation analyser." Berlin, ZKL Information, 1984.
17. C. Kaas-Petersen. User's guide. Centre for Nonl. Studies. Univ. of Leeds, Leeds, 1987.
18. A.I. Khibnik. "Computer oriented analysis of equilibrium and limit cycle bifurcations." Anstr. 10th Intern. Conf. on Nonlinear Oscillations. Sofia, Bulgarian Acad. Sci., 1984, p.299.
19. M. Kubiček and M. Marek. Computational methods in bifurcation theory and dissipative structures. Springer-Verlag, 1983 (Springer Series in Computational Physics).
20. W.C. Rheinboldt. Numerical analysis of parametrized nonlinear equations. A Wiley-Interscience Publication. New York, 1986. (The Univ. of Arkansas Lecture Notes in the Mathematical Sciences, v.7).
21. J. Rinzel and R.N. Miller. "Numerical calculation of stable and unstable solutions to the Hodgkin-Huxley equations." Math. Biosci., 49, 1980, 27-59.

Footnotes

*Nondegeneracy of a system of equations signifies that its Jacobi matrix is of the maximum rank.

**Or a principal system by the terminology accepted in [2].

***Very often but not always these functions are algebraic.

****Hereinafter we consider degeneracies occurring inside the family of limit cycles. The boundaries of this family should be studied separately by other methods.

*****Of great importance and interest are the so-called "strong" resonances q = 1, 2, 3, 4 [1]. Two such resonances, 1:1 ($\mu_{1,2} = 1$) and 1:2 ($\mu_{1,2} = -1$), have already been discussed.

****** In other words, the programs intended for numerous users studying different systems of differential equations.

Mathematics
and Modelling

ISBN 1-56072-104-9

STRUCTURE OF STRONGLY COUPLED POLARON

N.K. Balabaev and V.D. Lakhno

1. INTRODUCTION

The theory of polaron is the simplest example of the quantum field theory in solid state physics. At the same time, this example reveals many typical properties of a particle interacting with a quantum field. The results obtained for a polaron are of rather a general character and can be extended to such adjacent fields as the meson theory of nuclear forces, theory of motion of ions introduced into liquid helium, etc. The detailed description of the polaron problem can be found elsewhere [1].

The polaron problem is a problem of the behavior of a slow (nonrelativistic) conduction electron in a polar crystals. Considering a band structure one usually proceeds from the translational periodicity of a crystal, assuming that the Hamiltonian for electrons possesses the same periodicity. As a result, one arrives at the Bloch states distributed over the whole crystal with equal probabilities of electron being located in all the unit cells.

First, the idea on possible autolocation of a conduction electron in an ionic crystal was put forward by Landau [2]. He indicated that from the energy standpoint an electron in a system consisting of an excessive electron plus a perfect crystal can have two favorable locations - either as a free electron in the band of a non-deformed lattice or as a bound electron to be captured by a deformed lattice region. Pekar [3] has introduced the concept of a strongly coupled polaron according to which the conduction electron moving in a polar crystal polarizes its ionic lattice so strongly that the electron and the lattice form a self-consistent state. Pekar called such a state a polaron. Using the direct variational Ritz method, Pekar has found the state of an electron in a polar crystal that corresponds to the lowest energy [3].

Asymptotically exact equations for a strongly coupled polaron were derived by Bogolyubov [4]. In the present study based on the earlier authors' works [5] it is shown that the equations for the strongly coupled polaron admit a series of autolocalized spherically symmetric solutions. The first one, corresponding to the ground state, coincides with that found earlier by Pekar [3,6]. We also give the main characteristics of three next (energetically) solutions.

2. A NONRELATIVISTIC PARTICLE IN A QUANTUM FIELD. RESULTS OBTAINED IN THE ADIABATIC THEORY

The initial Hamiltonian for the description of a nonrelativistic particle in a quantum field has the form [4,7]:

$$H = H_p + H_v + H_{int}$$

$$H_p = -\frac{\hbar^2}{2\mu}\Delta r, \quad H_v = \frac{1}{2}\sum_f \hbar\omega_f(b_f b_f^+ + b_f^+ b_f), \tag{1}$$

$$H_{int} = \sum_f c_f e^{i\vec{f}r} b_f + c_f^* e^{-i\vec{f}r} b_f^+,$$

where H_p, H_v and H_{int} are Hamiltonians of a free particle, a field, and the interaction between the particle and the field, respecively. μ is the particle mass, r is its coordinate, b^+ and b are the generation and annihilation operators of field quanta with energy $\hbar\omega$, and c_f are certain constants characterizing the particle-field interaction.

The successive solution of the quantum problem (1) in the case of a strong particle-field interaction was given in [4,7] on the basis of the adiabatic approximation with due regard for the translational degeneracy of the Hamiltonian. In this case Hamiltonian (1) is represented in the form

$$\mathcal{H} = - \frac{\hbar^2}{2\mu}\Delta r + \sum_f A_f q_f e^{i\vec{f}r} +$$

$$+ \frac{1}{2} \sum \nu_f q_{-f} q_f - \frac{\varepsilon^4}{2} \sum_f \nu_f \frac{\partial}{\partial q_{-f}} \frac{\partial}{\partial q_f} , \quad (2)$$

where $q_f = (b_f + b_{-f}^+)/(\varepsilon \sqrt{2})$ and $-i\frac{\partial}{\partial_{q_f}} = i\varepsilon(b_f^+ - b_{-f})/\sqrt{2}$ define the complex field coordinates, $\hbar\omega_f = \varepsilon^2 \nu_f$, and $A_f = \sqrt{2}\, c_f$. The last term in (2) corresponds to the kinetic energy of the field and vanishes in the limit $\varepsilon \mid 0$ where the frequency of the field os- cillations tends to zero.

The successive procedure of the perturbation theory for Hamiltonian (2) by parameter provides the representation of the respective wave equation

$$(H - E)\, \Psi = 0$$

in the form

$$(H_0 + \varepsilon H_1 + \varepsilon^2 H_2 + \ldots - E)\, \Phi = 0$$

where $\Psi = e^{i\hat{p}\vec{q}}\, \Phi(\vec{\lambda}, Q_f)$, $\hat{p} = -i\hbar \frac{\partial}{\partial q}$. Here **q** is a part of r corresponding to the uniform rectilinear motion, λ is a fluctuating part of r, and $\{Q_f\}$ are the field coordinates. In the zeroth approximation, we have $\Phi(\vec{\lambda}, \{Q_f\}) = \varphi(\vec{\lambda})\, \theta(\{Q_f\})$.

For $\varphi(\lambda)$ the following equation is valid [4,7]:

$$\left(-\frac{\hbar^2}{2\mu}\Delta_{\vec{\lambda}} + U(\vec{\lambda}) - W_0\right)\varphi(\vec{\lambda}) = 0$$
$$U(\vec{\lambda}) = -\sum_f \frac{v_f A_f\ {}^2 \int e^{-i\vec{f}\lambda}\varphi(\vec{\lambda})^2 d^3\vec{\lambda}}{v_f^2 - (cf)^2} \tag{3}$$

where $c = v/\varepsilon^2$ and v is the average velocity of a forward motion of the particle and the field. The total energy is

$$E = W_0 + \frac{1}{2}\sum_f u_f\ {}^2\left(\nu_f + \frac{(fc)^2}{\nu_f}\right),$$
$$u_f = -\frac{A_f^* \nu_f}{\nu_f^2 - (cf)^2} \int e^{-i\vec{f}\lambda}\ \varphi(\vec{\lambda})^2\ d^3\vec{\lambda}. \tag{4}$$

In the limit of small velocities, using the expansion in v from (4), we arrive at the following formula for the particle mass:

$$E = E_0 + M v^2/2 + \dots \qquad M = \frac{1}{3}\frac{\hbar^2}{\varepsilon^4}\sum_f \frac{A_f^2 f^2}{\nu_f^3} \left|\int\varphi^* e^{i\vec{f}\lambda}\varphi\, d^3\vec{\lambda}\right|^2. \tag{5}$$

Relationships (3) -(5) completely specify the problem of the motion of a slow (nonrelativistic) particle in a quantum field in the limit of a strong particle-field interaction.

3. CASES OF DIFFERENT MEDIA

Hamiltonian (1) embraces numerous examples of the interaction between a quantum particle and a condensed medium. In the continuum approximation of the interaction between an excessive electron and a medium, i.e., under the assumption that the characteristic dimensions of the electron density distribution substantially exceed the lattice constant, the coefficients c_f can be written, for some cases, in an explicit form.

For an electron in an ionic crystal we have [1,6]:

$$c_f = \frac{e}{f}\sqrt{\frac{2\pi\hbar c\,\omega_f}{V}}, \qquad \omega_f \equiv \omega_0 \tag{6}$$

where e is the electron charge, V is the volume, $c = \tilde{\varepsilon}^{-1} = \varepsilon_{\infty}^{-1} - \varepsilon_{o}^{-1}$, and $\tilde{\varepsilon}$ is the effective dielectric constant.

In a piezoelectric conductor we have [8]:

$$c_f = -\frac{1}{f^{1/2}}\left(-\frac{4\pi\,\alpha_1}{V}\right)^{1/2}, \quad \omega_f = s\,f, \tag{7}$$

where $\alpha_1 = e^2(\hbar\varepsilon^2 sc_1)^{-1}\langle l_{ijk}^2\rangle$, s is the sound velocity, c_1 is the elasticity constant, and $\langle l_{ijk}^2\rangle$ is the averaged squared piezo electric tensor. For a homeopolar crystal [9]:

$$c = G\,f\sqrt{\frac{\hbar}{2\rho\omega_f V}}, \qquad \omega_f = sf, \tag{8}$$

where ρ is the crystal density and G is the deformation potential.

4. EQUATIONS FOR STRONGLY COUPLED POLARON

As was indicated earlier, the coefficients c_f for a polar medium are given by formula (6). Taking into account an additional equation

$$\frac{1}{\vec{\lambda}} = \frac{1}{2\hat{\pi}}\int\frac{e^{-i\vec{f}\lambda}}{\hat{f}}\,d^3f,$$

the integral-differential equation (3) at v = 0 can be reduced to the form

$$\begin{aligned}\frac{\hbar^2}{2\mu}\Delta\vec{\lambda}\,\varphi(\vec{\lambda}) + e\vec{\Pi}(\lambda)\vec{\varphi}(\lambda) + \vec{W}\varphi(\lambda) &= 0,\\ \Delta\vec{\lambda}\,\Pi(\vec{\lambda}) + 4\pi ce^2\vec{\varphi}(\lambda) &= 0,\end{aligned} \tag{9}$$

where $\Pi(\vec{\lambda})$ within the accuracy of the sign can be determined as a self-consistent potential field with an electron. Such a field and the wave function of an electron, $\varphi(\vec{\lambda})$, are simulataneously determined from (9).

We seek for spherically symmetric solutions of system (9). Let us use new variables:

$$\varphi(\vec{\lambda}) = \frac{W}{e\hbar}\sqrt{\frac{\mu}{2\pi c}}\, y(x), \qquad \Pi(\vec{\lambda}) = \frac{W}{e}\, z(x),$$
$$\vec{\lambda} = \frac{\hbar}{\sqrt{2\mu W}}\, x. \tag{10}$$

The substitution of (10) into (9) results in the system of differential equations in dimensionless variables:

$$y'' + \frac{2}{x} y' + zy - y = 0, \quad z'' + \frac{2}{x} z' + y^2 = 0. \tag{11}$$

On the variables y and z the following natural boundary conditions are imposed:

$$y'(0) = z'(0) = 0, \quad y(\infty) = z(\infty) = 0 \tag{11'}$$

Using the normalization conditions for the wave function

$$\int |\varphi(\vec{\lambda})|^2 d^3\lambda = 1,$$

we arrive at the expression for the electron energy:

$$W = -2\,\Gamma^{-2}\frac{e^4 c^2 \mu}{\hbar^2}, \qquad \Gamma = \int_0^\infty y^2 x^2 dx. \tag{12}$$

Thus the problem of the electron energy reduces to the computation of

the integral $\Gamma = \int_0^\infty y^2x^2dx$ where y(x) is the solution of the system of

equations (11)-(11′).

5. NUMERICAL DETERMINATION OF SOLUTIONS OF THE BOUNDARY PROBLEM

The solutions of the system of equations (11) with the boundary conditions (11′) can be computed numerically. It is convenient to change the variables:

$$\xi = xy, \quad \eta = xz,$$

In such variables the problem (11)-(11′) is rewritten in the form

$$\xi'' + \xi\left(\frac{\xi}{x} - 1\right) = 0, \qquad \eta'' + \frac{\xi^2}{x} = 0. \tag{12}$$

$$\xi(0) = \eta(0) = 0, \quad \xi(\infty) = \eta'(\infty) = 0. \tag{12'}$$

We search for the solution of the boundary problem (12)-(12′) by "ranging" proceeding from the left-hand boundary conditions. With this aim we represent the solutions of system (12) in the neighborhood of the point x = 0 as a series

$$\begin{aligned} \xi(x) &= a_1x + a_2x^2 + a_3x^3 + \ldots \\ \eta(x) &= b_1x + b_2x^2 + b_3x^3 + \ldots \end{aligned} \tag{13}$$

Thus the hand boundary conditions (12′) are automatically fulfilled. Substituting expansion (13) into equations (12), we arrive at the relationships for coefficients a_i and b_i. It turns out that they are uniquely expressed in terms of $a_1 \equiv a$ and $b_1 \equiv b$:

$$\xi(x) = ax + \frac{a(1-b)}{6}x^3 + \frac{a(1-2b+b^2+a^2)}{120}x^5 + \ldots$$

$$\eta(x) = bx + \frac{a^2}{6}x^3 + \frac{a^2(1-b)}{60}x^5 + \ldots$$

In other words, there exists a two-parametric family of the solutions to system (12) that satisfies the left-hand boundary condition (12′). The parameters a and b determine the slope of the curves $\xi(x)$ and $\eta(x)$ for $x \to 0$. Setting their values, it is possible to solve numerically the Cauchy problem on the given interval $[0, x_{fin}]$.

Now the solution of the boundary problem (12)-(12′) has reduced to the problem of finding such values of the parameters $a > 0$ and $b > 0$ that the solution of the respective Cauchy problem would satisfy the right-hand boundary conditions (12′) at $x_{fin} \to \infty$. The properties of differential equations (12) allow us to outline the algorithm for solving this problem. From the equation for η it follows that the function $\eta(x)$ is convex: $\eta''(x) = -\xi^2/x < 0$ at $x > 0$. For the solution of the problem (12)-(12′) we also have $\eta'(x) \geq 0$. If to take arbitrary values of the parameters $a > 0$ and $b > 0$, then the quantity $\eta(x;\, a,\, b)$ attains its maximum at a certain point $x = x^*$ and then goes into minus infinity. At the maximum $\eta'(x^*, a, b) = 0$. Let us put $x_{fin} = x^*$ in the Cauchy problem and determine the function of two variables

$$F(a,b) = \eta'(x_{fin};a,b).$$

The equation

$$F(a,b) = 0 \tag{14}$$

defines a certain dependence between the parameters a and b. Figure 1 shows this dependence at $x_{fin} = 10$. This curve was determined with the aid of the CURVE program complex [10]. The solutions of the Cauchy problem for different values of the parameters a and b on this curve in the motion along it are shown in Figure 2.

The analysis of the behavior of $\xi(x)$ allows one to put forward a hypothesis about the existence of some solutions of the boundary problem (12)-(12′). These solutions can be written in the following sequence: $\xi_n(x)$ $(n = 0, 1, 2, \ldots, n)$ intersects the x-axis n times and then

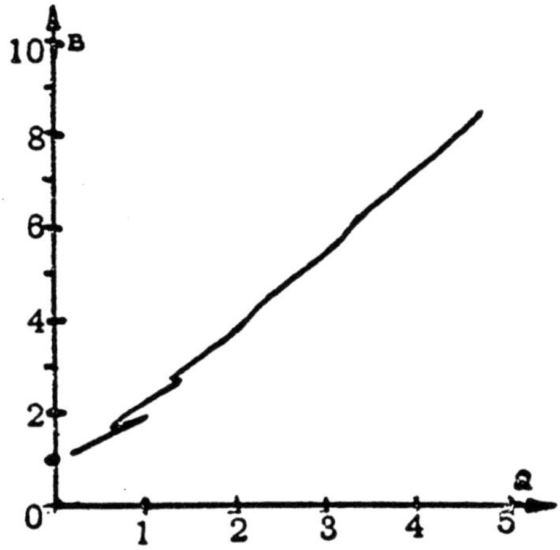

Figure 1. Dependence between the parameters, F(a,b) = 0, at $x_{fin} = 10$.
1 V

tends exponentially to zero; $\eta_n(x)$ increases monotonically approaching its extremum c_n values. In the following section it will be shown that the values of c_n coincide with the values of $\Gamma_n = \int_0^\infty y_n^2 x^2 dx \equiv \int_0^\infty \xi_n^2 dx$.

In order to determine several first solutions of the problem (12)-(12′) the following numerical procedure was used. Let us specify the functions:

$$F_1(a,b,x_{fin}) = \eta'(x_{fin}; a,b),$$
$$F_2(a,b,x_{fin}) = \xi'(x_{fin}; a,b).$$

Here $\xi(x; a, b)$ and $\eta(x; a,b)$ is the solution of the Cauchy problem on the segment $[0, x_{fin}]$. The system of equations

$$F_1(a,b,x_{fin}) = 0, \quad F_2(a,b,x_{fin}) = 0 \tag{15}$$

defines certain dependences between the variables a, b, and x_{fin}. These dependences were computed with the aid of the CURVE program complex. Various initial approximations providing different branches of curve (15) were chosen from the computation data illustrated by Figures 1 and 2.

Conditions (15) at $x_{fin} \to \infty$ are equivalent to the right-hand boundary conditions (12′). For the numerical determination of the different branches of curve (15) we had to move up to the value $x_{fin} \approx$ 15-25 to

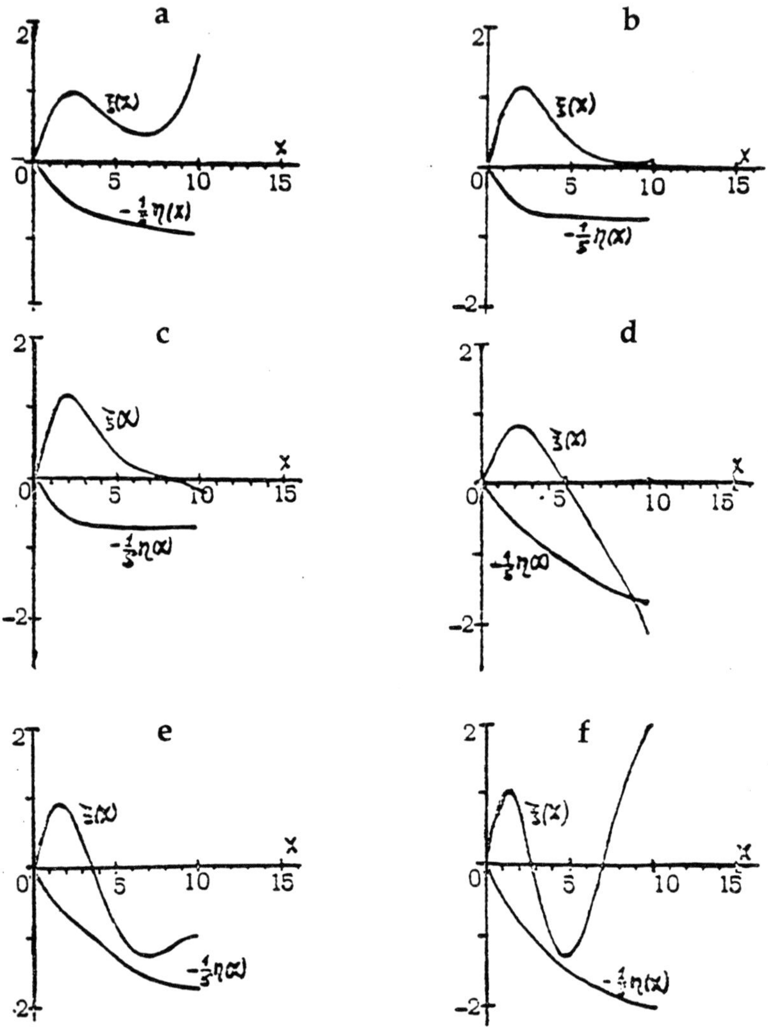

Figure 2. Functions $\xi(x;a,b)$ and $\eta(x;a,b)$ at different values of the parameters a and b on curve (14): (a) a = 0.733, b = 1.664; (b) a = 1.003, b = 1.921; (c) a = 1.025, b = 1.942; (d) a = 0.623, n = 1.662; (d) a = 0.900, b = 2.074; (e) a = 1.338. b = 2.727.

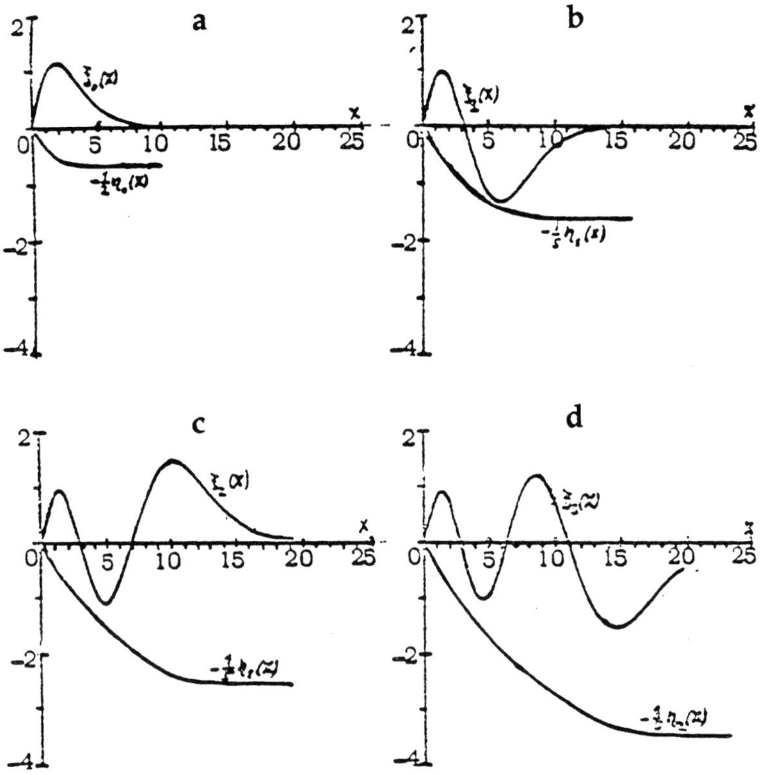

Figure 3. Various solutions to the system of equations (12)-(12′): (a) a = 1.021, b = 1.938; (b) a = 1.091, b = 2.320; (c) a = 1.118, b = 2.502; (d) a = 1.13, b = 2.61.

obtain the solution of the boundary problem (12)-(12′) with three-four reliable signs. Figure 3 shows the first four solutions for which we determined: $\Gamma_0 = 3.5052$, $\Gamma_1 = 8.060$, $\Gamma_2 = 12.66$, and $\Gamma_3 = 17.24$.

The solutions are readily obtained if the following conditions are given: $y_o(0) = 1.021$, $z_o(0) = 1.938$; $y_1(0) = 1.091$, $z_1(0) = 2.32$, $y = 0 = 0.118$; $z_2(0) = 2.502$. $y_3(0) - 1.13$, $z_3(0) = 2.61$.

6. PROPERTIES OF SOLUTIONS TO THE BOUNDARY PROBLEM

Let y(x) and z(x) be the solution of the boundary problem (11)-(11′). Let us introduce the notation

$$A_1 = \int_0^\infty x^2 y'^2 dx, \quad A_2 = \int_0^\infty x^2 y^2 dx, \quad A_3 = \int_0^\infty x^2 z\, y^2 dx,$$

$$A_4 = \int_0^\infty x^3 z' y^2 dx, \quad A_5 = \int_0^\infty x^2 z'^2 dx.$$

Then the following relationship is valid

$$A_1 : A_2 : A_3 : A_4 : A_5 = 1 : 3 : 4 : 2 : 4 \tag{16}$$

Relationship (16) can readily be derived from the linear dependences imposed on the quantities A_1, ..., A_5 obtained from (11) by multiplying the first equation by x^2y and x^2y' and the second equation by x^2z and x^3z' and then integrating them by parts.

The dependences (16) relating the integrals of the bounary problem solution can be used for checking the quality of the numerically determined solutions. It turned out that they are fulfilled up to the fourth decimal place, i.e., with the same accuracy with which the solutions of the boundary problem were determined.

One more important property of the solution to the boundary problem can be readily obtained from the second equation of system (12). We have

$$\int_0^\infty \xi^2 dx = -\int_0^\infty x\eta'' dx = -x\eta'\Big|_0^\infty + \int_0^\infty \eta' dx = \eta(\infty) - \eta(0) = \eta(\infty)$$

Taking into account that $\int_0^\infty \xi^2 dx = \int_0^\infty x^2 y^2 dx = A_2 \equiv \Gamma$, we obtain for the nth solution:

$$\eta_n(x) \equiv x z_n(x) \to \Gamma_n \text{ for } x \to \infty \tag{17}$$

The solution $y_o(x) = \xi_o(x)/x$ upon the transition to the initial variables by formulae (10)′ can be compared with the solution determined by Pekar [4] with the use of the Ritz direct variational method:

$$\varphi_0(\lambda) = 0,123\ \kappa^{3/2}(1 + \kappa\lambda + 0,451\ \kappa^2\lambda^2)e^{-\kappa\lambda}$$

$$\kappa = 0,66\ ce^2/\ \hbar^2, \qquad E_0 = -\ 0,0543\ e^4\mu\ c^2/\ \hbar^2$$

The numerical comparison shows that the deviation of these solutions from one another in the dimensionaless variables does not exceed $1 \cdot 10^{-3}$.

7. RADII AND EFFECTIVE MASSES OF PARTICLES

The determined solutions are self-consistent states of a nonrelativistic particle in a quantum field. Now let us determine some properties of such states.

A radius of a polaron is determined by the condition [6]:

$$V_n(\vec{\lambda} \leq R_n) = \frac{1}{2} V_n(\vec{\lambda} < \infty), \tag{18}$$

where V_n is the electron-field interaction energy. R_n corresponds to the crystal volume where half of the whole energy of the interaction between an electron and the crystal is contained. If to substitute into (18) the expression V_n in the form

$$V_n = \int \Pi_n(\vec{\lambda})\varphi_n^2(\vec{\lambda})d^3\vec{\lambda} \quad \int z_n^2 y_n^2 x\ dx, \tag{19}$$

then this relationship determines the characteristic dimensions of the region of the charge distribution of an electron captured by a distorted region of the lattice, i.e., the dimensions of the electron autolocalization region. If to substitute into (18) the expression for V_n in the form

$$V_n \int \left(\frac{dz_n}{dx} \right)^2 x^2 dx, \tag{20}$$

then relationship (18) determines the characteristic dimensions of the region of the lattice distortion created by an electron. These two characteristic dimensions do not coincide. In both cases (18) yields:

$$R_n = \frac{1}{2} \xi_n \Gamma_n \frac{\hbar^2}{\mu c e^2} \tag{21}$$

Also, if V_n is taken in the form (19), then for the first three states we have : $\xi_0 = 2.03$, $\xi_1 = 5.41$, $\xi_2 = 9.01$. If V_n is chosen in form (20), then $\xi_0 = 5.03$, $\xi_1 = 12.0$, $\xi_2 = 18.9$. For example, for the ground state the dimensions of the region of electron autolocalization are almost three times smaller than the characteristic size of the region of the lattice distortion.

Following [6], we shall call the polaron radius a value obtained from (18) if V_n is chosen in form (20). The value $\xi_0 = 5.3$ leads to the polaron radius coinciding with the value determined by Pekar, $R_c \approx 10\hbar^2/\mu c e^2$. With an increase of the state number the polaron radius also increases (Table 1).

The effective mass is given by expression (5). With due regard for (6) the effective mass in the nth state is:

$$M = \frac{4\pi c e^2}{3\omega^2} \int \varphi^4(\vec{\lambda}) d^3\lambda \tag{22}$$

In dimensionless variables, instead of (22) we obtain the following expression:

Table 1. Polaron radius (R_n), effective mass (M_n), and energy (E_n) of various self-consistent states of electron

n	0	1	2	3
$E_n / \frac{e^4 \mu c^2}{\hbar^2}$	-0.05426	-0.0103	-0.00416	-0.00223
$R_n / \frac{\hbar^2}{\mu c e^2}$	9.29	48.4	120	216
$M_n / \mu\alpha^4$	0.0227	3.0 10^{-4}	2.9 10^{-5}	5.0 10^{-6}

$$M_n = 32\ \mu\ \alpha^4 L_n /\ 3\Gamma_n^5, \qquad L_n = \int_0^\infty y_n^4 x^2 dx, \tag{23}$$

where $\alpha = \frac{e^2c}{\hbar} \sqrt{\frac{\mu}{2\hbar\omega}}$ is the constant of the electron-phonon coupling. For the first four states we have: $L_0 = 1.126$; $L_1 = 0.969$; $L_2 = 0.873$; $L_3 = 0.807$. The effective mass of a a particle in the ground state determined from (23) is: $\mathbf{M} = 5.68 \cdot 10^{-3}(\mu e^2 c/\hbar^3 \cdot e^2 c/\omega^2$, i.e., is close to the value obtained in [6]. From (23) it follows that the effective masses $\mathbf{M}_n$ decrease with an increase in number n.

The obtained results are summarized in Table 1 listing the numerical values of polaron radii, effective masses, and energies of different self-consistent states of an electron.

Concluding the article, we should like to note that for a pieso-electric medium, where the particle interaction with the field is set by equation (7), we arrive again at equations (11)-(11′) obtained earlier for a polar medium. For a nonpolar medium with the deformational mechanism of interaction (8) equation (3) leads to a nonlinear Schrödinger equation with cubic nonlinearity

$$y'' + \frac{2}{x} y' + y^3 - y = 0$$
$$y'(0) = y(\infty) = 0$$

that has been analyzed in detail elsewhere [11].

REFERENCES

1. Polarons (ed. Yu.A. Firsov) Moscow: Nauka, 1975 (in Russian).
2. L.D. Landau. Physik Z. Sowjetunion, 1933, 3, 644. In: L.D.Landau, Collection of works, v.1, Moscow: Nauka, 1969, p.90 (in Russian).
3. S.I. Pekar, ZhETF, 1946, 16, 341; 1946, 16, 335.
4. N.N. Bogolyubov. Ukr. Matem. Zh., 1950, 2, 3; Selected works. Vol 2. Kiev: Naukova Dumka, 1970, 499 (in Russian).
5. N.K. Balabaev and V.D. Lakhno. Teor. i Mat. Fiz., 1980, 45, No.1, 139; Preprint. Pushchino, ONTI NTsBI AN SSSR, 1979, 23pp. (in Russian).
6. S.I. Pekar, "Studies in the electronic theory of crystals." Gostekhizdat, 1951 (in Russian).
7. S.V. Tyablikov. ZhETF, 1951, 21, 377.
8. G. Whitfield, I. Gerstner, and K. Tharmalingam. Phys. Rev., 1968, 165, 993.
9. Ch.Kittel. Quantum theory of solids. Moscow: Nauka, 1967 (in Russian).
10. N.K. Balabaev and L.V. Lunevskaya. "Motion along a curve in the n-dimensional space. Algorithms and programs in FORTRAN." Pushchino: ONTI NTsBI AN SSSR, 1978 (in Russian).
11. N.K. Balabaev, V.D. Lakhno, and A.M. Molchanov. Preprint, Pushchino: ONTI NTsBI AN SSSR, 1983, 16 (in Russian).

Mathematics
and Modelling

ISBN 1-56072-104-9

PULSES OF COMPLICATED SHAPE IN MODELS OF NERVE CONDUCTION

Yu.A. Kuznetsov

INTRODUCTION

In 1952 Hodgkin and Huxley [21] suggested a model describing the behavior of nerve pulses over the axon membrane of a squid. A Hodgkin-Huxley system had the form of a "reaction-diffusion" type system with the components (V, n, m, h) on a line with diffusion occurring only along the first component. The Hodgkin-Huxley model is phenomenological but nevertheless it has played the key role in biophysics of nerve fibers and is an initial point for the mathematical analysis of the excitation pulse propagation along a nerve fiber.

To such a pulse there corresponds the automodel solution of the Hodgkin-Huxley equations of the form

$$V(x,t) = V(x + ct) \equiv V(\xi)$$

etc. for all the remaining components. In this case $V_{\xi \to \pm\infty}(\xi)$ is equal to the potential of rest. To such a pulse there corresponds a homoclinic solution of the automodel system for the Hodgkin-Huxley equations. Hodgkin and Huxley determined it with the aid of an arithmometer in 1952. The determined pulse shape and the propagation velocity agree quite well with the experimental data for the squid axon [21].

The initial Hodgkin-Huxley system is too complicated for the mathematical analysis. Therefore the attempts were undertaken to reduce it to simpler models preserving all its main qualitative properties. Then one tries to get results obtained on such "caricature" models also on the complete Hodgkin-Huxley equations. By present such a program has been implemented to a large extent.

Separating the motions in the Hodgkin-Huxley system into fast and slow components, it is possible to reduce it to a system of two equations of the form

$$\begin{cases} V_t = DV_{xx} + P(V,n) \\ \dot{n} = Q(V,\ n) \end{cases} \tag{1}$$

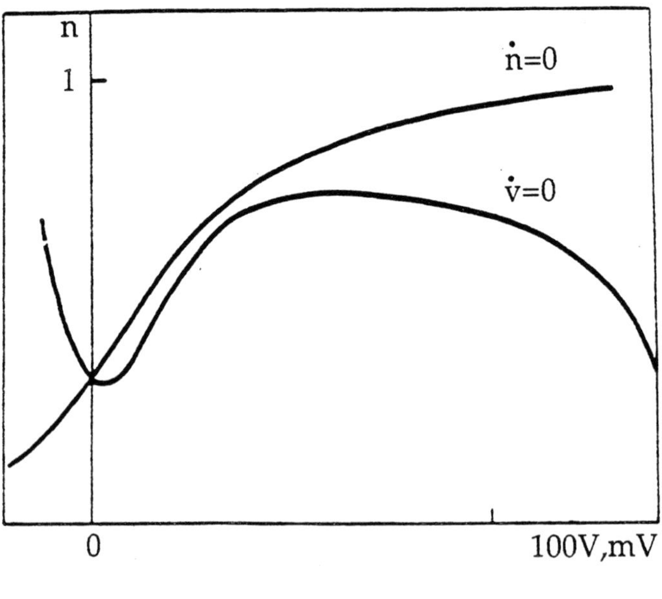

Figure 1. Zero-isoclines for the Hodgkin-Huxley didffusion-free model reduced to a second-order system.

(1) 100 mV

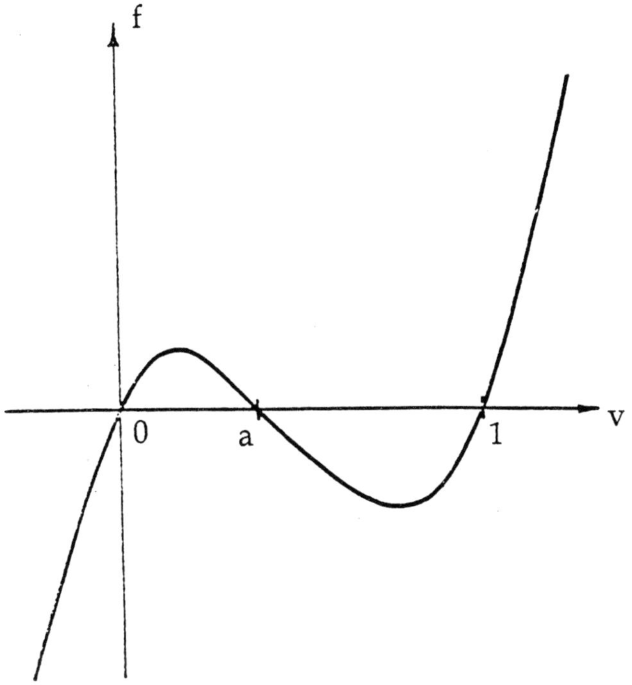

Figure 2. The shape of the f(v) function.x

for which the zero-isoclines of the space-homogeneous solutions have the form shown in Figure 1. Such a reduction was made in [5]. The zero-isoclines can be then approximated by simpler functions. Thus upon the appropriate scaling of variables and translation of their origin, we proceed from system (1) to the following model:

$$\begin{cases} v_t = v_{xx} - f(v) - \omega \\ w_t = b(v - \gamma\omega) \end{cases} \tag{2}$$

The form of the function f(v) is shown in Figure 2. It is usually set by a cubic polynomial $f(v) = v(v - a)(v - 1)$, $0 < a < 1$, $b > 0$, $0 \leq \gamma < 1$. The zero-isoclines corresponding to system (2) are depicted in Figure 3. The analysis of system (2) shows that its qualitative properties are preserved even if to take $\gamma \equiv 0$ (below we shall take just this value). System (2) was suggested

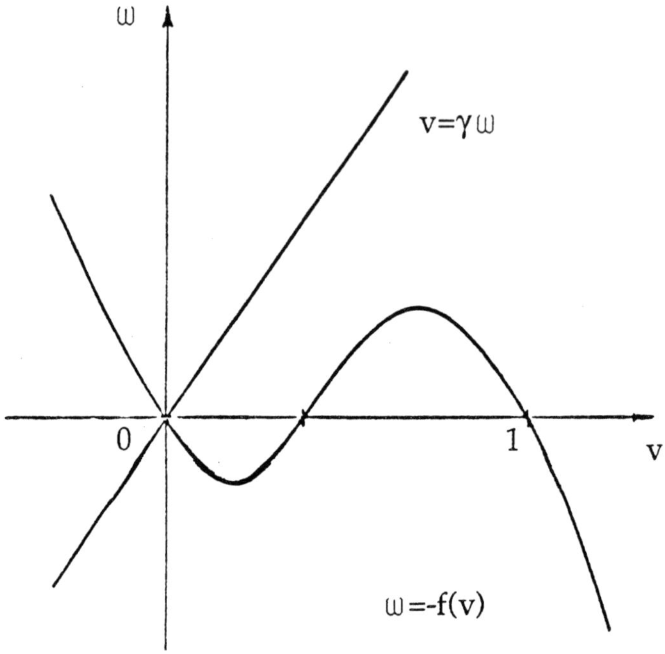

Figure 3. Zero-isoclines for the diffusion-free FitzHugh-Nagumo system.

for space-homogeneous solutions by FitzHugh [19] in 1961. System (2) with partial derivatives was suggested by Nagumo, Arimoto, and Yoshizawa [26]. At present it is usually called the FitzHugh-Nagumo system. In [19], system (2) was suggested proceeding from pure phenomenological considerations. Taking into account the reduction made in [5], it is possible to establish the corrrespondence between the variables of the FitzHugh-Nagumo and Hodgkin-Huxley systems, and also to compare the effect of variations in parameters. The FitzHugh-Nagumo model is the basic model in those problems of the theory of excitable media where of essence is only the spatial variable [10].

The solution of system (2) at $t > 0$ is determined by the initial conditions $v(x,0) = v_0(x)$ and $w(x,0) = w_0(x)$ where $x \in \mathbf{R}$. This Cauchy problem is well posed in many functional spaces [27], e.g., in the $\mathbf{BC}(\mathbf{R})$ space of limited and homogeneously-continuous vector-functions on a line $\mathbf{R}$ or in the Sobolev space $\mathbf{W}_2^2(\mathbf{R})$. The existence of solutions of system (2) at all $t > 0$ follows from the existence for it of a positively invariant region [15].

The main questions that arise about system (2) are: when has the system autowave solutions and what properties may it posses? The simplest autowaves in the FitzHugh-Nagumo system are solitary travelling pulses and periodic travelling waves. They were studied by munerous authors (see References in [3,4,7,11]). No other types of autowaves in system (2) have been discovered until recently. On the basis of the experience collected at the NIVTs AN SSSR*, we shall consider the problem on the existence and geometric properties of travelling waves in system (2) as a problem of the bifurcations of the phase portrait of the respective automodel system. Such an approach provides the establishment of new properties for this widely used model. As will be shown later on, the automodel system has a nonlocal bifurcation of codimensionality two due to the formation of a loop of the saddle separatrix with an additional degeneration. This allows us to predict a possible existence for system (2) of travelling waves of a complicated form that look like solitary "pulses" with different numbers of maxima ("bound pulses") or like waves with stochastic profiles, and to explain parametrically their generation. Below we shall discuss the relation between the theoretical predictions and the published experimental data.

The method and the style of the present article were directly influenced by the A.M. Molchanov and E.E. Shnol ideas about the mathematical analysis of the problems of natural sciences.

Some results were obtained together with A.B. Panfilov (Institute of Biophysics of the USSR Academy of Sciences) and were published as a preprint [6].

1. AUTOMODEL SYSTEM

Stationary travelling waves are the solutions of system (2) in the form $v(x,t) = V(\xi)$, $w(x,t) = W(\xi)$ where $\xi = x + ct$. Here c is an unknown velocity of the wave propagation; for definiteness we shall take that $c > 0$. The $V(\xi)$ and $W(\xi)$ profiles of the travelling wave satisfy the system of ordinary differential equations:

$$\begin{cases} \dot{V} = U \\ \dot{U} = cU + f(V) + W \\ \dot{W} = \frac{b}{c} V \end{cases} \tag{3}$$

where the point denotes differentiation with respect to ξ and $U \equiv dV/d\xi$. The systtem (3) is called an "automodel system", it depends on three positive parameters (a, b, c). The qualitative analysis of automodel system (3) allows us to make some important statements about the existence and geometrical properties of travelling waves in the FitzHugh-Nagumo system (2).

2. PROPERTIES OF EQUILIBRIA OF AUTOMODEL SYSTEM

System (3) at all the positive values of the parameters (a, b, c) has the only equilibrium 0 = (0, 0, 0). The linearization matrix of system (3) at point 0 has the form

$$A = \begin{pmatrix} 0 & 1 & 0 \\ a & c & 1 \\ b/c & 0 & 0 \end{pmatrix}$$

The eigenvalues of the equilibrium 0 satisfy the characteristic equation

$$\lambda^3 - c\lambda^2 - a\lambda - \frac{b}{c} = 0 \tag{4}$$

For any positive values of the parameters (a, b, c) equation (4) has no roots on the imaginary axis. If all the roots of equation (4) are real,

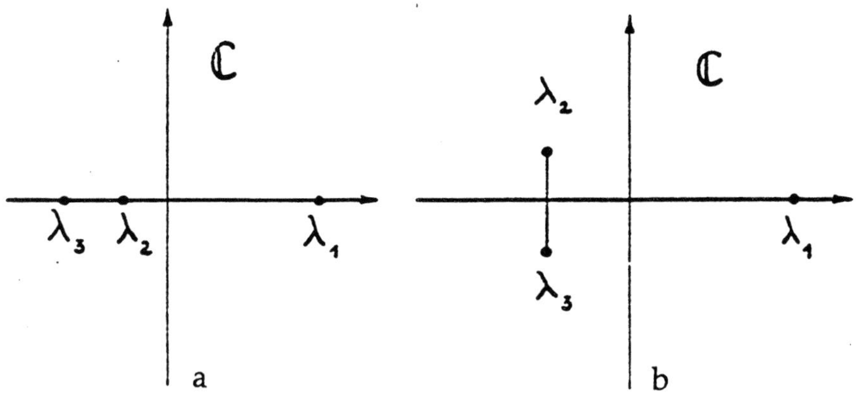

Figure 4. Typical location of the equilibrium eigenfunctions in an automodel FitzHugh-Nagumo system.

then two roots are negative and one root is positive. Since the eigenvalues cannot intersect the imaginary axis, then at all positive (a, b, c) two eigenvalues lie in the left-hand half-plane, and one eigenvalue is in the right-hand half-plane (Figure 4). Therefore the equilibrium O is a topological saddle with the one-dimensional unstable invariant manifold, $\mathbf{W}^{(u)}(0)$, and a two-dimensional stable invariant manifold, $\mathbf{W}^{(s)}(0)$. At zero, the unstable manifold $\mathbf{W}^{(u)}$ is tangential to the vector

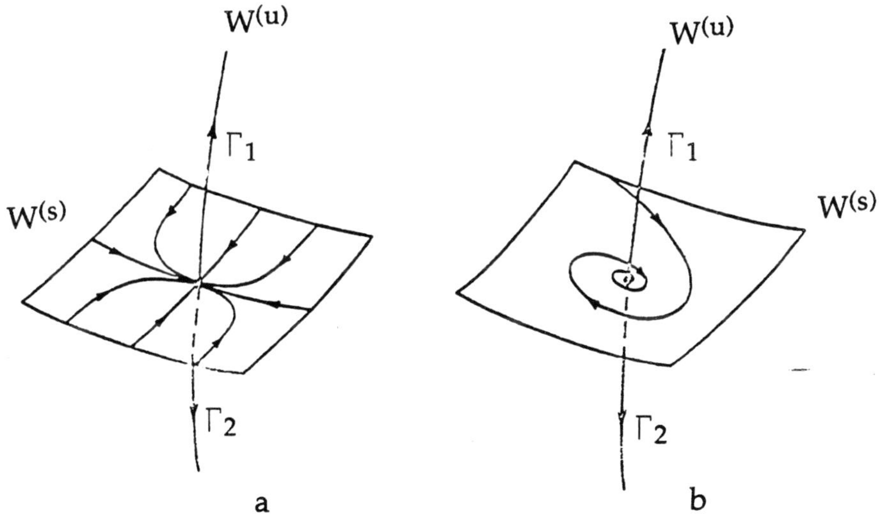

Figure 5. A saddle equilibrium in an automodel system. Saddle (a) and saddle-focus (b).

$$Z_1 = (1,\ \lambda_1,\ \frac{b}{c\lambda_1}),$$

which is the eigenvector of the matrix| A corresponding to the eigenvalue $\lambda_1 > 0$. The manifold $W^{(u)}(0)$ consists of the equilibrium O and two outgoing trajectories $\Gamma_1(0)$ and $\Gamma_2(0)$ ("separatrices"). We shall take that $\Gamma_1(0)$ leaves the equilibrium O along the vector Z_1. The saddle position of the equilibrium O can be either a saddle or a saddle-focus (Figure 5). The boundary separating these two cases corresponds to the equilibrium with double real eigenvalue $\lambda_2 = \lambda_3 < 0$. In the space of parameters (a, b, c) this boundary is set by the equation R(a, b, c) = 0, where R is the resultant of the characteristic polynomial and its derivative. We have:

$$R = c^2(4b - a^2) + 2a(9b - 2a^2) + 27\frac{b^2}{c^2}$$

Thus, at a fixed value of $b > 0$, the saddle - saddle-focus transformation occurs in the (a,c) plane in system (3) on the line

$$D_b = \{(a,c) : c^4(4b - a^2) + 2ac^2(9b - 2a^2) + 27b^2 = 0\}$$

On this line, the A matrix is equivalent to the Jordan form:

$$\begin{pmatrix} \lambda_1 & 0 & 0 \\ 0 & \lambda_2 & 1 \\ 0 & 0 & \lambda_3 \end{pmatrix}$$

where $\lambda_1 > 0$ and $\lambda_2 = \lambda_3 < 0$ are its eigenvalues. Several D_b lines at different values of $b > 0$ are shown in Figure 6. All these lines have vertical asymptotic lines $a^2 = 4b$. The region in which the equilibrium 0 is a saddle-focus is adjacent to the coordinate axes on the (a,c)-plane and vanishes at $b \to 0$. Up to the recent time, in the analysis of system (3) one has considered only the case $b < 1$ taking that the point O is a saddle. (As a matter of fact, the possible saddle-focus behavior was first indicated in [24]).

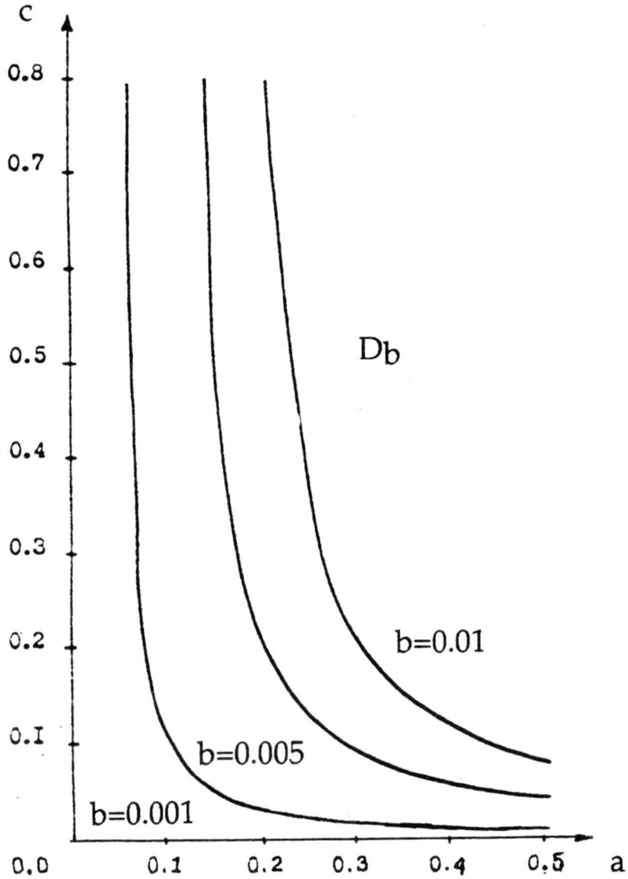

Figure 6. The saddle - saddle-focus transformation lines in an automodel FitzHugh-Nagumo system at different values of the parameter b.

3. SINGLE TRAVELLING PULSES

The separatrix $\Gamma_1(0)$ entering a positive octant along the vector Z_1 at certain parameter values can return back to point O and form a loop [26, 20, 22]. To such a loop there corresponds a single travelling pulse in system (2). The bifurcation values of the parameter values at which a separatrix loop exists can be determined only numerically. Figure 7 shows several bifurcation curves $P_b^{(1)}$ of the separatrix loop $\Gamma_1(0)$ at different values of the parameter b computed by the algorithms suggested in [8]. Each curve $P_b^{(1)}$ is ambiguously projected onto the abscissa which corresponds to the well known fact of the existence of fast and slow pulses in system (2). Although the stability

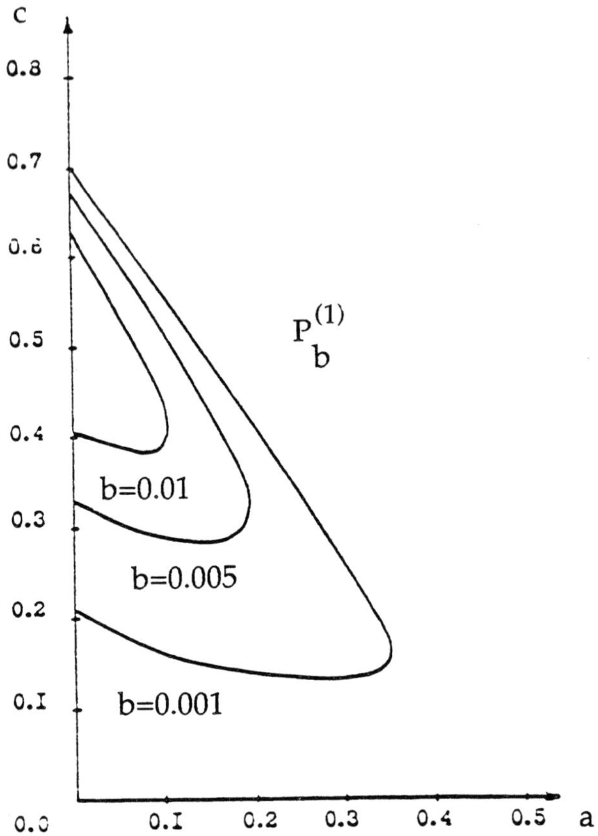

Figure 7. The lines of saddle-equilibrium loops in an automodel FitzHugh-Nagumo system at different values of the parameter b.

problems are beyond the present study, we should like to note that at small b a fast pulse is stable, whereas a slow one is unstable [4, 11, 23, 22].

Note that there is such $a^*(b)$ that at $a > a^*$ the stationary propagation of an individual pulse is impossible. At $b \to 0$ the curve $P_b^{(1)}$ tends to unite the segment of the abscissa $\{(a,c):c = 0, 0 < a < 1/2$ and a segment of the line $\{(a,c):c = (1/)(1-2a), 0 < a < 1/2\}$. This can be shown with the aid of the theory of singular perturbations. Thus for a > 1/2, pulses cannot exist at any values of b.

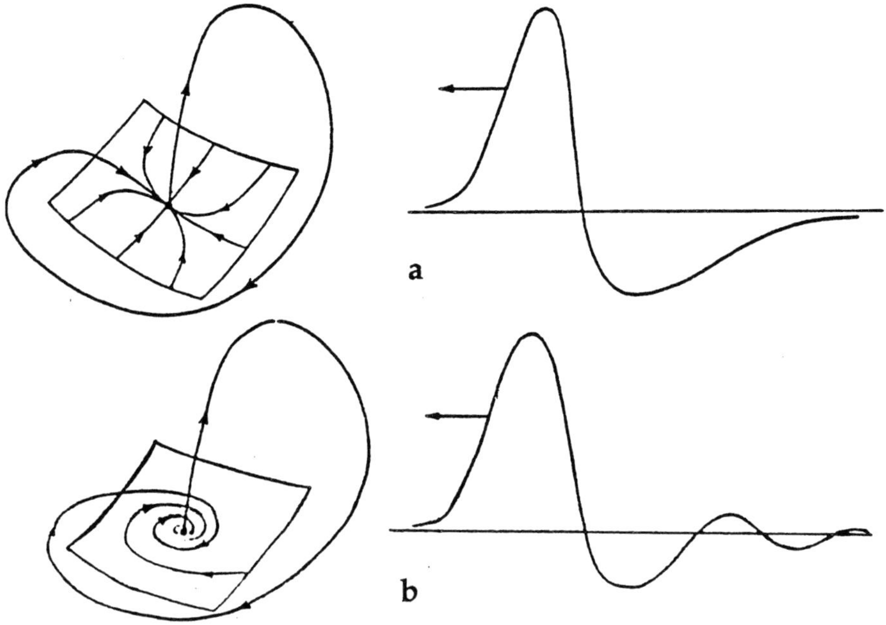

Figure 8. To a saddle loop there corresponds a pulse with a monotonous tail (a) and to a saddle-focus loop there corresponds a pulse with the oscillating tail.

The shape of a pulse corresponding to the separtrix loop is strongly dependent on the character of the equilibrium 0. If 0 is a saddle, then the pulse is of a conventional shape if it is a saddle-focus, then it is a single pulse with the "oscillating tail" (Figure 8). In system (3) both pulse types can exist. Comparing Figs. 6 and 7, it is easy to see that for all $b > 0$ the loop curve $P_b^{(1)}$ passes through the region where the point 0 is a saddle focus, and at $b \geq 0.01$ it lies completely in this region. Figure 9 shows the mutual position of curves $P_b^{(1)}$ and D_b at $b = 0.001$.

4. LONG-PERIOD TRAVELLING WAVES

Now let us consider bifurcations of the phase portrait of system (3) occurring in the intersection of the loop line $P_b^{(1)}$. These bifurcations are described by the Shil'nikov theorems [12, 13, 14]. We begin our consideration with the case where the point O is a saddle (all λ_i are real and different). In accordance with the results obtained by Shil'nikov, in

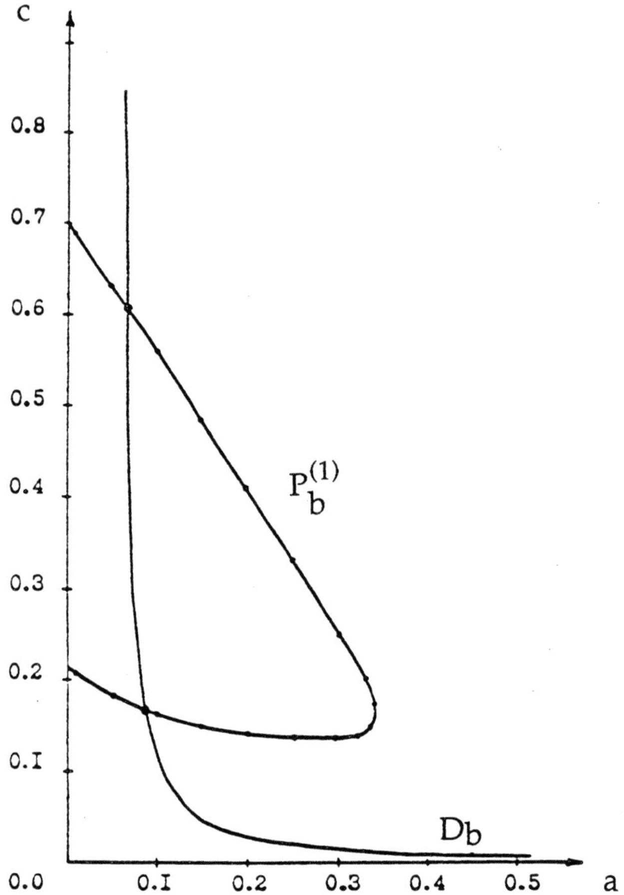

Figure 9. The mutual location of the loop line $P_b^{(1)}$ and the saddle - saddle-focus transformation line D_b at b = 0.001.

case of the general position the destruction of the separatrix loop of the saddle O should be accompanied by a generation of a single limit cycle. In order to apply the respective Shil'nikov theorem [12] one should check whether these conditions of the general position are fulfilled in the system under consideration. As far as we know, these conditions for system (3) have not been checked as yet. Thus one should check whether on the line of the saddle loop

(1) the saddle quantity $\sigma \neq 0$;

(2) the separatrix Γ_1 enters the saddle along its leading direction;

(3) at $\sigma > 0$ the loop has a certain orientation.

It is most easy to check the fulfillment of the first nondegeneracy condition. Using the equality $\lambda_1 + \lambda_2 + \lambda_3 = c$ (see (4)), it is easy to show that always $\sigma = \lambda_1 + \max \operatorname{Re} \lambda_{2,3} > 0$.

For verification of two other nondegeneracy conditions one has to know with sufficient accuracy the solution $(V, U, W)(\xi)$ corresponding to the separatrix loop of the saddle O. The attempt to determine this solution by the numerical integration of the separtrix $\Gamma_1(0)$ faces substantial difficulty - an accumulation of the error of numerical integration due to trajectory divergence in the direction of unstable manifold (along the vector $\pm Z_1$) in the vicinity of the saddle. This does not allow the determination by this method of that loop portion that corresponds to slowly attenuating "tail" of a travelling pulse. This difficulty was overcome in a way similar to that used in [9,25] by the reduction of the problem of loop determination to the solution of a boundary problem. In this case our aim is not finding a loop or a pulse as such but rather the checking of the fulfillment of the conditions of the Shil'nikov theorem.

At the bifurcation values of the parameters (a, b, c) that can be rather accurately determined by integrating the separatrix $\Gamma_1(0)$, the solution to be determined that corresponds to the separtrix loop satisfies a nonlinear boundary problem on an infinite straight line

$$\begin{cases} V_{\xi\xi} - cV_{\xi} - f(V) - W = 0 \\ W_{\xi} - \dfrac{b}{c}V = 0 \\ V(-\infty) = 0 \\ W(-\infty) = 0 \\ V(\infty) = 0 \end{cases} \tag{5}$$

This boundary problem was approximated by a discrete difference scheme on a finite interval [-T, T], where $T > 0$ is a sufficiently large number chosen in such a way that 2T would be larger than the characteristic pulse duration. The obtained system of nonlinear algebraic equations was solved by the Newton method. As an initial approximation we used the solution obtained by numerical integration of the separatrix with the arbitrarily approximated "tail". Thus obtained solution was stable to the increase in T and the variation in the system parameters (a, b, c) and boundary conditions.

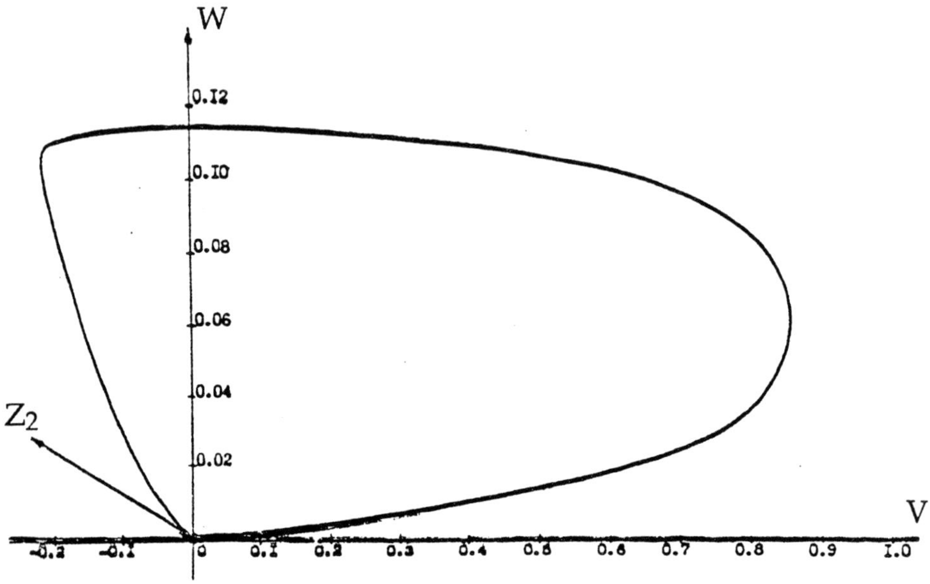

Figure 10. A saddle separatrix loop constructed by solving a nonlinear boundary problem a = 0.175, b = 0.005, and c = 0.376828; Z_2 is the leading direction on a stable saddle monifold.

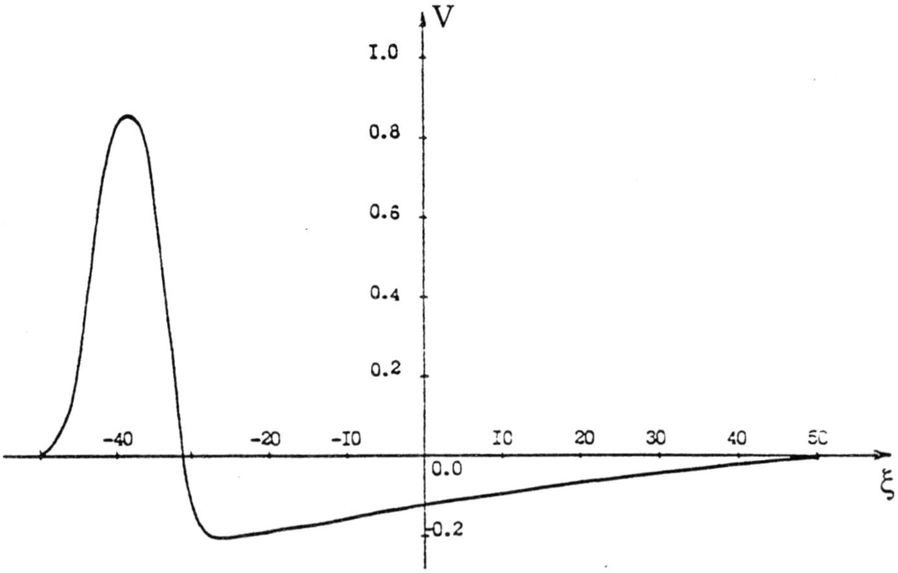

Figure 11. The potential pulse profile corresponding to a separatrix loop depicted in **Figure** 10.

The construction of a curve in the phase space (V, U, W) corresponding to the detemined solution of the boundary problem at different bifurcation parameters showed that the separatrix $\Gamma_1(0)$ enters the saddle 0 along the leading direction. This direction is set by the vector

$$Z_2 = (1, \lambda_2, \frac{b}{c\lambda_2}),$$

where $\lambda_2 < 0$ is the negative eigenvalue closest to the negative axis of the saddle 0. Figure 10 shows a separatrix loop constructed by solving the boudnary problem, and Figure 11 shows the V-component profile of the pulse corresponding to this loop.

Thus the fulfillment of the second nondegeneracy condition according to Shil'nikov has been checked. Since $\sigma > 0$, one has to prove the validity of the third condition for the general position, i.e., to determine the loop orientation. A loop is called orientable if a two-dimensional manifold obtained by closure of the continuation of the stable invariant manifold of the saddle O along the loop in its vicinity is orientable. Otherwise the loop is non-orientable. The loop orientation can be established by calculating a sign of a certain derivative of the correspondence mapping for the global loop portion [12, 14]. We calculated this derivative numerically by solving a certain approximating boundary problem. For all the bifurcation values of the parameters the loop of the saddle separatrix in system (3) turned out to be orientable.

Thus, we have varified numerically the fulfillment of all the conditions for the generaal position of the loop : $\sigma > 0$, the loop is closed along the leading direction and is orientable. Hence, the intersectin of the loop portion $P_b^{(1)}$ corresponding to the saddle loop in the motion on the plane of the parameters (a,c) from outside into the region limited by the loop line and the ordinate axis is accompanied by the generation of the only saddle limit cycle with orientable invariable manifolds. In system (3), to this limit cycle there correspond periodic travelling waves in the FitzHugh-Nagumo system (2). The existence of such waves can be proved by the singular perturbation technique [16] valid only for small b. Almost nothing is known about the stability of periodic travelling waves.

5. BIFURCATION MECHANISM OF BOUND-PULSE GENERATION [6]

In the previous section we have studied the generation of periodic solutions of automodel system (3) in the vicinity of the portion of the loop $P_b^{(1)}$ of the separatrix corresponding to the saddle loop. But, as has already been noted, the line of a loop $P_b^{(1)}$ for all $b > 0$ passes through the region where the equilibrium O is a saddle-focus with $\sigma > 0$. On these portions in the vicinity of the loop there exists a nontrivial invariant set including an infinite number of saddle limit cycles. This is established with the aid of the

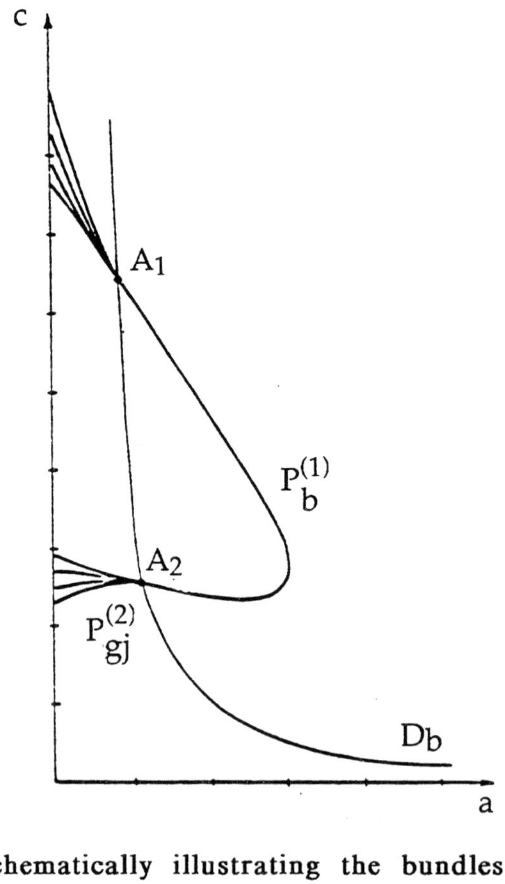

Figure 12. Schematically illustrating the bundles of lines of the double-turn separatrix loops emerging from the bifurcation points A_1 and A_2 of codimensionality two.

Shil'nikov theorem [13]. Thus in system (2) there exist an infinite number of long-period and nonperiodic travelling waves of a complicated profile.

The intersection of the loop line of the saddle-focus is accompanied by an infinite number of various bifurcations. The complete bifurcation pattern is unknown. But an essential progress in the investigation of automodel system (3) for such parameter values can be achieved by the use of the bifurcation points of codimensinality two. In system (3), the (a,c)-plane at sufficiently small fixed n > 0 has two points of nonlocal bifurcations of codimensionality two - these are points A_1 and A_2 of the intersection of the line of the loop $P_b^{(1)}$ with the line of the saddle - saddle-focus transformation (Figure 12). As was established by L.A. Belyakov [1,2], in case of the general position the region of the saddle-focus is entered by an infinite number of lines $P_{bj}^{(2)}$, j = 1, 2, ... emerging from points $A_{1,2}$. In system (3), there exist double-tracing loops of the separatrix Γ_1 of the saddle-focus O with different number of "turns" in the vicinity of the saddle-focus (Figure 13).

To be able to apply the Belyakov theorem, one should check the fulfillment of the conditions for the general position. They are fulfilled. Indeed, on the saddle portions of the line $P_b^{(1)}$ the nondegeneracy conditions for the respective bifurcation of codimensionality one are fulfilled; numerically it was checked by us earlier. At points A_1 and A_2 themselves the linearization matrix has eigenvalues $\lambda_1 > 0$ and $\lambda_2 = \lambda_3 < 0$, in this case $\sigma = \lambda_1 + \lambda_2 > 0$, and to the multiple eigenvalue $\lambda_{2,3}$ there corresponds a second-order Jordan cell.

To double-tracing in automodel system (3) there correspond travelling pulses with two large maxima and different number of small oscillations in between (Figure 13). Such a travelling wave looks as a bound system of two "conventional" pulses travelling as a whole. We call such pulses bound or double pulses. From points $A_{1,2}$ the bundles of lines of the loops of any "tracing multiplicity" emerge, whence it follows that there exist travelling pulses with any number of large maxima and different numbers of small oscillations in between.

Thus using the theory of bifurcations of codimensinality two, we have established, that in the FitzHugh-Nagumo system there exist bound travelling pulses. We indicated also those values of the system parameters (a,b) at which one can expect the generation of such pulses.

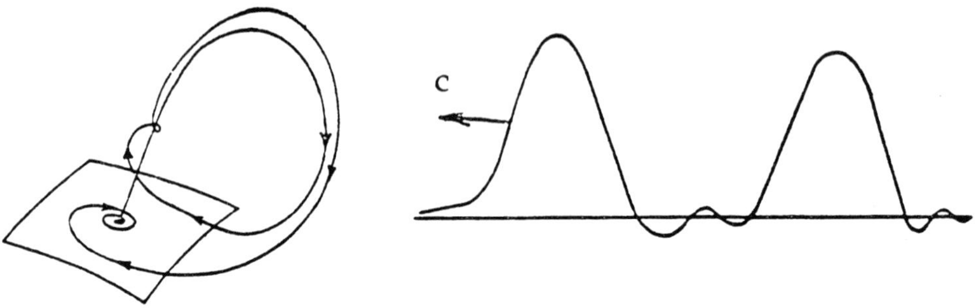

Figure 13. To a double-turn saddle-focus separatrix there corresponds a travelling double (bound) pulse.

Independently, the possible existence of bound pulses in nerve conduction models was indicated later on by Evans, Feroe, and Fenichel [17, 18].

The stability problem of bound pulses requires a separate consideration. The results of numerical experiments with system (2) and the results stated in [22] allow us to assume that stable pulses correpond to the loop lines emerging from point A_1.

6. DISCUSSION

At first glance, the possible existence of bound pulses contradicts the widespread opinion that the second pulse, being triggered after the first one, is always characterized by a lower velocity and therefore has a lag. Such a standpoint is usually supported by the fact that the second pulse propagates along the "decelerated" membrane with a residual negative potential ($v < 0$). This standpoint is also expressed in such a form that periodic waves in system (2) are always characterized by a velocity lower than that of a single pulse. The numerical experiments with the piecewise-linear approximation $f(v)$ confirm this standpoint for a stable pulse [4]. But recent numerical experiments on periodic travelling waves in the Hodgkin-Huxley system carried out by Rinzel and Miller [28] have shown that under certain conditions periodic travelling waves can propagate faster than a single pulse. At the same time the question about possible propagation of bound pulses remained open.

The above presented analysis of the FitzHugh-Nagumo system (2) with the use the bifurcation theory methods has shown a possible existence of bound pulses propagating as a whole. A heuristic explanation of such an effect consists in the following. Double pulses arise at such values of the parameters of system (2) for which a single pulse has the oscillating tail (a saddle-focus loop). Thus in this case the main action potential is followed by a series of regions with the positive ($v > 0$) residual potential. If the second pulse is triggered after the first one with such an interval that it would reach such a pre-excited region, then it is expected that it would catch the first pulse and then both pulses would propagate further on as a whole. The condition $\sigma > 0$ corresponds to the requirement that such pre-excited zones were sufficiently extended in comparison with the characteristic length of the positive pulse phase. Of course, it is only a post factum explanation but it gives the grounds to hope that the phenomenon under consideration is rather universal. Recently, this standpoint has been confirmed experimentally.

In 1982 Scott and Pinardi [29] published the results of their experiments carried out on a gigantic squid axon in which the R-ratio of the velocities of two subsequently triggered pulses was measured as a function of the interval T between them. The experiments showed that there are several intervals T_i such that both pulses have the same velocity ($R_i= 1$) and move as a whole. Such a formation can naturally be identified with the above described theoretical double pulse. The ambiguity of the T_i values and their magnitude qualitatively agree with the "saddle-focus explanation" of the bound pulse generation.

Thus the substantially simplified FitzHugh-Nagumo model reproduces qualitatively correctly such a complicated autowave effect as the propagation of a system of bound pulses observed in real systems. Thus it is expedient to use this modes as a basic model of excitable media.

The above developed methods seem to be useful for the analysis of travelling waves in various models of distributed kinetic systems.

REFERENCES

1. L.A. Belyakov. "On a bifurcation set in the system with homoclinic saddle curve." Matem. Zametki, 1980, 28, No.6, 911-922.
2. L.A. Belyakov. "On the structure of bifurcation sets in systems with a separatrix loop of the saddle-focus." In: Qualitative methods

of the theory of nonlinear oscillations. Trans. 9th Internat. Conf. on Nonlinear Oscillations. Vol.2, Kiev, 1984.52-55.

3. A.S. Dikanskii|. "Diffusion equations with nonlinear kinetics." Pushchino, 1979. Deposited at VINITI, No.1405-80, 1980, p.89 (in Russian).
4. V.S. Zykov. Simulation of the wave processes in excitable media. Moscow: Nauka, 1984, 165 pp.(in Russian).
5. G.R. Ivanitskii|, V.I. Krinskii|, and E.E. Sel'kov. Methematical biophysics of a cell. Moscow: Nauka, 1978, pp. 310 (in Russian).
6. Yu.A. Kuznetsov and A.V. Panfilov. "Stochastic waves in the FitzHugh-Nagumo system." (Preprint NTsBI AN SSSR) Pushchino, 1981, 8pp. (in Russian).
7. Yu.A. Kuznetsov. "Existence and stability of travelling waves in "reaction-diffusion" systems with one spatial variable." (Preprint, NTsBI AN SSSR) Pushchino, 1982, 37 pp. (in Russian).
8. Yu.A. Kuznetsov. "One-dimensional separatrices of parameter-dependent ordinary differential equations. Algorithms and Programs in FORTRAN. Materials on computer software. Issue 8. Pushchino, NTsBI AN SSSR, 1983, 47 pp (in Russian).
9. V.S. Posvyanskii| and E.E. Shnol'. "On the simplest model of cold flame." Preprint IPM AN SSSR No.25. Moscow, 1974. 39pp. (in Russian).
10. Yu.M. Romanovskii|, N.V.Stepanova, and D.S. Chernavskii|. Mathematical biophysics. Moscow: Nauka 1984, 304 pp (in Russian).
11. E. Scott. Waves in active and nonlinear media in electronic applications. Moscow: Sovetskoe Radio, 1977, 368 pp (in Russian).
12. L.P. Shil'nikov. "On generation of periodic motion from a trajectory dually asymptotic with respect to the saddle-type equilibrium." Matem. Sb., 1968, 77(119), 461-472 (in Russian).
13. L.P. Shil'nikov. "To the structure of an extended neighborhood of a rough saddle-focus type equilibrium." Matem. Sb., 1970, 81(123), pp.92-103 (in Russian).
14. L.P. Shil'nikov. "Bifurcation theory and the Lorentz model." In: J. Marsden and M. McCraken. Bifurcation of the cycle generation and its applications. Appendix II. Moscow: Mir, 1980, 317-335 (in Russian).

15. K.N. Chuen, C.C. Conley, and J.A. Smoller. Positively invariant regions for systems of nonlinear parabolic equations. - Indiana Univ. Math. J., 1977, 26, 373-392.
16. C.C. Conley. "On travelling wave solutions of nonlinear equations." In: Lect. Notes in Phys. Berlin etc. Springer-Verlag, 1975, 38, 498-510.
17. J.W. Evans, N. Fenichel, and J.A. Feroe. "Double impulse solutions in nerve axon equations." SIAM J. Appl. Math., 1982, 42, No.2, 219-234.
18. J.A. Feroe. "Existence and stability of multiple impulse solutions of a nerve equation." Siam J. Appl. Math., 1982, 42, No.2, 235-246.
19. R. FitzHugh R. "Impulses and physiological states in theoretical models of nerve membrane." Biophys. J., 1961, No.3, 445-446.
20. S.P. Hastings. "On the existence of homoclinic and periodic orbits for the FitzHugh-Nagumo equations." Quart. J. Math. Oxford (2), 1976, 27, 123-134.
21. A.L. Hodgkin and A.F. Huxley. "A quantitative description of membrane current and its application to conduction and excitation in nerve." J. Physiol., 1952, 117, 500-544.
22. C.K.R.T. Jones. "Stability of the travelling wave solution of the Fitz-Nagumo system." Trans. Amer. Math. Soc., 1984, 286, No.2, 421-469.
23. D.A. Larson. "On the stability of certain wave solutions to Nagumo's equation." Quart. J. Math. Oxford (2), 1977, 28, 339-352.
24. H.P. Jr.McKean. "Nagumo's equation." Adv. in Math., 1970, 4, 209-223.
25. R.M. Miura. "Accurate computation of the stable solitary wave for the FitzHugh-Nagumo equations." J. Math. Biol., 1982, 13, 247-269.
26. J. Nagumo, S. Arimoto, and S. Yoshizawa. "An active pulse transmission line simulating nerve axon." Proc. IRE, 1962. 50, 2061-2070.
27. J. Rauch and J. Smoller. "Qualitative theory of the FitzHugh-Nagumo equations." Adv. in Math., 1978, 27, 12-44.
28. J. Rinzel and R.N. Miller. "Numerical calculation of stable and unstable periodic solutions to the Hodgkin-Huxley equations." Math. Biosci., 1980, 49, No.1/2, 27-59.

29. A. Scott and U.V. Pinardi. "Nerve pulse interactions." In: Biomathematics in 1980. Eds. L.M. Riccardi and A.C. Scott. Amsterdam etc.: North-Holland Publ. Comp., 1982, 49-72

Footnote
[*]Research Computational Center of the USSR Academy of Sciences.

Mathematics
and Modelling

ISBN 1-56072-104-9

THEORETICAL AND MATHEMATICAL ECOLOGY: DANGEROUS BOUNDARIES AND CRITERIA OF APPROACHING THEM

A.D. Basykin

Mathematical modeling in ecology is characterized by some specific features in comparison with mathematical modeling in biology which, in its extreme form, reflects the difficulties encountered in application of mathematics to biology. All the biological objects are unique, but the natural ecological systems differ from one another more drastically than, e.g., different individuals of one species, nerve fibers of representatives of different species, etc. Therefore in mathematical modeling of ecologic systems, of special importance is the contradiction between the practicability and accuracy of the model, on the one hand, and its generality, on the other. This contradiction is clearly reflected by the development of mathematical ecology along two main directions. The aim of the imitational or portrait direction is the attainment of the maximum accuracy and closeness to the object under consideration, whereas the aim of the second, mathematical modeling proper, is the detection, mathematical description, and the analysis of the properties inherent in the maximum number of ecological systems.

One of the most important property, from both theoretical and practical standpoints, common to the majority of ecological communities is the so-called "buffer action" or "elasticity". These somewhat misnomers denote the following characteristic feature of the ecological system reaction on the action of external factors: unless the intensity of the action does not exceed a certain threshold value, the systems, so to say, "adsorb" this action but qualitatively they remain unchanged. But upon the attainment of a certain critical value, the system undergoes a certain drastic jumpwise change and transits into a qualitatively new state.

It should be noted that the difference between the qualitative and quantitative, jump-wise and gradual changes is, of course, very conventional: it is first and foremost the problem of time scales on which these changes occur in comparison with the time scale of the system itself. There are two scales of characteristic times of ecological systems - small succession times of the order of hundreds of years and large evolution and geological times of the order of tens of thousands of years and longer. Therefore for ecological systems, jumpwise qualitative changes are those occurring at times of the order of many years on several decades. From the standpoint of human life span, such changes seem to be gradual. It is here that the psychological difficulty of their consideration as jumpwise processes lies.

Thus, there exist two main differences between the imitational and mathematical models:

(a) the former describe the unique properties characteristic of concrete objects, whereas the latter describe the properties of a wide class of objects;

(b) the former describe the specific features of a normal functioning mode of system and small, quantitative changes, occurring there under the action of comparatively weak external factors; the latter should forecast qualitative ("disastrous") restructuring in functioning of ecological systems under the action of strong ("post-critical") factors.

Both problems are equally well-timed for practice. The main practical "user" of the mathematical modeling in ecology is the developing system of ecological monitoring: models should forecast what, where, when, how often and with what accuracy should be measured, what changes occurring in ecological systems should be payed attention to. The problems arising here are also of a dual character [2]:

(a) optimization (or at least some rationalization) of the exploitation of natural systems with the preservation of their qualitative features;

(b) forecasting of qualitative (disastrous) restructuring of the natural systems under the action of anthropogeneous factors. The latter is especially important for global monitoring and monitoring of unique especially valuable natural objects.

The first problem should be solved with the aid of imitational models, the second one with the aid of mathematical ones.

Thus one of the key problems of mathematical modeling proper is the development of the theoretical basis for forecasting qualitative restructuring in natural ecological systems under the action of external antropogeneous loading and the development of criteria for approaching to dangerous loading intensities. An adequate mathematical appratus for the solution of such a problem seems to be the bifurcation theory of dynamical systems.

Within the framework of the idealization used in the theory of dynamical systems, an ecological system to be modeled should be considered as isolated from any external factors, i.e., as a system functioning under unvariable conditions. In this case, upon the completion of a transient process the system attains a stationary functioning mode of one of three possible types:

(i) equilibrium,

(ii) periodic oscillations,

(iii) aperiodic oscillations (the quasistochastic mode).

Note also that the study of the third mode in mathematical ecology is at its very beginning.

Under the fixed external conditions the system can have several "attraction modes" of functioning, that, in turn, may be of several types. Depending on the initial conditions and the completion of the transient process, the system starts functioning in one of such modes.

As to the external factors that act onto ecological systems, two types of factors should be distinguished:

(a) single, one-time, pulse factors under invariable conditions of the system existence, the so-called "phase" factors. Examples here are killing of a part of the population of a certain species that enters the ecological system, a single volley-like pollutants discharge, etc;

(b) the factors that change the conditions of the system existence, the so-called "parameteric" factors. Examples are the establishment of a constant annual bugging quota for a population which is a hunting object, constant discharge of pullutants in certain concentrations, etc. The action of the first-type (phase) factors can be

(a) weak; then the system remains inside the attraction region of the earlier functioning mode; upon the completion of a certain transient process the system start functioning again in its initial mode;

(b) strong; then the system is brought outside the attraction boundary of the initial mode. Under such an action, the system transits into the region of attraction of a new qualitatively different functioning mode established in the system after completion of a certain transient process.

It should be emphasized that the concept of a strong and weak phase action is formulated only for a concrete attraction mode under the fixed external conditions. A mathematical model cannot and should not pretend to describe accurately the boundaries of attraction model (i.e. dangerous boundaries of phase factors) in a real system and, therefore, to forecast quantitatively the system response to concrete phase actions. The best possible result that can be achieved in this case is the qualitative description of a shape of such a boundary and, what is the most important, of the characteristic features of transient mode evolution with an approach to the dangerous boundary of the phase action intensity.

The second-type factors (parametric) can also be weak and strong:

(a) weak with respect to a certain attraction mode are those parametric factors under the action of which the given mode remains unchanged or undergoes weak, i.e., quantitative changes;

(b) strong are those parameteric factors under the action of which a system transits in a jump-wise manner into a qualitatively new functioning mode.

As in the analysis of phase factors, the mathematical model in this case cannot provide an exact quantitative description of the location of a boundary separating weak and strong parametric factors (i.e. of a dangerous parametric boundary). The model however should indicate the qualitative character of the location of various dangerous boundaries and forecast the specific features of the system behavior indicating its approach to a dangerous boundary.

Let us consider consistently some characteristic features of the system behavior that indicates the approach of the external loads to their dangerous level. In this case the main criterion of the approaching a dangerous boundary is the manifestation of specific features of a transient process upon the completion of which the system restores its initial functioning mode. Let us consider first equilibrium functioning modes. A transient process in this case is either monotonous or has the form of attenuating oscillations. In a monotonous transient process, the approach to a dangerous phase boundary is indicated by a decelerating process of returning of the system from a perturbed to the initial state with an increase of the load. The well known example here is the dynamics of a population after single killing when its number turned out to be close to a critical (i.e., dangerous) value.

In case of an oscillating transient process the approach to a dangerous phase boundary is indicated by two facts:

(a) with an increase of the loading intensity the period of attenuating oscillations following directly the load application drastically decreases, their shape acquiring a more and more prononced relaxation form. Figuratively, it can be said that the system "stays longer and longer" in the vicinity of a certain state;

(b) the shape and the period of attenuating oscillations does not substantially change with loading intensity but the decrement of oscillation attenuation decreases immediately upon the load application.

The above formulated criteria of approaching dangerous phase boundaries in oscillating transient processes remain valid if the normal functioning mode of a system is not equilibrium but oscillating one. It should be emphasized that we mentioned here only some simplest criteria of approaching dangerous phase boundaries. There can exist some other more complicated and qualitatively different criteria. We can indicate here some general features of the system response to the phase action. They reduce to the fact that weak perturbations allow a system to return rapidly to the attraction mode. Perturbations close to dangerous induce in the system a transient mode close to metastable. Such a mode can be either stationary or oscillating (even quasistochastic). With an approach of the intensity of a phase factor to a dangerous value, the most time of the transient process the system is in the vicinity of the metastable mode attained due to the action of external factors.

The criteria of approaching dangerous phase boundaries can be different, but one feature is common for all of them - they manifest themselves in the dynamics of the system outgoing from a perturbed state. In other words, the approach to a dengerous phase boundary is always indicated by the initial stages of a transient process arising immediately upon perturbation, whereas the concluding stages of the transient process remain to be unchanged independently of an increase of the external factor intensity.

The parametric action changes, as a rule, also the attraction mode of the system. If a parametric factor is weak, i.e., if it does not pass through a dangerous boundary, the attraction mode of the system changes only feebly. In the general case the approach of the factor intensity to a dangerous boundary can be indicated by either a new attraction mode attained because of the action of this factor or by the specific features of the transient mode manifested in the process of attainment of a new attraction mode. Let us consider the cases where an attraction mode reduces either to equilibrium or autooscillations.

In case of an equilibrium attraction mode, the character of its change due to the action of a parametric factor cannot be a criterion of approaching a dangerous boundary. Moreover, the variation in the attration mode can mask the approach to such a boundary: thus if a parametric load gradually increases, the density of some populations entering a certain ecological system can gradually increase and then, upon the attainment of a dangerous boundary of the factor intensity it can drop to zero.

The approach of a dangerous boundary in this case can be indicated by the specific features of the transient process alone:

(a) in a monotonous transient process such a criterion is a decelerating (with an increase of the parametric load) attainment by the system of a new equilibrium. One should pay attention to the fact that, at first glance, the criteria of approaching the phase and parametric dangerous boundaries seem to be opposite: in case of a phase perturbation, the danger is indicated by a gradually decelerating process of the system outgoing from the perturbed state, and in the case of a parametric perturbation it is indicated by a decelerating process of system enter into a new equilibrium state.

(b) In an oscillating transient proceess, the approach to the dangerous boundary is indicated by a decrease in the attentuation decrement with the attainment by a system of a new equilibrium state due to an increase of a parametric load. Note also that in distinction from the above formulated criteria, the latter one is not unconditioned: a decrease in the attenuation decrement can indicate the approach not only to a dangerous boundary but also to a smooth (with an increase in loading) transition from the equilibrium attraction mode to a small--amplitude autooscillating one.

It is known that there are four types of dangerous parametric boundaries for oscillating modes [4]. In the practice of ecological modeling only two of them have been encountered, namely:

(1) merge together and "annihilation" of stable and unstable limit cycles and

(2) destruction of a stable limit cycle on the separtrix contour.

In the first case the criterion of the approach to a dangerous boundary is only feebly marked. It consists in decelerating slow reduction of oscillations to the attraction ones with an approach to a dangerous boundary. In the second case, it is not the specific features of the transient process that serve as a criterion but the evolution of the attraction mode itself. With an approach to a dangerous boundry the period of oscillations increases and their shape acquires more and more relaxation character, the system passes most time in the vicinity of a certain state.

In addition to the above described rigorous criteria of the approach to a dangerous boundary, there are some other features of the system behavior that seem to be "suspicious" from the standpoint of the approach to a dangerous boundary. These are the generation in the system of oscillations with an increase of the intensity of parametric loading and the transition from regular oscillations to stochastic ones (which cannot be distinguished from random ones) with a further loading. Such a type of a dangerous parametric boundary was observed in the study of a model of three populations interacting according to the "one predator - two victims" principle [5].

Returning to the above stated problem on the relationship between the imitative and mathematical modeling proper, one can say that the unbounded enthusiasm about the prospectives of imitative modeling reigned at late sixties and early seventies has changed by the middle seventies into a more sober understanding of the limited character of this approach and the necessity of constructing along with imitative also very simplified models oriented to the analysis of the main general laws of functioning of ecological systems. At present, we are at the threshold of the synthesis of these two directions developed until recently rather independently. The future of the mathematical ecology seems to lie not on the path of constructing models of "optimum" complexity, but in creating a hierarchy of "embedded" models of increasing complexity that should answer different problems related to the same systems.

REFERENCES

1. A.D. Basykin. "Mathematics and ecology." In: Problems of modern theoretical biology. 1983, 12-24. Pushchino: ONTI NTsBI AN SSSR (in Russian).
2. Yu.A. Izrael', L.M. Filippova, G.E. Pisarev, et al. "Ecological monitoring and regulation of the state of a natural medium." In: Problems of ecological monitoring and modeling of ecological systems. v.4, 6-19. Leningrad: Gidrometeoizdat 1981 (in Russian).
3. A.D. Basykin. Mathematical biophysics of interacting populations. Moscow: Nauka, 1985 (in Russian).
4. N.K. Bautin and L.P. Shil'nikov. "Behavior of dynamical systems in the vicinity of the stability region boundaries of equilibria for periodic motions ("dangerous" and "safe" boundaries)." In: G. Marsden and M. McCraken, Bifurcation of the cycle generation and its applications. Addition 1., 294-316. Moscow: Mir (in Russian).
5. E.A. Aponina, Yu.M. Aponin, and A.D. Basykin. "Analysis of complicated dynamical behavior in the "one predator - two victims" model." In: Problems of ecological monitoring and modeling of ecological systems. V.4, 163-180. Leningrad: Gidrometeoizdat, 1982 (in Russian).

Mathematics
and Modelling

ISBN 1-56072-104-9

ASYMPTOTIC BEHAVIOR OF POPULATION MODELS WITH AGE AND INTERNAL STRUCTURES

G.P. Karev

INTRODUCTION

Below a population model is considered in which each individual of a species is characterized by an age and a certain set of variables X = (x_1, ..., x_n) (internal with respect to the population as a whole) such as size, biological mass of an individual, etc. Such models are called population models with the age and internal structures or, in a more concise form, structured models [2,14]. The dynamics of such a model is described by a function l(X, t, a), the density of a number of individuals of age a at the moment t with a set of values of internal variables X hereinafter called the "state" of an individual. The main processes occurring in a population is death and reproduction depending the individual age and state.

The generally accepted but not the only possible way of the mathematical description of this model is the following system of equations:

$$
\left.\begin{aligned}
&\dot{x} = F(a,X) \\
&\dot{a} = 1 \\
&\frac{\partial l}{\partial t} + \frac{\partial l}{\partial a} + \operatorname{div}(lF) = -\mu l \\
&l(X,0,a) = \varphi(X,a) \\
&\beta(X,t) \equiv l(X,t,0) = \int_E \int_0^\infty \beta(X,Y,a)\, l(Y,t,a)\, dY\, da
\end{aligned}\right\} \quad (A)
$$

Here $F = (f_1(a,X), \ldots f_n(a,X))$ is the vector-function of the variation rate of the internal variables X (so that the equations A_1 and A_2 describe the dynamics of the state of an individual), $\mu(a,X)$ is the death density, $\beta(X, Y, a)$ is the density of individual birth in the state X from parents of age a in the state Y; $\varphi(X,a)$ is the initial distribution, and $E = \{X = (x_1, \ldots, x_n)\}$ is a set of possible values of variables X; as a rule, E is a region of the positive octant of R_n^+.

The population models possessing only the age structure are considered in numerous publications (see a special issue of Computers & Math. with Appl., 1983, 9, No.3). At the same time it was repeatedly indicated that various applications dictate the necessity of considering also models possessing both the age and the internal structure (Ch.8 in [24], 12].

The asymptotic behavior of a population model with the age structure (the Lotka and Von Förster models) are described by the Sharp-Lotka theorem rigorously proved by Feller [9] on the basis of the analysis of the renewal equation satisfied by the density of newly-born individuals. As far as the author knows, by present an analogue of the Sharp-Lotka theorem has been obtained only for a particular case of the (A) model: Webb [14] has studied a model with "two ages" by the semigroup methods and indicated how the obtained results can be extended to the case of a more general second internal variable.

The present article is devoted to a complete asymptotic description of the (A) model based on the Marcovian renewal theory.

1. INTEGRAL FORM OF A MODEL OF STRUCTURED POPULATION

Let us reformulate the problem (A) in such a way that it could be studied by the methods of the renewal theory. With this aim we write the equation for the function B(X, t) equal to the number of newly born (per unit time) in the state X at the moment t.

Obviously that

$$B(X,t) = B_1(X,t) + B_2(X,t),$$

where $B_1(X,t)$ is the rate of the appearance of the first generation, i.e., the number of direct descendants (per unit time) of that part of the population which existed at the initial moment of time and had the distribution $\varphi(X,a)$; $B_2(X,t)$ is the rate of the appearance of descendants of individuals appeared in the population after the moment t = 0, i.e., of the second and the following generations.

Let X(a,Y) be a solution of the Cauchy problem for the system

$$\left.\begin{aligned} \dot{X} &= F(a,X) \\ \dot{a} &= 1 \end{aligned}\right\} \qquad (1.1)$$

under the initial condition X(0) = Y.

We denote by n(a,Y) the probability of an individual born in the state Y at the moment t > 0 living to the age a and by t ≤ a the probability of an individual born in the state Y being at the initial moment t = 0 of the age a - t, so that the number of individuals born in the state Y and being at moment t of the age a is equal to n(t, a, Y)φ(X(a-t, y), a-t).

Therefore

$$B_1(Z,t) = \int_E \int_t^{\infty} \beta(Z,X(a,Y),a)\ n(t,a,Y)\ \varphi(X(a-t,Y),a-t)\ da\ dY \tag{1.2}$$

Further on, the quantity

$$K(Z,Y,a) = \beta(Z,X(a,Y),a)\ n(a,Y) \tag{1.3}$$

is equal to the number of newly born in the state Z (per unit time) for one parental individual with the initial state Y attained the age a. Therefore

$$B_2(Z,t) = \int_E dY \int_0^t da(K(Z,Y,a)\ B(Y,t-a)) \tag{1.4}$$

Relationships (1.2-1.4) provide the desired renewal equation for B(Z,t).

A formal derivation of the renewal equation is based on the following. Let us assume that the Cauchy problem for system (1.1) has the only solution X(a,Y) for all a ≥0. Let us denote by C(a,X) = $\{c_1(a,X), \ldots, c_n(a,X)\}$ a set of n first integrals of the system (1.1.) chosen in such a way that $c_i(0,X) = x_i$ for all i = 1,...,n. We intoroduce also the following notation

$$J(t,a,Y) = \exp\left(- \int_{(a-t)^+}^{a} \sum_{i=1}^{n} \frac{df_i}{dX_i} (X(u,C(a,Y)))du\right) \tag{1.5}$$

$$n(t,a,Y) = \exp\left(- \int_{(a-t)^+}^{a} \mu(u,X(u,Y))\ du\right) \tag{1.6}$$

Note that at $t \geq a$ the quantities $J(t,a,Y) \equiv J(a,Y)$ and $n(t,a,Y) \equiv n(a,Y)$ are t-independent. In virtue of the formula on page 143 in [7], $J(a,Y)$ is equal to the Jacobian of the system of functions $C(a, Y)$, and $J(t, a, Y)$ is equal to the Jacobian of the system of functions $\{x_1(a-t, C(a,Y)),\ldots, x_n(a-t, C(a,y))\}$, $a \geq t$.

It can be proven that the solution of the problem (A) has the form

$$l(Z,t,a) = \begin{cases} \varphi(X(a-t,C(a,Z))\ J(t,a,Z)\ n(t,a,C(a,Z)), & a \geq t \\ B(C(a,Z),t-a)\ J(a,Z)\ n(a,C(a,Z)), & a \leq t \end{cases} \tag{1.7}$$

whence, in particular, we arrive at the "transparent" formula

$$l(X(a,Y),t,a) = \begin{cases} \varphi(X(a-t,Y),a-t)\ n(t,a,Y) \\ B(Y,t-a)\ n(a,Y) \end{cases} \tag{1.8}$$

The technical details of the relationships derivation are given elsewhere [2]. From (1.7) it follows that for any function (for which the respective integrals have sense) the following equality is valid

$$\int g(X) l(X,t,a) dt = \begin{cases} \int_E g(X(a,Y))\varphi(X(a-t),Y),a-t) n(t,a,Y) dY, & a \geq t \\ \int_E g(X(a,Y)) B(Y,t-a) n(a,Y) dY, & a \leq t. \end{cases}$$

This formula and equations (A_5) yield directly the renewal equations for $B(X,t)$.

Now let us formulate the "integral" variant of the model of the structured population. Let X(a, Y) be a function describing the dynamics of the state of an individual depending on age a and the initial state Y (it is unimportant whether X(a,Y) is the solution of the Cauchy problem for the system (1.1) or not). Then the population dynamics is determined by the system

$$\left.\begin{aligned} &l(X(a,Y),t,a) = \begin{cases} \varphi(X(a-t,Y),a-t)n(t,a,Y), & a\geq t \\ B(Y,t-a)n(a,Y), & t\geq a \end{cases} \\ &B(Y,t) = B_1(Y,t) + \int_E \int_0^t K(Y,Z,u)B(Z,t-u)\, dZ\, du \end{aligned}\right\} \tag{B}$$

The functions B_1, K, n are defined by equalities (1.2, 1.3, and 1.6), respectively. The functions φ, η, and β are set and have the sense as earlier.

The integral form of the model (B) has some advantages over the "differential" form of the model (A), e.g., it allows us to avoid the artificial requirements of sufficient smoothness of functions φ and β, which allows one, in particular, to consider not only continuous but also discrete reproduction. Note that the integral model (B) of the structured population has as clear "biological" sense as the differential model and therefore can also be taken as a formal initial population description instead of being derived from the system (A).

2. RENEWAL THEOREMS FOR EQUATIONS WITH DECOMPOSABLE KERNELS

The asymptotic behavior of the solutions of the equations in the form (B_2) is studied by the theory of the Markovian renewal. Let us introduce necessary definitions. Let on a measurable space (E x $[0,\infty)$, A x B) /where (E, A) is a measurable space with the enumerably generated σ-algebra A and B is a Borel σ-algebra on $[0, \infty)$/ a non-negative kernel M(x, dy, dt) be given, i.e., an A- measurable measure with respect to x on (A x B) such that

$$\sup_{x \in E'} \mathcal{M}(x, E, [0, \infty)) < \infty . \tag{2.1}$$

The following equation

$$Z(x,t) = f(x,t) + \int_E \int_0^t \mathcal{M}(x, dy, du)\, Z(y, t-u), \tag{2.2}$$

is called the renewal equation with kernel M. Here f and Z are the known and unknown functins, respectively.

The kernel:

$$R(x,dy,dt) = \sum_{k \geq 0} M^{k*}\ (x,dy,dt), \tag{2.3}$$

is called a resolvent of the kernel M. Here

$$\mathcal{M}^{0*}(x, A, [0, t]) = X_A(x) \equiv \begin{cases} 1, & x \in A \in \mathcal{A} \\ 0, & x \in A \end{cases}$$

$$\mathcal{M}^{(k+1)*}(x, A, [0, t] = \int_E \int_0^t \mathcal{M}(x, dy, du)\ \mathcal{M}^{k*}(y, A, [0, t-u]).$$

The function

$$(\mathcal{M} * q)(x,t) = \int_E \int_0^t \mathcal{M}(x,dy,du)\ g(y,t-u).$$

is called a a convolution of the kernel **M** and the function g. The solution of equation (2.2) can be represented in the form

$$Z(x,t) = (R^*f)\ (x,t) \tag{2.4}$$

The asymptotic behavior of convolutions (2.4) depends on the spectral properties of the operator **M** acting in the Banach space of bounded A-measurable functions by the formula

$$(\mathcal{M}\ f)(x) = \int_E \mathcal{M}(x,dy)\ f(y) \tag{2.5}$$

$$\mathcal{M}(x,dy) \equiv \mathcal{M}(x,dy,[0,\infty)). \tag{2.5}$$

The classical renewal theory (Ch.11 in [6]) deals with the asymptotic behavior of the solutions to the equations

$$Z(t) = f(t) + \int_0^t \mathcal{M}(du)\ Z(t-u) \tag{2.7}$$

(i.e., in this case E $\neq \emptyset$), the| theory of the Markovian renewal (Sect.9 in Ch.8 [5], Sect.6 in Ch.12 [4]) considers the matrix equation

$$Z_i(t) = f_i(t) + \sum_{j=1}^{N} \int_0^t \mathcal{M}_{ij}(du)\ Z_j(t-u), \qquad i = 1,...N \tag{2.8}$$

(in this case E = {1, 2, ..., N}).

In [9], the main theorem of the renewal theory is derived for general equation (2.2). This theorem is rather complicated from the standpoint of both its proof and formulation. In particular, the check of the non-decomposability conditions and recurrency by Harris of the operator M| (2.5) and also a search for an invariant vector and the measure of this operator are, in the general case, rather a complicated problem. Therefore, of interest is the fact that the renewal theorems for rather a wide class of kernels M (hereinafter called decomposable) reduce to the classical (one-dimensional) renewal theory for equation (2.7).

Definition. A kernel M(x, dy, dt) is called decomposable, if it can be represented in the form

$$M(x,dy,dt,) = \sum_{k=1}^{\infty} \gamma (x) M_k(dy,dt), \tag{2.9}$$

where $\gamma_i(x) \geq 0$ is a measurable function, $M_k(dy, dt)$ is a nonnegative measure on (E x [0,∞, A x B) for all k.

In virtue of condition (2.1), we have

$$\sup_x \sum_i \gamma_i(x)\ M_i(E,[0,\infty)) < \infty \tag{2.10}$$

Later on we shall show that the renewal equation with the decomposable kernel reduces to the equation in the form (2.8) in which E = {1,2, ...}, i.e., to the equation of the Markovian renewal with the enumerable space of states. In this case, if the sum in (2.9) consists of final number of N nonzero terms, then the respective equation (2.2) reduces directly to equation (2.8). In turn, the renewal theory with the enumerable space E reduces to the classical theory for equation (2.7). This approach allows us to apply some statements of the classical theory to the case of the renewal equations with decomposable kernels.

Let us introduce the notation

$$F_{ij}(du) = \int_E M_i(dZ,du)\ \gamma_i(Z). \tag{2.10}$$

Proposition 2.1. For convolutions of a decomposable kernel the following representation is valid for all k = 1, 2, ...

$$M^{k*}(x,A\,[0,t]) = \sum_{i=1}^{\infty} \gamma_i\ (x)\ b_i^{(k)}(A,t),$$

where $b^{(1)}(A,t) = M_i(A,[0,t])$

$$b_i^{(k+1)}(A,t) = \sum_{j=1}^{\infty} \int_0^t F_{ij}(du)\ b_j^{(k)}(A,t-u).$$

Proposition 2.2. The resolvent R(x, A, [O t]) of a decomposable kernel **M** (2.9) satisfies the equations

$$R(x,A,[0,t]) = X_A(x) + \sum \gamma_i\ (x)\ R_i(A,t), \tag{2.11}$$

where

$$R_i(A,t) = N_i(A,t) + \sum_j \int_0^t F_{ij}(du)\ R_j(t-u). \tag{2.12}$$

The proof of Proposition 2.1 is carried out by induction with respect to k:

$$M^{(k+1)*}(x,A[0,t]) = \int_E \int_0^t \left(\sum_i \gamma_i(x) M_i(dy,du)\right)\left(\sum_j \gamma_i(y) b_j^{(k)}(A,t-u)\right)$$

$$= \int_0^t \sum_{i,j} \gamma_i(x) F_{ij}(du)\ b_j^{(k)}(A,t-u) = \sum_i \gamma_i(x) b_i^{(k+1)}(A,T)$$

Introduce the notation

$$R_i(A,t) = \sum_{k=1}^{\infty} b_i^{(k)}\ (A,t).$$

Then equalities (2.11) and (2.12) follow directly from the definition of resolvent (2.3) and Proposition 2.1.

From these propositions it follows that the problem of the search for the resolvent R, and hence, of the solution of equation (2.2) with the decomposable kernel reduces to the solution of renewal equation (2.8) for the case where the space E is a set of natural numbers.

The asymptotic properties of the resolvent R and convolutions (R * g)(x,t) are dependent on the spectral radius of the operator M (2.5). Let us establish the correspondence between the spectral properties of the operator M and the matrix $\|F_{ij}(\infty)\|$.

Introduce the notation $\Gamma_i(A) = M_i(A, [0,\infty))$. Then kernel (2.6) of the operator $\mathcal{M}$ (2 .5) can be represented in the form

$$\mathcal{M}(x, dy) \equiv \mathcal{M}(x, dy, [0, \infty)) = S\sum_i \gamma_i(x)\Gamma_i(dy)$$

Proposition 2.3. A function $\gamma(x)$ is an eigenfunction of the operator M that corresponds to the eigenvalue λ then and only then when $\gamma(x) = \sum_i u_i\gamma_i(x)$, where $\bar{u} = (u_1, u_2, \ldots)$ is the right-hand eigen vector of the matrix $\|F_{ij}(\infty)\|$, i.e., $\sum_i F_{ij}(\infty)u_j = \lambda u_i$ at all i.

The measure $\Gamma(A)$ is the eigenmeasure of the operator $\mathcal{M}$, i.e.,

$$\int_E \Gamma(dx)\, \mathcal{M}(x, A) = \lambda\, \Gamma(A)$$

then and only then when $\Gamma(A) = \sum_i v_i\Gamma_i(A)$, where $v = (v_1, v_2, \ldots)$ is the left-hand eigenvector of the matrix $\|F_{ij}(\infty)\|$, i.e.,

$$\sum_i v_i F_{ij}\ (\infty) = \lambda v.$$

Proof. Let $\gamma(x)$ be an eigenfunction of the operator $\mathcal{M}$. Then

$$\lambda\gamma(x) = (\mathcal{M}\gamma)(x) = \sum_i \gamma_i(x) \int_E \Gamma_i(dy)\ \gamma(y) = u_i\gamma_i(x)I$$

where

$$u_i \equiv \int_E \Gamma_i(dy)\ \gamma(y) = \sum_j \int_E \Gamma_i(dy) \frac{u_i}{\lambda} \gamma_i(y) = \frac{1}{\lambda} \sum_j F_{ij}(\infty)u_j$$

i.e., $\bar{u} = (u_1, u_2, \ldots)$ is an eigenvector of the matrix $\|F_{ij}(\infty)\|$. The same consideration carried out in the reverse sequence shows that if

$$\sum_j F_{ij}\ (\infty)u_j = \lambda u_i$$

then

$$\mathcal{M}(\sum_i u_i\gamma_i)(x) = \lambda\sum_i u_i\gamma_i(x) \equiv \lambda\gamma(x)$$

The second statement is proved in a similar way.

Thus, we have established the one-to-one correspondence between the eigenvalues, eigenvectors, and the measures of the operator $\mathcal{M}$ and the eigenvalues and eigenvectors (right- and left-hand, respectively) of the matrix $\|F_{ij}(\infty)\|$. In particular, the spectral

radius of the operator $\mathcal{M}$ is equal to the spectral radius R (Perron's root) [11] of the matrix $\|F_{ij}(\infty)\|$.
Hereinafter we suggest that the following is valid.

Condition 1. The spectral radius R = 1 and the matrix $\|F_{ij}(\infty)\|$ has the only left-hand, $\bar{u}$, and right-hand, $\bar{v}$, invariant vectors (within the accuracy of the constant cofactor), or, what is equivalent, the operator M has the only invariant function $\gamma(x) = \sum_i u_i \gamma_i(x)$ and the invariant measure $\Gamma(dy) = \sum_i u_i v_i \equiv \int \gamma(x)\Gamma(dx) = 1$. This condition is fulfilled if the operator M is decomposable and recurrent (according to Harris) and fulfilled equivalently if the matrix $\|F_{ij}(\infty)\|$ is decomposable and recurrent (see [9]).

Now let us remind the main theorems of the Markovian renewal for the case where E is a set of natural numbers.

Let the matrix $\|F_{ij}(dt)\|$, $ij \in E$ be such that $\|F_{ij}(dt)\|$ are finite nonnegative measures on $[0, \infty)$ and the matrix $\|F_{ij}(\infty)\|$ satisfy the condition I, and in this case $\sup_i \sum_i F_{ij}(\infty) < \infty$.

Let $\bar{u} = (u_1, u_2, \ldots)$ and $\bar{v} = (v_1, v_2, \ldots)$ be the left- and right-hand eigenvectors of the matrix $\|F_{ij}(\infty)\|$ corresponding to the eigenvalue $\lambda = 1$ (the only eigenvalues within the constant factor, in accordance with the condition (1), and $\sum_i u_i v_i = 1$.

Definition. The renewal matrix is called the matrix

$$H(t) = \sum_{n\geq 1} F^{n*}(t)$$

where

$$F^{1*} = F, \quad F^{(n+1)*}(t) = \int_0^t F(dn)F^{n*}(t-n)$$

<u>Definition</u>. The matrix $\|F_{ij}(t)\|$, $i,j \in E$ is called a lattice matrix with the step δ and the shift function $\rho(i)$ if for all $i, j \in E$ all the points of the growth of the function $F_{ij}(t)$ are concentrated on the shifted lattice $\{\rho(j) - \rho(i) + n\delta$, n is an integer$\}$ and δ is the maximum number possessing the above indicated property.

<u>Definition</u>. A function $f(i,t)$, $i \in E$, $t \in [0,\infty)$ is called directly integrable according to Riemann (or (R,v)-integrable) if

$$\sum_i v_i \sum_{n\geq 1} \sup_{T_{n,1}} f(i,t) < \infty$$

$$\lim_{\Delta\downarrow 0} \sum_i \sum_n v_i \left(\sup_{T_{n,\Delta}} f(i,t) - \inf_{T_{n,\Delta}} f(i,t) \right) = 0$$

where

$$T_{n,\Delta} = \{ t: (n-1)\Delta \leq t < n\Delta \}$$

Let us introduce the notation

$$m = \sum_{i,j=1}^{\infty} v_i u_j \int_0^{\infty} t \, F_{ij}(dt).$$

<u>Theorem 2.1</u>. Let $m < \infty$. If the matrix $\|F_{ij}(t)\|$ is a nonlattice matrix, then for any $h > 0$ the following equation is valid

$$\lim_{t\to 0} (H_{ij}(t+h) - H_{ij}(t)) = \frac{u_i v_j}{m} h$$

If the matrix $\|F_{ij}(t)\|$ is a lattice matrix with the step δ and the shift function ρ, then

$$\lim_{n\to\infty} (H_{ij} (\rho(j) - \rho(i) + (n+1)\delta) - H_{ij} (\rho(j) - \rho(i) + n\delta) = \frac{\delta u_i v_j}{m}$$

The proof of the first statement almost completely repeats the proof of the main theorem of the article [8] for the finite set E with due regard for conditions (2.13). Namely, it is proved that at all i, j = 1, 2, ...

$$H_{ij}(t) = P_{ij}(t) + \int_0^t P_{ij}(du)\, H_{jj}(t-u), \tag{2.14}$$

where

$$P_{ij}(t) = F_{ij}(t) + \sum_{k\neq j} \int_0^t F_{ik}(ds)\, P_{kj}(t-s),$$

with

$$P_{ij}(\infty) = F_{ij}(\infty) + \sum_{k\neq j} F_{ik}(\infty)\, P_{kj}(\infty).$$

If to introduce the notation

$$\mu_{ij} = \int_0^\infty tP_{ij}(dt), \qquad a_{ij} = \int^\infty tF_{ij}(dt),$$

then

$$\mu_{ij} = \sum_k a_{ik} \frac{u_k}{u_j} + \sum_{k \neq j} F_{ik}(\infty)\mu_{ik}$$

and

$$\mu_{jj} = m/(u_j v_j)$$

where $m = \sum_{i,k} v_i a_{ik} u_k$.

Now, u|s|i|ng the Blackwell renewal theorem (therem 12.2 in [3]), we obtain from equation (2.14) at i = j:

$$H_{jj}(t+h) - H_{jj}(t) \rightarrow \frac{h\, u_j v_j}{m}, \qquad t \rightarrow \infty,$$

and at $i \neq j$

$$H_{ij}(t+h) - H_{ij}(t) \rightarrow \frac{h\, u_i v_j}{m}, \qquad t \rightarrow \infty,$$

The second statement follows from the Erdesh-Feller-Pollard theorem (therem 12.3 in [3]).

In a way similar to that used for proving therem 1.2 from Ch.11 in [5], it is possible to establish the equivalence of the theorem 2.1 and the following statement.

Theorem 2.2 (nodal renewal theorem). Let $m < \infty$. If the matrix $\|F_{ij}(t)\|$ is a non-lattice matrix and if the non-negative function f(i,t) is (R,v)-integrable, then

$$Z_i(t) \equiv \sum_j \int_0^t H_{ij}(du)\, f(j,t-u) \underset{t \rightarrow \infty}{\rightarrow} \frac{u_i}{m} \sum_{j=1}^{\infty} v_j \int_0^{\infty} f(j,t)\, dt$$

If the matrix $\|F_{ij}(t)\|$ is a lattice matrix with the step δ and the translation function ρ and if the nonnegative function $f(i,t)$ is such that

$$\sum_i \sum_n v_i \, f(i, n\delta + \rho(i)) < \infty ,$$

then

$$Z_i(\rho(i) + n\delta) \underset{n \to \infty}{\to} \frac{u_i \delta}{m} \sum_j v_j \sum_{k \geq 0} f(j, \rho(j) + k\delta).$$

Now return back to the renewal theorems for equations with decomposable kernels. Since the γ-algebra, by condition, is enumerably generated, then there exists the representation ([9])

$$M(x,dy,dt) = M(x,dy)\ \Phi(x,y,dt)$$

where $\Phi(x, y, dt)$ is the probability distribution on $[0, \infty)$ which is A x A-measurable with respect to x and y.

Definition. A kernel $M(x, dy, dt)$ is called a lattice kernel, if there exist $\delta > 0$ and a nonnegative funtion $\rho(x)$ such that almost all $\Phi(x, y, dt)$ distributions in measure $M(x, dy)$ are concentrated on the translated lattices $\{\rho(x) - \rho(y) + n\delta, n = 0, \pm 1, \ldots\}$. The maximum δ is called a step and ρ is called a translation function.

Let the lattice kernel M be decomposable and the distribution $\Phi_i(x, y, dt)$ be defined by the equality

$$\gamma_i(x)\ M_i(dy,dt) = M(x,dy)\ \Phi_i(x,y,dt)C$$

One can readily see that for a decomposable kernel M the lattice on which the distributions $\Phi_i(x, y, dt)$ is concentrated at all i, and, hence, also the distribution Φ (x, y, dt) are x-independent, and therefore $\rho(x) = \text{const}$. Thus, the decomposable kernel is a lattice kernel with the step δ if almost all the distributions $\Phi_i(x, y, dt)$ at all i are concentrated on the lattice $\{n\delta, n = 0, 1, \ldots\}$.

In particular, the decomposable kernel M is a lattice kernel with a step δ then and only then when the matrix $\|F_{ij}(t)\|$ (2.10) is a lattice matrix with step δ and the translation function $\rho \equiv 0$.

Definition. A nonnegative (A x B)-measurable function g(x,t) is called a (R,Γ)-integrable function if

$$\int_E \Gamma(dx) \sum_{n\geq 0} \sup_{T_{n,1}} g(x,t) < \infty ,$$

$$\lim_{\Delta\downarrow 0} \Delta \int_E \Gamma(dx) \sum_{n\geq 0} \{ \sup_{T_{n,\Delta}} g(x,t) - \inf_{T_{n,\Delta}} g(x,t) \} = 0$$

The following theorem establishes the asymptotic behavior of the solution to the renewal equation

$$Z(x,t) = g(x,t) + \int_E \int_0^t \mathcal{M}(x,dy,du)\ Z(y,t-u).$$

Theorem 2.3. (The nodal renewal theorem for a decomposable kernel). Let a decomposable kernel M satisfy the condition 1 and

$$m \equiv \int_E \Gamma(dx) \int_E \int_0^\infty \mathcal{M}(x,dy,dt)\ t\gamma(y) < \infty .$$

Then, if a kernel M is a nonlattice kernel and a function g(x,t) is (R,Γ)-integrable, then

$$Z(x,t) \equiv (R * g)(x,t) \underset{t\to\infty}{\longrightarrow} \frac{\gamma(x)}{m} \int_E \Gamma(dy) \int_0^\infty g(y,t)dt .$$

If the kernel M is a lattice kernel with the step δ, then

$$Z(x, n\delta) \xrightarrow[n\to\infty]{} \frac{\delta\gamma(x)}{m} \sum_E \int_0 \Gamma(dy)\; g(y, k\delta) \;.$$

Proof. According to Proposition 2.2, we have

$$(R * g)(x,t) = \sum_i \gamma_i\; (x) G_i(t)$$

where

$$G_i(t) = \int_E \int_0^t R_i(dy, du)\; g(y, t-u).$$

The next statement follows from the rule of integration by parts.

Lemma. Let r(t) be a monotonically increasing function, r(0) = 0, f(t) be a nonnegative function, and F(dt) be a finite measure on $[0,\infty)$. We introduce the notation

$$\Phi(t) = \int_0^t F(du)\; r(t-u), \quad V(t) = \int_0^t r(du)\; f(t-u).$$

Then

$$\int_0^t \Phi(du)\; f(t-u) = \int_0^t F(du)\; V(t-u).$$

From the above Lemma and equation (2.12) we have

$$G_i(t) = f(i,t) + \sum_j \int_0^t F_{ij}(du)\; G_j(t-u) \qquad (2.17)$$

where

$$f(i,t) = \int_E \int_0^t \mathcal{M}_i(dy,du)\; g(y,t-u).$$

It can readily be seen that from the (R,Γ)-integrability of the g(x,t) function it follows that f(i,t) is a (R,v)-integrable function. Note that

$$m \equiv \int_E \Gamma(du) \int_E \int_0^\infty \mathcal{M}(x,dy,dt)\; \gamma(y)t = \sum_{i,j} v_i u_j \int_0^\infty + F_{ij}(dt).$$

If the kernel M, and hence the matrix $\|F_{ij}(t)\|$ are nonlattice, then, applying theorem 2.2 to equation (2.17), we have

$$G_i(t) = \frac{u_i}{\bar{m}} \sum_j v_j \int_0^\infty f(j,t)dt.$$

Changing the order of integration over dt and $N_j(dy,dt)$ we have

$$G_i(t) = \frac{u_i}{\bar{m}} \sum_j v_j \int_E \Gamma_j(dy) \int_0^\infty g(y,t)dt =$$

$$= \frac{u_i}{\bar{m}} \int_E \Gamma(dy) \int_0^\infty g(y,t)dt.$$

Now the first statement of the theorem follows from equation (2.15); the second statement is proved in a similar way.

3. GENERALIZATION OF THE SHARP-LOTKA THEOREM

The results obtained in the previous section allows us to study a the behavior of the model (B) and thus of the model (A).
We assume that the following conditions are met:

1. The functions β, γ, μ, and φ are nonnegative and measurable.

2. The function $X(a,Y)$ defined at all $a \geq 0$, $Y \in E_o \subseteq E$ is nonnegative and bounded. Here E_o is a bounded region of allowed initial states. If $X(a,Y)$ is set by system (1.1), then this condition means that the Cauchy problem for system (1.1) has the unique solution for all $a > 0$ and $X(0) \in E_o$.

Let us determine the renewal kernel density

$$K(Z,Y,t) = \beta(Z,X(t,Y),t)\, n(t,Y)$$

and the renewal kernel

$$\mathcal{M}(Z,A,[0,t]) = \int_A \int_0^t K(Z,Y,a)\, dY\, da.$$

Since the distribution set by the renewal kernel is not necessarily absolutely continuous with respect to the Lebesgue measure, then the quantity K can be a generalized function of the arguments y,t. Let us assume that

3. $\beta(Z,Y,a) = \sum_{i=1}^{\infty} \gamma_i\ (Z)\ \beta_i(Y,a)$

so that $K(Z,Y,a) = \sum_i \gamma_i\ (Z)\ \beta_i(Y,a)$,

where $K_i(Y,a) = \beta_i(Y,a)n(a,Y)$

4. There exists a number λ such that Perron's root of the matrix

$$\| \hat{F}_{ij} \| = \| \int_0^\infty e^{-\lambda t} F_{ij}(dt) \|$$

is equal to unity.

Here

$$F_{ij}(t) = \int_E \mathcal{M}_i(dX,[0,t])\ \gamma_j(X)$$

$$\mathcal{M}_i(A,[0.t]) = \int_A \int_0^t K_i(Y,a)dY\ da.$$

If to introduce the notation

$$\hat{\mathcal{M}}_i(dY,dt) = e^{-\lambda t}\mathcal{M}_i(dY,dt),$$

$$\hat{\mathcal{M}}(dX,dY,dt) = \sum_i \gamma_i(X)\hat{\mathcal{M}}_i(dY,dt),$$

then the spectral radius of the operator $\mathcal{T}$ where

$$(\mathcal{T}f)(X) = \int_E \hat{\mathcal{M}}(X,dY,[0,\infty))\ f(Y)$$

is equal to unity.

5. $$\sup_X \hat{\mathcal{M}}(X,E,[0,\infty)) = \sup_X \sum_i \gamma_i X) \int_E \int_0^\infty e^{-\lambda t}\mathcal{M}_i(dY,dt) < \infty .$$

∞ .

6. There exist the unique (within the constant factor) right-nand $\bar{u} = (u_1, u_2, ...)$ and left-hand $\bar{v} = (v_1, v_2, ...)$ invariant vectors of the matrix $\|\hat{F}_{ij}\|$ or, what is equivalent, unique (within a constant factor) invariant function $\gamma(x)$ and the invariant measure $\Gamma(dy)$ of the operator $\mathcal{T}$. \In this case (see Proposition 2.3):

$$\gamma(X) = \sum_i u_i \gamma_i(X), \qquad \Gamma(dY) = \sum_i v_i \hat{\Gamma}_i(dY),$$
$$\hat{\Gamma}_i(dY) = \hat{\mathcal{M}}_i(dY,[0,\infty)).$$

Let us also assume that

$$\sum_i u_i v_i = \left(= \int_E \gamma(X)\ \Gamma(dX)\right) = 1$$

Condition 6 is fulfilled if the matrix $\|\hat{F}_{ij}\|$ is nondecomposable and positively recurrent.

7. $m \equiv \sum_{i,j} v_i u_j \int_0^\infty t\hat{F}_{ij}(dt) = \int_E \Gamma(dX) \int_E \int_0^\infty e^{-\lambda t}\mathcal{M}(X,dY,dt)\ \gamma(Y)t < \infty i$

8. Let us introduce the notation

$$g(i,t) = e^{-\lambda t}\int_E \int_t^\infty \beta_i(X(a,Y),a)\ \varphi(X(a-t,Y),a-t)\ n(t,a,Y)\ da\ dY$$

The function g(i,t) is (R,v)-integrable. For this, it is sufficient that either $\beta_i(X,a) = 0$ at all i and $a > a_o$
or the functions β_i be bounded in the set and inf $\mu = \mu > 0$.

9. $V(\varphi) \equiv \sum_i v_i \int_0^\infty g(i,t) < \infty$

<u>Theorem</u> <u>3.1</u>. Let conditions 1-9 be fulfilled and the matrix $\|F_{ij}(t)\|$ be a nonlattice matrix. Then the functions B and l satisfying the equations of problem (B) exist, are unique, and

$$\lim_{t\to\infty} e^{-\lambda t}B(x,t) = \frac{\gamma(x)\ V(\varphi)}{m} \tag{3.2}$$

$$\lim_{t\to\infty} e^{-\lambda t}l(X(a,y),t,a) = \frac{\gamma(y)\ V(\varphi)}{m}e^{-\lambda a}n(a,y). \tag{3.3}$$

Proof. The existence and uniqueness of the function **B** (and hence also l) follows from the general properties of the renewal equation. Let us put

$$\hat{g}(x,t) = e^{-\lambda t}B_1(x,t), \qquad \hat{B}(x,t) = e^{-\lambda t}B(x,t). \tag{3.4}$$

Then equation (B_2) is equivalent to the equation

$$\hat{B}(x,t) = \hat{g}(x,t) + (\hat{\mathcal{M}} * \hat{B})\ (x,t) \tag{3.5}$$

Then, in virtue of condition 4, the spectral radius of the operator **M** (or the matrix $\|\hat{F}_{ij}\|$) is unity. Moreover, it is obvious that

$$\hat{g}(x,t) = \sum_i \gamma_i(x)\ g(i,t).$$

Since $\int_E g(dz) \sum_i \gamma(x)g(i,t) = \sum_i \sum_k v_k g(i,t)\ \hat{F}_{ki} = \sum v_i g(i,t),$

then, as can readily be seen, from condition 8 it follows that the function $\hat{g}(X,t)$ is (**R**,**Γ**)-integrable and also

$$\sum_E g(dx) \int_0^\infty \hat{g}(x,t)\ dt = V(\varphi).$$

Therefore, applying Theorem 2.2 to equations (3.5), we arrive at the limiting relationship (3.1). Equality (3.3) follows from (3.2) and equation (B_1).

Note. If the Cauchy problem for system (1.1) has the unique solution X(a,Y) at all $a \geq 0$ and $Y \in E_o$, then the functions B and l are the solutions of the problem (A).

Proposition 3.2. The initial distribution

$$\overline{\varphi}\ (X(a,Y),a)\ \equiv e^{-\lambda a}\gamma(Y)\ n(a,Y)$$

is a "quasistationary" in the sense that for such an initial distribution

$$l(X(a,Y),ta)\ =\ e^{\lambda t}\ \overline{\varphi}(X(a,Y),a). \tag{3.6}$$

In this case $V(\overline{\varphi}) =$ m.

Proof. Suffice it to show that for the initial distribution φ we have B(Z,t) = e γ(Z).

Note also that n(a-t,Y) n(t,a,Y) = n(a,Y) at a≥ t. Substituting (3.6) into the right-hand side of the equation (B_2) we have, in accordance with condition 6:

$$B(Z,t)\ =\ e^{\lambda t}\int_E dy\ \int_0^\infty \beta(x(a,y),Z,a)\ e^{\lambda a}\gamma(y)\ n(a,y)\ da\ =$$

$$=\ e^{\lambda t}\int_E \hat{\mathcal{M}}(Z,dy,[0,\infty])\gamma(y)\ =\ e^{\lambda t}\gamma(Z).$$

This proves the equality (3.7). With due regard for (1.8) the equality (3.7) then yields equation (3.6). The equality $V(\overline{\varphi})\ =$ m is either checked directly or follows from (3.6) and Theorem 3.1.

Thus Theorem 3.1, similar to the classical Sharp-Lotka theorem, states that the population density l(X,t,a) asymptotically converges to "stable" distribution (3.3) (see [13]) possessing the property that a relative size of any age-structured group of the population with such a distribution is time-independent. Such a distribution differs from the "quasistationary" distirubion of Proposition 3.2 only by a constant factor.

Now, let us consider the case of a season or pulse reproduction modeled by the reproduction density β(X, Y, a) of the form

$$\beta(x,y,a) = \sum \beta\,(x,y)\,\delta(a-\sigma n)$$

where δ(a) is a Dirac δ-function, $0 \leq \beta_n(X,Y) < c$ = const for all n, σ = const > 0. Restricting ourselves to the case of a decomposable kernel, we put

$$B_n(x,y) = \sum_i \gamma_i\,(x)\,\beta_i(y,n)$$

so that

$$\beta(x,y,a) = \sum_i \gamma_i\,(x)\,\sum_n \beta_i\,(y,n)\,\delta(a-\sigma n) \tag{3.8}$$

Introduce the notation

$$W(\varphi) = \sum_{k=0}^{\infty} \int_E \Gamma(dy)\,\hat{g}(y,k\sigma) = \sum_{k=0}^{\infty}\,\sum_{i=1}^{\infty} v_i g(i,k\sigma),$$

where $\hat{g}(y,t)$ and g(i,t) are defined by (3.4) and (3.1), respectively.

<u>Theorem</u> <u>3.3</u>. Let the density of birth β be defined by equation (3.8), conditions 1-7 be fulfilled, and W(φ) < ∞. Then

$$\lim_{n\to\infty} e^{-\lambda n\sigma} B(X, n\sigma) = \frac{\sigma}{m} \gamma(x) W(\varphi),$$

$$\lim_{n\to\infty} e^{-\lambda n\sigma} l(X(a,y), a+n\sigma, a) = \frac{\sigma}{m} \gamma(y)\, W(\varphi)\, n(a,y).$$

The proof follows from the second part of Theorem 3.1.

REFERENCES

1. A.L. Gimmel'farb, L.R. Ginzburg, R.A. Popuektov, et al. Dynamical theory of biological populations. Moscow: Nauka, 1974 (in Russian).
2. G.P. Karev. "Dynamics and asymptotic behavior of the models of structured populations." In: Simulation of the processes of biological development. Moscow: All-Union Research Institute of System Studies, 1987 (in Russian).
3. I.N. Kovalenko, N.Yu. Kuznetsov, and V.M. Shurenkov. Random processes. Kiev: Naukova Dumka, 1983 (in Russian).
4. R.A. Poluektov, Yu.A. Pykh, and I.A. Shvytov. Dynamical models of ecological systems. Moscow: Nauka, 1980 (in Russian).
5. B.A. Sevast'yanov. Branching processes. Moscow: Nauka, 1971 (in Russian).
6. V. Feller. Introduction into the probability theory and its applications. v.2. Moscow: Mir, 1967 (in Russian).
7. F. Hartman. Ordinary differential equations. Moscow: Mir, 1970 (in Russian).
8. V.M. Shurenkov. "A note about the equation of multidimensional renewal." Probability theory and its applications, v.22, 4, 1975 (in Russian).
9. V.M. Shurenkov. "To the theory of the Markovian renewal." Ibid, v.29, 2, 1984 (in Russian).
10. W. Feller. "On the integral equation of renewal theory." Ann. Math. Stat. 1941, 12.
11. J. Greiner. "A typical Perron-Frobenius theorem with application to an age-dependent population equation. Infinite-dimensional systems." Proceedings. Springer, 1984.

12. G. Oster. "Lectures in population dynamics." In: Lecture in Appl. Math., v.16, Providence, 1977.
13. G.F. Webb. "A semigroup proof of the Sharp-Lotka theorem. Infinite-dimensional systems." Proceedings, Springer, 1984.
14. G.F. Webb. "Dynamics of population structured by internal variables." Math. Z., 189, H.3, 1985.

Mathematics
and Modelling

ISBN 1-56072-104-9

MATHEMATICAL SIMULATION OF PERIODIC PROCESSES OF MICROORGANISM CULTIVATION WITH ALLOWANCE FOR DYNAMICS OF DISSOLVED GASES

Yu.M. Aponin, E.N. Vanyakin, and V.V. Osipov

One of essential factors determining growth and development of microorganisms is the content of oxygen and carbon dioxide in the culture liquid. Low oxygen solubility and therefore high blowing and stirring rates result in a low-inertia O_2- and CO_2- concentrations which therefore become quick-response variables. In the process of cultivation, against the background of slow microbe biomass growth, one can observe drastic changes and fast oscillations of O_2- and CO_2-concentrations [1-4].

Fast motions of low-inertia variables arise as a reaction to the action of externmal pulse perturbations produced deliberately in order to determine the current state of the process. Fast motions can also be self-excited [1-3]. In this case they are a kind of signals indicating the completion of some stages of the cultivation process. The study of the mechanism of fast motions and their classification are very important for an operator who needs such data for the prognosis of the further process and also for the development of efficient computer-oriented controlling algorithms. Therefore the analysis of mathematical models describing both slow and fast processes occurring in the periodic microorganism cultivation is both important and well-timed.

In the present study a mathematical model of periodic microorganism cultivation is considered that takes into account the complex effect of dissolved oxygen and carbon dioxide on the specific growth rate. The model describes the evolution of a slow variable (the biomass concentration) and also fast transient processes and oscillations related to oxygen and carbon dioxide.

The study of the model was carried out by the methods of the qualitative theory of differential equations and the theories of stability and bifurcations [5-7]. The main attention was payed to the analysis of fast motions in the cultivation process and to the study of the partition of the parameter space of the model into the regions with qualitatively different types of the dynamical behavior of low-inertia variables.

Model assumptions. The equations of the mathematical model were derived from the following material balance relationships with respect to oxygen, carbon dioxide, and the biomass (see, e.g., [8]):

$$\begin{cases} \dot{c}_1 = k_1 \cdot (a_1 - c_1) - g_1 \cdot N \\ \dot{c}_2 = k_2 \cdot (a_2 - c_2) + g_2 \cdot N \\ \dot{N} = \mu \cdot N \end{cases} \tag{1}$$

where dots above the letters indicate the time (t) derivative,

c_1, c_2, and N are the concentrations of dissolved oxygen, carbon dioxide, and the biomass;

q_1, q_2, and μ are the specific rates of oxygen consumption by microorganisms, evolution of oxygen dioxide, and the specific rate of biomass growth;

a_1 and a_2 are the saturated oxygen and oxygen dioxide concentrations; and

k_1 and k_2 are the volumetric mass-transfer coefficients.

The relationships of the material balance (1) determine a sufficiently general class of the mathematical models of periodic microorganism cultivation, the concrete form of the equations being dependent on the additional assumptions about the specific rates q_1, q_2, and μ. Below we shall consider a model derived from (1) under the following assumptions.

1. There exists a linear dependence between the rate of oxygen consumption and the rate of carbon dioxide evolution

$$g_2 = r \bullet g_1 + d,$$

where r and d are empirical constants.

2. The specific growth rate μ depends on the concentrations of dissolved oxygen and carbon dioxide as

$$\mu = \frac{c_1}{k_1 + c_1} \cdot \left[\mu_1 + \frac{(\mu_2 - \mu_1) \cdot c_2^{\gamma}}{k_2^{\gamma} + c_2^{\gamma}} \right], \tag{3}$$

where k_1, k_2, μ_1, μ_2, and γ are empiric constants.

3. The following relationship is valid

$$q_1 = l \cdot \mu + \frac{m \cdot c_1}{k_1 + c_1}, \tag{4}$$

where l is the stoichiometric coefficient with respect to oxygen; the second term in equation (4) characterizes the oxygen consumption necessary for maintaining microorganisms in the viable state; and m is the maximum value of the maintenance coefficient [9,10].

Note that the first term in the right-hand side of equation (3) is the well known Mono formula that describes quite reliably the dependence of μ on c_1 for various microorganism species [11]. Some literature data also indicate the inhibiting and stimulating action of oxygen dioxide on microorganism growth [12-19]. The second term in (3) describes various decreasing (if $\mu_2 < \mu_1$) and increasing (if $\mu_2 > \mu_1$) dependences of μ on c_2 that can be used for approximating experimental data.

Equations of mathematical model. With due regard for the above assumptions, the material-balance equations (1) can be made dimensionless:

$$
\begin{cases}
\delta \cdot y = y_0 - y + \alpha \cdot Z \cdot \frac{x}{1+x} \cdot \left(\kappa + \frac{v + y^{\gamma}}{1 + y^{\gamma}}\right) + \beta \cdot Z \\
x = x_0 - x - Z \cdot \frac{x}{1+x} \cdot \left(\kappa + \frac{v + y^{\gamma}}{1 + y^{\gamma}}\right) \\
Z = \varepsilon \cdot Z \cdot \frac{x}{1+x} \cdot \frac{v + y^{\gamma}}{1 + y^{\gamma}}
\end{cases}
\qquad (5)
$$

where x, y, z are dimensionless oxygen, carbon dioxide, and biomass concentrations expressed in terms of the respective dimensional concentrations by the formulae:

$$
x = \frac{c_1}{K_1}\ ; \quad y = \frac{c_2}{K_2}\ ; \quad Z = \frac{\mu_2 l}{k_1 \cdot K_1} \cdot N\ ; \quad \tau = k_1 \cdot t\ , \qquad (6)
$$

where τ is dimensionless time. The parameters entering equations (5) are also dimensionless and can be expressed in terms of the initial dimensional parameters of the model in such a way

(7)

$$\delta = \frac{k_1}{K_2}; \quad \varepsilon = \frac{\mu_2}{K_1}; \quad \nu = \frac{\mu_1}{\mu_2}; \quad \alpha = r \cdot \frac{k_1 \cdot K_1}{K_2 \cdot K_2};$$

$$x_0 = \frac{a_1}{K_1}; \quad y_0 = \frac{a_2}{K_2}; \quad \kappa = \frac{m}{1 \cdot \mu_2}; \quad \beta = \frac{d}{1 \cdot \mu_2} \cdot \frac{k_1 \cdot K_1}{K_2 \cdot K_2}. \tag{7}$$

Small parameter. The dimensionless form of the equations describing the mathematical model simplifies the study of the various types of model behavior as a function of the system parameters, since the number of parameters in system (5) is smaller than the number of the initial dimensional parameters. Moreover, on the basis of dimensionless equations (5) the condition of low-inertia for O_2- and CO_2-concentrations can be represented in the form $\varepsilon < 1$. This inequality is characteristic of many periodic cultivation processes. Indeed, the volumetric coefficient of the oxygen mass transfer k_1, depending on the conditions of aeration and stirring, and also on the fermenter design, ranges from several tens up to several thousands of reciprocal hours [20]. At the same time, the value of μ_2 characterizing the specific growth rate does not exceed $\mu_2 = 1$ (1/r). Therefore $\varepsilon = \mu_2/k_1$ is a small parameter indicating that the characteristic time scales of variables x and y substantially differ from that of variable Z.

System of equations of fast motions. In the limiting case, equa tions (5) at $\varepsilon = 0$ yield Z = const, i.e., the motion trajectory of system (5) lies in a certain plane M parallel to the coordinate (x,y) plane (Figure 1). If the value of ε is small but non-zero, then the variable Z will increase (although slowly), i.e., the plane M will slowly move towards the coordinate Z-axis. Therefore at small $\varepsilon > 0$ the motion of the system (5) can be considered as the super position of two motions - fast in the plane M and slow translatio nal motion towards the X-zxis in this plane.

When the biomass concentration attains the Z-value, the fast motion of low-inertia variables x and y are described by the following system of differential equations

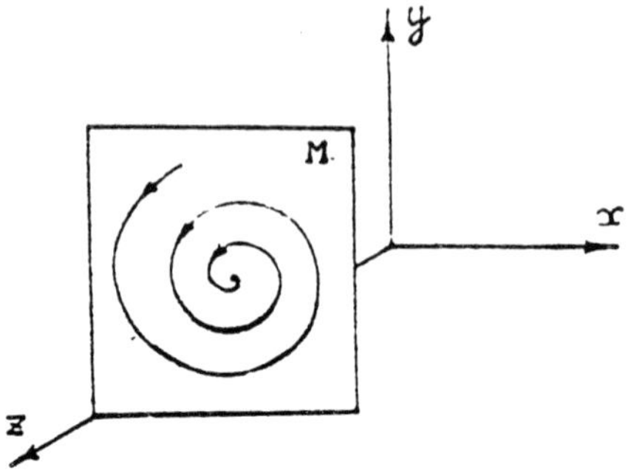

Figure 1. Plane of fast motions **M** in the space of the mathematical model states.

$$\begin{cases} \dot{x} = x_0 - x - \zeta \cdot \frac{x}{1+x} \cdot \left(\kappa + \frac{\nu + y^{\gamma}}{1+y^{\gamma}}\right) \\ \delta\dot{y} = \eta - y + \alpha\zeta \cdot \frac{x}{1+x} \cdot \left(\kappa + \frac{\nu + y^{\gamma}}{1+y^{\gamma}}\right), \end{cases} \quad (8)$$

where the parameters ζ and η are expressed in terms of the biomass concentration by the formulae

$$\zeta = Z, \quad \eta = y_0 + \beta Z. \quad (9)$$

The phase portrait of system (8), which is naturally drawn in the plane of fast motions M, provides a sufficiently complete description of self-excited oscillations of variables x, y, and fast transient processes arising as a response to the external pulse perturbations. In this process, the equilibria of system (8) correspond to quasistationary states with respect to oxygen and carbon dioxide.

Qualitative changes in the character of fast motions in the periodic cultivation process. One of possible phase portraits of system (8) is shown in Figure 1. There is the only stable quasistationary state. In other words, under the action of external pulse perturbation such a state is restored upon the completion of a fast transient process which, in this case, has the form of gradually attenuating oscillations.

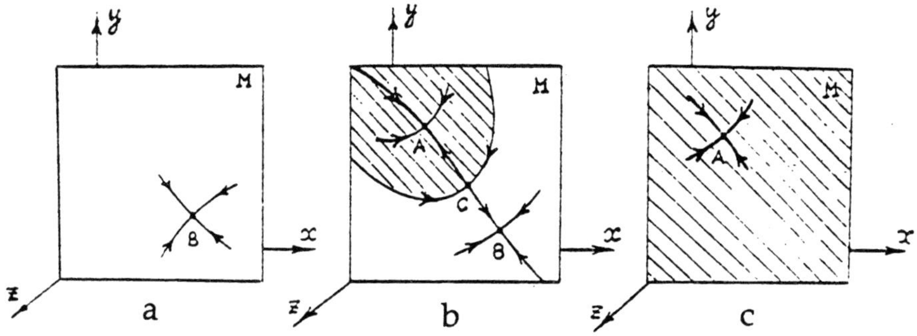

Figure 2. Restructuring of the phase portrait for systems of equations of fast motions occurring in the process of periodic cultivation. Such a type of the phase portrait evolution is possible, e.g., for the parameter values indicated in for Figure 3.

In the cultivation process, an increase of the biomass concentration results in slow displacement of the plane of fast motions M towards the coordinate Z-axis. The phase portrait in this plane also changes, the qualitative restructuring of the phase portrait can occur upon the attainment of a certain threshold (critical) values of the biomass concentration.

One of possible types of the phase portrait evolution of the system of fast motion equations is shown in Figure 2. There are two critical values of the biomass concentration, Z_1 and Z_2 with $Z_1 < Z_2$. If the biomass concentration Z is lower than the low critical value Z_1, then there exists the only stable quasistationary state B (fig.2a). Upon the attainment of the first critical value of the biomass concentration Z_1, the phase portrait experiences qualitative restructuring: there appear two more quasistationary states A and C - the state C is unstable but the state A is stable (its attraction region is dashed in Figure 2b). In the cultivation process, with an approach to the biomass concentration of its second critical value, the attraction region of the state A becomes wider, whereas the attraction region of the state B becomes narrower. Finally, in the passage through the value $Z = Z_2$ both states, B and C, merge together and disappear. At $Z > Z_2$ there exists again the only quasistationary state with respect to oxygen and carbon dioxide (Figure c).

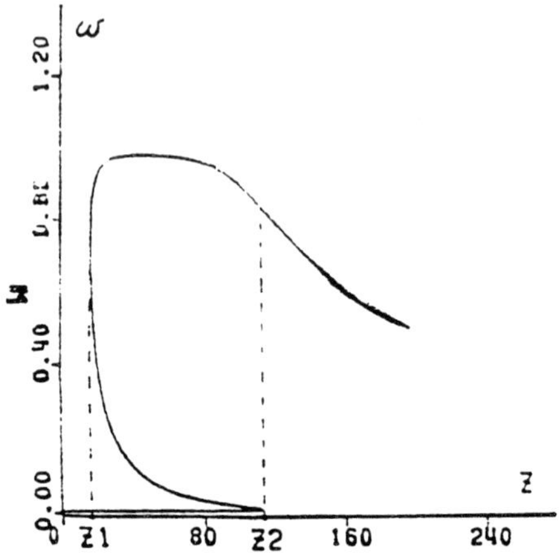

Figure 3. The normalized specific growth rate ω as a functiion of the biomass concentration z at $x_0 = 100$, $y_0 = 0.1$, $\alpha = 0.1$, $\beta = 0$, $\gamma = 4$, $\kappa = 0$, $\nu = 0.01$. Between z_1 and z_2 there are three branches of w values. The upper (lower) branch corresponds to such values of the specific growth rate ω that are attained in the stable quasistationary state A(B) (Figure 2). The middle branch corresponds to the unstable state C.

The trigger-like behavior of low-inertia variables and instability of microbiological processes. Let $\bar{x}$ and $\bar{y}$ be quasistationary values of variables x and y, in other words, the coordinates of the equilibrium of system (8). Then the quantity $w = \frac{x}{1+x} \cdot \frac{\nu+y^\gamma}{1+y^\gamma}$ characterizes the specific rate of microorganism growth which is attained in the respective quasistationary state (see the last equation in (5)).

As follows from computations (Figure 3), the A and B states have substantially different specific rates of biomass growth. Biomass in state B is almost constant (does not increase) and the state A is characterized by a sufficiently high growth rate. In case of external pulse perturbations, one state can switch into another one. Such a trigger-like behavior can be considered as an instability of the microbiological process in which the growth kinetics substantially ranges in a series of cultivations carried out under the same controllable conditions

- in some cases the biomass grows quite well and in some other cases growth almost ceases.

The hysteresis character of the dependence of the specific growth rate on the biomass concentration is also of interest in connection with choosing an appropriate sowing dose. If a sowing dose (i.e. the initial biomass concentration) is chosen to the left of the critical value Z_1, then the biomass almost does not grow (Figure 3). If the sowing dose is chosen to the right of the value $Z = Z_2$, then a high growth rate is attained almost immediately. But it is clear that the sowing dose can be substantially reduced by choosing it to the right of the value $Z = Z_1$ but rather close to it. But in this case it is necessary to have a controlling factor bringing the system into a quasistationary state A with a high growth rate. This can be provided by a pulse action onto the system bringing it to the attraction region of the state A if initially it was in the state B.

Self-excitation oscillations. As has been shown by the study of the model, there exist such critical values of the biomass concentration that their attainment in the system of equations of fast motion (8) results in generation of stable longitudinal cycles. This is accompanied by self-excited oscillations of the concentrations of dissolved O_2 and CO_2. For example, the numerical experiments on the mathematical model in a wide parameter range showed the existence of three growth stages (Figure 2). The first stage is characterized by a low growth rate, the biomass almost does not increase, in other words, this stage can naturally be identified with the lag-phase. The transition into the second stage is accompanied by self-excitation of O_2 and CO_2 concentrational oscillations and an increase in the growth rate. Finally in the transition to the third stage the oscillations attenuate, and the growth rate is stabilized at a new higher level. Thus, in this case self-excited oscillations of low-inertia variables can be considered as a sign indicating a subsequent acceleration of microorganism growth.

The parametric analysis of the model. The phase portrait of the system of equations of fast motion depends on eight parameters. As follows from (7), in this case the parameters

$$\alpha,\ x_0,\ \kappa,\ \nu,\ \gamma,\ \text{and}\ \delta \qquad (10)$$

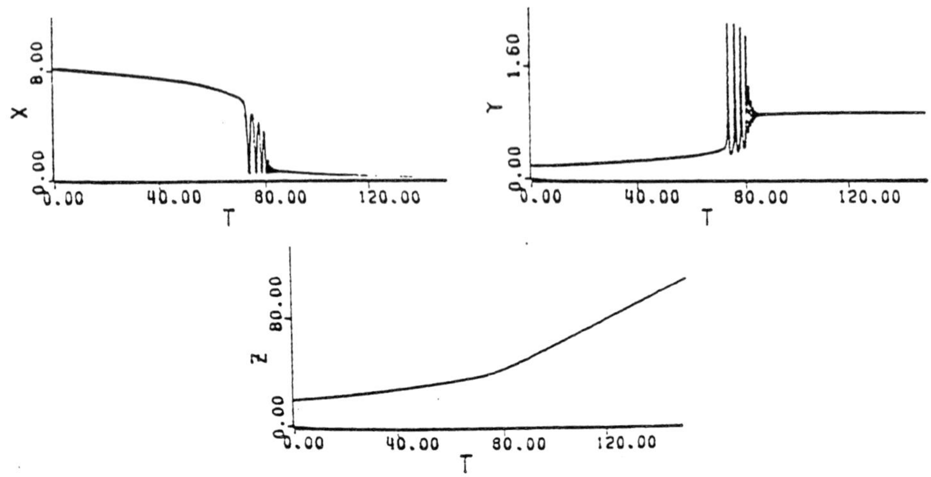

Figure 4. The results of numerical integration of the system of equations (5) that describes the variation of the oxygen, x, carbon dioxide, y, and biomass, z, concentrations in the process of periodic cultivation; $T = \tau$ is dimensionless time, $\alpha = 0.1$; $\beta = 0$; $\gamma = 4$; $\delta = 0.1$; $x = 0$, $\nu = 0.1$, $\varepsilon = 0.1$, $x_o = 10$, $y_o = 0$.

can vary only with the change of the physical-chemical conditions in the fermenter, the aeration or stirring mode, or metabolism swithching. The values of parameters ζ and η continuously vary in the cultivation process at the expense of the variation in the biomass concentration Z (see (9)). Therefore it is interesting to study the partition of the plane of parameters (ζ,η) into the regions with qualitatively different phase portraits of system (8). Such a partition, i.e., the parametric portrait, depends, of course, on the values of parameters (10).

The most important elements of the parametric portrait are the multiplicity line (K) and netrality line (N). The K line is a boundary of the region of parameters ζ and η at values providing three equilibria in system (8). Let us denote this region by G (Figure 5a). If the point (ζ, η) lies outside the region G, then in the system of fast motions (8) there exists the only equilibrium. In the process of periodic cultivation the passage of point (ζ,η) through the line K gives rise either to generation or annihilation of a pair of equilibria depending on the fact whether the point (ζ,η) enters or leaves the region G. If in the process of cultivation the point (ζ, η) passes through the line N then the stability of one of the

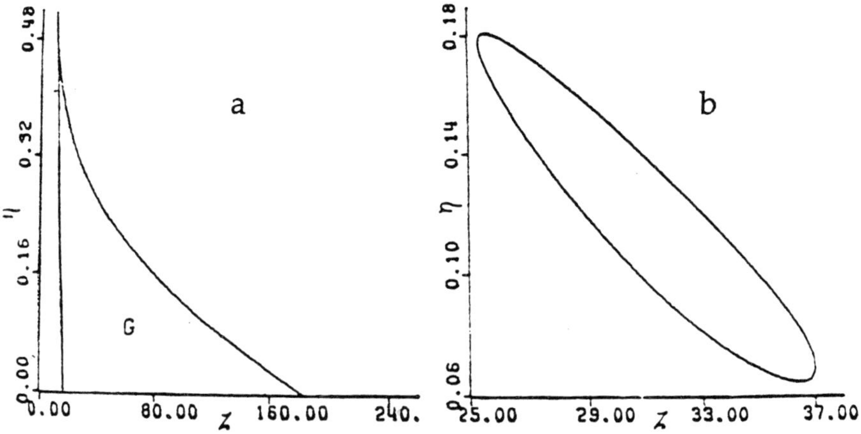

Figure 5. Examples of the arrangement of lines K and N in the plane of the evolution parameters (ζ, η): (a) the line K, $\alpha = 0.1$; $x_o = 100$, $\kappa = 0$; $\nu = 0.01$; $\gamma = 4$. The K line is δ-independent; (b) Line N, $\alpha = 0.01$; $x_o = 7$, $\kappa = 0.01$; $\nu = 0.155$, $\Gamma = 4$; $\delta = 0.1$.

equilibria of system (8) changes. Thus the lines K and N determine the partition of the (ζ,η) plane into the regions differing by either the number or the character of stability of equilibria of the system of fast motion equations.

As has shown the analysis of the model, the following situations are distinguished depending on the values of parameters (10) in the (ζ,η) plane.

1. The lines K and N are absent. Then the phase portrait of system (8) for any parameter values consists of the only stable equilibrium and the trajectories entering it (Figure 1 or Figure 2a).

2. There exists only the multiplicity line (Figure 5a). Then in the region G the phase portrait of system (8) is shown in Figure 2b. Outside the region G, the phase portrait is as in the previous case.

3. There exists only the neutrality line (Figure 5b). Then system (8) has the only equilibrium independently of the ζ, η values. But if in the process of cultivation the point (ζ, η) enters the region limited by the line N, this equilibrium looses its stability. This is accompanied by self-excitation of O_2 and CO_2 concentration oscillations.

4. There exist two lines, K and N. In this case the line N can have self-intersection points, and also the points of the tangency with the multiplicity line, which do not disappear with small variations of parameters (10). This is accompanied by a substantial complication of the parametric portrait. The study of the mathematical model is developed up to writing respective prog rams for computer drawing of K and N lines at any values of para meters (10). There were also developed the programs for constructingseparatrices and quasistationary states (with the indication of the character of their stability) in the plane of fast motion M. These programs provide sufficiently complete data on qualitative restructuring in the dynamical behavior of low-inertia variables in the process of the periodic cultivation for any range of the values of parameters y, β_o, and (10).

The mathematical model (5) uses the minimum set of variables necessary for the description of the growth and respiration processes of microorganisms. Therefore the model corresponds to the stages of the periodic cultivation where growth has not been limited by a substrate or inhibited by the accumulated metabolism products as yet. But the obtained result can also be used as the initial approximation for the investigation of multiparametric mathematical models with wider sets of fast and slow variables if to use the standard programs based on the concept of continuation by parameter [21].

The variety of the types of the dynamical behavior of the model under consideration, the fast self-excited changes, and the oscillations of low-inertia variables provide a rich material for testing, relocating of adaptive identification algorithms, and predicting microbiological processes [8,22].

REFERENCES

1. E. Mochan and E.K. Pye. Nature new biology, 1973, 242, No.119, 177-179.
2. D.E.F. Harrison. In: Biological and Biochemical Oscillations. Ed.B. Chance et al. New York: Academic Press, 1973, 256-260.
3. L.K. Nyiri, G.M. Toth, and C. Marvin. Biotechnology and Bioengineering, 1975, 17, No.1.1, 1663-1678.
4. D.W. Ribbons. Applied Microbiology, 1969, 18, No.3, 438-443.

5. A.A. Andronov, E.A. Leontovich, I.I. Gordon, and A.G. Maer. Qualitative theory of dynamical systems on a plane. Moscow: Nauka, 1966 (in Russian).
6. A.A. Andronov, E.A. Leontovich, I.I. Gordon, and A.G. Maer. Bifurcation theory of dynamical systems on a plane. Moscow: Nauka, 1967 (in Russian).
7. V.I. Arnol'd. Additional chapters to the theory of ordinary differential equations. Moscow: Nauka, 1978 (in Russian).
8. G. Stephanopoulas and Ka-Yiu-San. Biotechnology and Bioengineering, 1984, 26, No.10, 1176-1188.
9. S. J. Perth. Fundamentals of microorganism and cell cultivation. Moscow: Mir, 1978 (in Russian).
10. N.S. Panikov. Microbiologiya, 1984, 53, No.4, 586-594.
11. I.S. Langmuir. The Biochemical Journal, 1954, 57, No.1, 81-87.
12. A.P.I. Trinci and C. Whittaker. Trans. Brit. Micol. Soc., 1968, 51, No.3, 594-596.
13. I.I. Hodgkiss and K. Harvey. Trans. Brit. Micol. Soc., 1972, 59, No.3, 409-416.
14. M. Smith and D. Noble. Trans. Brit. Micol. Soc., 1972, 58, No.3, 499-506.
15. R. Repaske, A.C. Renashe, and R.D. Mayer. J. Bacteriol., 1974, 117, No.2, 652-661.
16. R. Repaske and M.A. Clayton. J. Bacteriol., 1978, 135, No.3, 1162-1164.
17. T.S. Nikishina, S.S. Tarasova, V.M. Nikolushkina, et al. Priklladnaya Biokhimiya i Mikrobiologiya, 1974, 10, No.1, 68-73.
18. V.N. Ivanov. Energy aspect of microorganism growth. - Kiev: Naukova Dumka, 1981, 79-88 (in Russian).
19. R.P. Jones and P.F. Greenfield. Enzyme Microb. Technol., 1982, 4, 210-223.
20. K.P. Gaponov. Processes and apparatus of microbiological industries. Moscow: Legkaya i Pishchevaya Promyshlennost', 1981, 54-57 (in Russian).
21. A.I. Khibnik and E.E. Shnol'. Programs for qualitative investigation of differential equations. Informational material. Pushchino: ONTI NTsBI AN SSSR, 1982 (in Russian).
22. Yu. Stanishkis. Optimum control of biotechnologic processes. Vilnius: MOKSLAS, 1984 (in Russian).

Subject Index